FISH

Life, Environment and Diversity

N. B. Marshall

AGROBIOS (INDIA)

Published by:
AGROBIOS (INDIA)
Agro House, Behind Nasrani Cinema
Chopasani Road, Jodhpur 342 002
Phone: 91-0291-2642319, Fax: 2643993
E. mail: agrobios@sify.com

AGROBIOS (INDIA)

9 788177 540017
Rs. 2500.00 / US$ .00
Published by:
AGROBIOS (INDIA)

Published by: Dr. Updesh Purohit for Agrobios (India), Jodhpur
Lasertypeset at: Shriya Computers and Printers, Jodhpur
Cover Design by: Reena
Printed at: Hinglas Offset Printers, Jodhpur

To Edmund H. Watts,
designer of ships,
with diverse designs on fishes

Contents

Contents

List of Illustrations

Acknowledgements

The following were used as sources in preparing illustrations: Hoogland, Morris and Tinbergen (108) for Plate 2; Gunther Sterber 1962, *Fresh Water Fishes of the World* for Plates 4 and 5; Bigelow and Schroeder (177) for Plates 6; E. Bertelsen (104) for Plate 7; Rudolph Freund 1961, *Scientific American* for Plate 11.

The author also wishes to thank the following for supplying photographs for use in this volume: George Lower, Plates 14, 15, 20, 21, 26, 27, 29, 35, 37 and 40; Peter M. David, Plates 16, 17, 18, 19 and 41; Douglas P. Wilson, F.R.P.S., Plates 22, 24, 25, 30, 31, 39, 42 and 43; Mrs P. D. Wilson, Plate 23; Chuck Kehoe, Plate 28; Paul Popper, Plates 32 and 33; Walter Starck, Plate 34, Frank W. Lane, Plate 36; Mike Davis, Plate 38.

Preface

The plan of this book will be found in Chapter 1. This preface can thus be confined to expressions of gratitude.

First, I owe much to the British Museum (Natural History), where I have been given freedom to study fishes – and enough extramural time to write *The Life of Fishes*.

Even so, this book was finished during a visit to the USA. Here I have found a friendly and congenial atmosphere at the Woods Hole Oceanographic Institution and the Museum of Comparative Zoology, Harvard University.

Lastly, I am delighted once more to thank my wife for the care and patience she gave to the half-tone plates and line-drawings in the book. A man, as Rebecca West wrote, cannot describe the life of a fish, but good illustrations help the readers' imagination.

Chapter 1

An Introductory Guide

FISHES and their element move over most of the earth. Nearly all of this element is salty; for the ocean not only covers seven-tenths of the globe, but has a mean depth of about 4,000 metres. Visible fresh waters, which average a few metres in depth, stand or flow over little more than one-hundredth of the earth's surface.

Fishes are almost ubiquitous and virtually numberless. In the ocean they live at all levels from boisterous, well-lit intertidal regions down to the cold, sunless waters that flow over the sediments of the abyssal plains. In fresh waters they range from the sparkling torrents of tropical hill-streams to the dark and cool recesses of the deepest lakes. Some find a living in pitch-dark underground streams. A good many species, whether by breathing close to the surface, or through the acquisition of air-breathing organs, exist in stagnant, tropical swamps. By many means, which are central concerns of this book, fishes have mastered their element.

Life, as Professor C. H. Waddington [1] reminds us so cogently, proceeds along three time scales: long, medium and short. The long scale, which may involve many millions of years, embraces the evolutionary history of living organisms. Critical study of the comparative anatomy and classification of modern fishes may often provide insights to some of their main evolutionary radiations, but proper consideration of the past must involve the study of fossils. This book is confined to living fishes.* Nor do I propose to consider the multiplication of species. Even a short review of species formation would take more space than is available here. For those interested, Professor Ernst Mayr [2] has given considerable thought to fish speciation in his masterly book, *Animal species and evolution.*

Life from day to day – which is the pace of the short time scale – and early life histories – the first phase of the medium one – are two

* Professor Alfred S. Romer will be dealing with the palaeontology of fishes and other animals in another book in this series.

main concerns of this book. The first scale is marked by the physiological and behavioural features of everyday life. If this life is maintained, then life may be multiplied. A full life-history, which is a rare achievement for most kinds of fishes, begins with a fertilized egg. If this egg starts a life that plays one or more parts in reproducing its kind before death, then the life history is biologically complete.

Life from day to day

Whatever move they make, whether in exploring their surroundings, taking food, evading enemies, migrating, courting, mating, and so forth, fishes cleave through a medium that is eight hundred times denser than air. The more streamlined they are – and many fishes have elegant lines – the less energy is lost in overcoming the resistance of water. Form, motion and fin patterns are considered in Chapters 2 and 3. Here we see that fishes have adopted many different, though essentially similar, ways of using their fins and bodies. Even more astonishing is the great variety of body and fin forms that have been evolved in so stringent a medium as water.

Any healthy adult fish maintains its form, despite all the constructive and destructive sides of its metabolism. The mechanical retention of this form largely resides in the skin, which usually produces a scaly skeleton. The framework of form is provided by the inner skeleton, which moves to the pull of the many attached muscles that activate the head, body and fins. The dynamic rôle of the skin and skeleton in the everyday life of fishes is elaborated in Chapter 4.

A fish breathes and feeds to acquire the energy needed for its daily activities. Much of this energy goes to the muscles used in swimming. Through the evolution of a gas-filled swimbladder, which keeps them weightless in water, most of the bony fishes, called teleosts, have acquired a means of 'saving their breath'. They can hover or swim at one level without using energy to stay at this level. The significance of the swimbladder in the life and living spaces of teleosts is the subject of Chapter 5.

But all kinds of fishes have to work hard for their oxygen. Throughout their life they must move a dense liquid over their gills, and water contains relatively little dissolved oxygen. The mastery of these physical problems is discussed in Chapter 6, which also deals with air-breathing species. Some consideration of food and growth follows in Chapter 7.

The everyday life of fishes is dependent also on physiological processes that steady the water and salt balance of the body. Chapter 8 deals with these and related aspects.

Though water is much less transparent than air, fishes are long-sighted. Many deep-sea species have eyes that are superbly designed to see in dim surroundings, but some abyssal forms and cave-fishes are perfectly able to live without eyes. But no fish ever loses its ears, which may remind us that comparable sounds travel further in water than in air. Fishes also have gained a distant 'water-touch' sense through the evolution of lateral line organs. The main use of these is in prompt, short-range detection and location of disturbances in the water, such as those stirred by moving prey, predators, associates and so forth. A fish may take more time to scent its food or enemies, for odours diffuse slowly in water. Even so, fishes of the order Ostariophysi, which includes carps and catfishes, have evolved a way of olfactory signalling through the production of 'alarm substances' (p. 148). In water, as in air, taste and touch require contact with a sensed object. These and other aspects of the sensory life of fishes are discussed in Chapter 9.

Many teleosts, including some that live near the deep-sea floor, are sound producers. It almost looks as though they have come to know, so to speak, the good acoustical properties of their medium. At all events, sound signals are a good means of communication in dim and sunless waters. Fishes have also risen above such visual restrictions through the evolution of light organs and electric organs. The former are most highly developed in teleosts that live in the twilight zone of the deep ocean. But light organs may be adapted to lure prey or confuse enemies as well as to flash signals. Through the discharges of their electric organs, fishes may not only keep in touch with each other, but also have a means for navigation and the location of prey or predators in turbid, tropical fresh waters. The reader will see why Chapter 10 is called 'Physical' fishes.

Besides striving to maintain their health and vigour, fishes have ways of concealing and protecting their lives. Discussion of the forms of camouflage, such as countershading, dazzle-markings, eye-spots, background matching, deceptive resemblances and mimicry, takes up most of Chapter 11, which also deals with means of defence and offence.

Except when courting and mating, many fishes lead separate lives. Yet the schooling habit, whereby the same-sized members of a species swim together in orderly and peaceable ways, is widespread among marine and freshwater teleosts. Perhaps the individuals of one out of every four species schools during some part of their life. The first part of Chapter 12 is given to this prominent side of fish behaviour. The rest is concerned with other kinds of association: symbiotic (living together to mutual advantage), commensal (living together to the advantage of one and without serious harm to the other) and inquiline (lodging in some bodily space of another organism without doing harm to it).

The last chapter (13) in the section on life from day to day is concerned with wider behavioural aspects. Much of this chapter, which deals with the courting and mating 'language' of fishes, is a prelude to the section on early life-histories. Each species seems to have its own language, one that in removing specific sexual reluctances preserves the genetic identity of the species. 'Each species', to cite Ernst Mayr, 'is a delicately integrated genetic system that has been selected through many generations to fit into a definite niche in its environment. Hybridization would lead to the breakdown of the system and would result in the production of disharmonious types.' [2] Chapter 13 also contains a review of how fishes know their surroundings and make for home, or migrate to spawning grounds over long distances. Since fishes come to learn the features of their territories and home ranges, it is appropriate that the chapter should also contain a wider treatment of learning.

Early life histories

Life histories, which make up the medium time-scale of existence, last from three to fifteen years in most species of fishes. Here, after dealing with the everyday life of the adults, we need only consider early life histories; these are first described in Chapter 14 on reproductive aspects. Before reviewing the ways of reproduction, some attention is given to the formation of the germ cells. The gonads also produce hormones, which, in concert with others released by the pituitary gland, are intimately involved in the reproductive life of fishes. Proper levels and rhythms of temperature and light are also concerned in guiding their breeding cycles. Chapter 14 is followed by two chapters on the early life-histories of marine and freshwater fishes. One of the main aims of these chapters is to trace how particular forms of early life-history are directed to maintaining local populations. In comparing and contrasting conditions in salt and fresh waters, a key fact is that the ocean has favoured the evolution of floating eggs. Grounded eggs are almost an invariable rule in freshwater habitats.

Living spaces

To breathe, feed and grow, a fish needs a proper share of aquatic space. Fishes, as Robert E. Coker says, '. . . cannot live on water alone, but they need it in great quantity. Why ? Not just to provide room for lots of fish; chiefly because only a large volume of water can carry adequate amounts of essential respiratory gases, along with other nutrients.' [3] The members of each local population also require ample space for mating and spawning. If the populations of a species continue to produce young that spread into favourable surroundings, this species will keep, or even extend, its holding on the earth. But conditions may be such that the holding becomes smaller. If it becomes too small for too long, the species will cease to exist.

Marine fishes are four times more diverse than those of fresh waters, but the latter easily outnumber the species of deep-sea fishes, which have the largest living space on earth. Chapter 17 is concerned with the deep-sea fauna. Most kinds of fishes live in coastal and coral seas; and in reviewing them in Chapter 18, the more constant environments of polar and tropical regions are considered before the more changeable conditions of the temperate zones. Freshwater fishes are dealt with in Chapter 19.

These chapters on living spaces are centred on two aspects: the adaptations of the fishes to their environment; and on their zoogeography. Treatment of the first aspect involves both a drawing together and an extension of ideas and facts given in the chapters on daily life and early life histories. There is still much to be learned of the zoogeography of fishes, but certain outstanding features are becoming clearer. Discussion of these emerging ideas forms the other main part of these concluding chapters on the life of fishes.

The *dramatis personae* are generally listed before the play, but names do not become characters until the last curtain falls. This is one reason for making the diversity of fishes the last chapter in this book. The classification of fishes, which seeks to express their diversity, is also quite complex. This is why much of Chapter 20 is given in small print and need only be used for reference.

The diversity of fishes sharply reflects their mastery of water. There are some twenty thousand living species, which are arranged in many families and larger groups. But Chapter 20 is meant to be more than a systematic list of fishes. Inevitably, since systematics is the study of living diversity from all possible aspects, one is led to insistent and difficult questions. How, for instance, has the teleost sort of bony fish been so much more successful than all other sorts combined? One is led to consider the size spectrum of teleosts; the evolution of buoyant eggs and bodies; their precise ways of moving; their adaptable jaws, and so forth. In fresh waters, moreover, one type of teleost has flourished greatly: two out of three species belong to the order Ostariophysi containing carps, characins and catfishes. The ordinal name alludes to the chains of small bones that connect the ears to the swimbladder, a linkage that endows the Ostariophysi with enhanced powers of hearing and many are also sound producers. Fishes of this order also seem to be unique in developing cells in the skin that secrete alarm substances. If the skin of a fish is injured, these substances diffuse into the water, and, on being scented by members of the same or related species, lead to general fright and retreat. One shrewdly suspects that very good hearing and alarm substances have helped the Ostariophysi to flourish in fresh waters.

Study of fish classification raises a third question. One teleost in three is a perch-like fish, of the orders Percomorphi and Scleroparei; and species of these two orders are almost everywhere dominant in the seas that move over the continental shelves. Is this dominance related to their mobile jaws, their spiny fin rays, adaptable fin patterns, their visual powers, or to what? In these three instances, and there are others, one cannot think of the classification of fishes without thinking of their ways of life and living spaces. We thus return to the main concern of this book.

Life from Day to Day

Chapter 2

Form and Motion (1)

We fish are upheld and supported to all sides. We lean confidently and harmoniously upon our element. We move in all dimensions and whatever course we take, the mighty waters out of reverence for our virtue change shape accordingly.

ISAK DINESEN, Seven Gothic Tales

'WHY is the fish in the water swifter than the bird in the air . . .?' Leonardo da Vinci, who spent much of his life being charmed by motion in water, followed his question by adding: '. . . it ought to be the contrary since the water is heavier and thicker than the air . . .' Today, some fifteen generations after Leonardo's times, we are well aware that machines move very much faster through air than through water. Water is, of course, heavier and thicker than air, which is why motion through the denser medium is so much more impeded. To return to the fish and the bird, a darting one-pound trout – making an effort that cannot be long sustained – travels at little more than five miles an hour. A racing pigeon of the same weight can fly for hours and with ease at ten times the speed of the fish.

This is not all. The trout, having a buoyant, gas-filled swimbladder, has no weight in water: nearly all of its swimming energy can thus be given to forward motion. Some of the effort made by the bird's flight muscles must go to keeping its altitude. And even if the fish had no swimbladder, it would still not be much heavier than its environment. To sum up: while the fish must shoulder its way through a medium that is eight hundred times as dense as air, this medium is accordingly so buoyant that it almost neutralizes the pull of gravity.

The diverse forms of fishes reflect both their ways of motion and their means of making a living. A deep-sea angler fish, who lures her prey to a trap-like mouth, has no need for the fine shape of a herring,

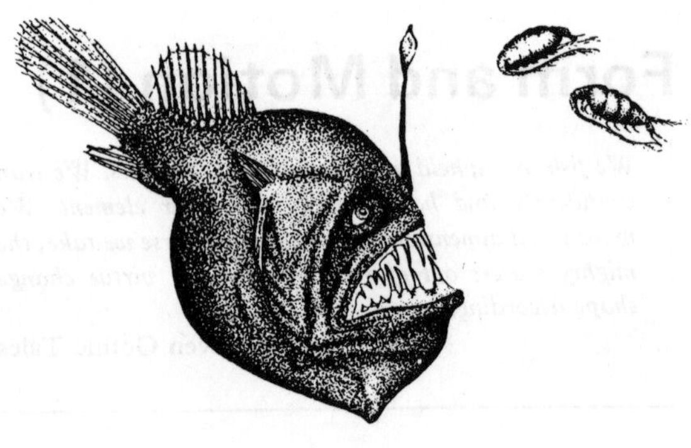

Figure 1. Body form and way of making a living. The shapely, fusiform herring swims in search of copepods (shown over its head) and other organisms. The globular female angler-fish (*Melanocetus*) probably hovers in one place and flashes her light, so luring copepods, etc.

much of whose life is spent in a restless search for its food, the plankton (figure 1). The more active the fish, the more this shape must come to terms with motion through water and approach a streamlined form. The convergence of form between two very different kinds of swift predatory fishes, a mackerel shark and a bluefin tunny, is a striking instance of the shaping influence of water, expressed, no doubt, after fruitful evolutionary experiment. Figure 2 expresses this much better than words, but certain similarities are worth stressing: (1) the compact, immensely muscular body, shaped rather like the hull of an airship; (2) the slim tail stalk or caudal peduncle, with a keel on either side; (3) the sickle-shaped or lunate tail fin.

The Mesozoic marine reptiles known as ichthyosaurs also had a lunate tail fin, and a body shape rather like that of a swordfish. Again, dolphins and porpoises, belonging to yet another group, the mammals, have twin tail flukes of lunate form, preceded by an elegant, compressed tail which merges with a compact, airship-shaped body. Regardless of whether the tail is moved from side to side, as in fishes and ichthyosaurs, or up and down, as in dolphins and other cetaceans, the tail fin has what aircraft designers would call a high aspect ratio, which can be expressed as the square of the fin-span divided by the fin area. The fin and tail drive a body that looks to be nicely shaped for motion in water.

Relatively few fishes have these special adaptations for swift motion, but most species are shaped something like a mackerel,

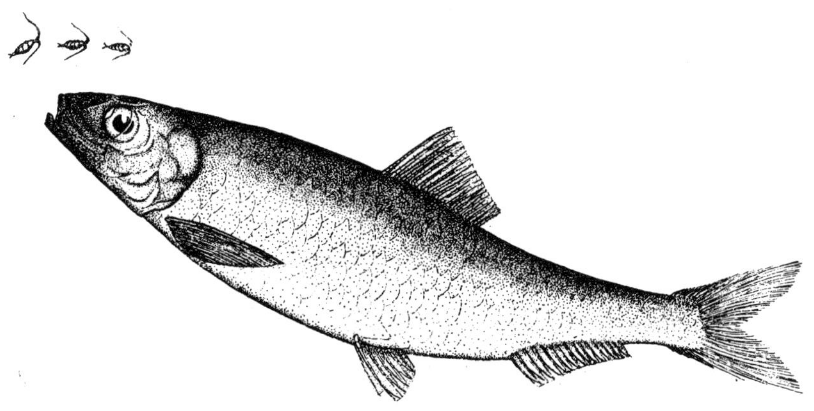

shark or tunny. Their body shapes are reminiscent of a handspinner's spindle: they are fusiform. Towards the end of the last century, a number of ship designers studied the lines of certain spindle-shaped fishes, so as to compare fish and ship hull forms. Like a ship, the fish was analyzed into two main parts; the forebody, or entrance, and the afterbody, or run, the division coming at the position of greatest cross-sectional area of the body. This transition was found to lie one third to two-fifths of the way along the span of the fish. The taper of the afterbody is thus more gradual than that of the forebody. After studying these and other ship-like features of fish form, such as the fineness coefficient, the designers found there were certain close similarities between man's design of high speed vessels and the natural design of some fishes. But this kind of convergence is not too surprising in so demanding a medium as water.

The earliest known sharks and bony fishes also had a fusiform body, which is, so to say, the thematic shape of modern jawed fishes. Even so, there has been much evolutionary scope for variation on this theme. Of all aquatic environments, such variation is best displayed in a coral reef or atoll (see also Chapter 17 on deep-sea fishes). Many reef fishes have deep bodies, much compressed from side to side, for instance file-fishes, butterfly-fishes and angel-fishes; others are elongated and slim in form, for example trumpet-fishes, cornet-fishes and pipe-fishes; and there are many serpentine species in the eel order, such as moray-eels and snake-eels. Away from the corals, on sandy bottoms, live certain kinds of flounders (Bothidae). The body is deep and compressed, but they rest, as do all flatfishes, on one side, the 'blind one' of the body. Flatheads (Platycephalidae), like many flatfishes, lie half buried in the sand, but their broad heads are depressed and splayed outwards, not compressed like those of deep-bodied fishes. Lastly, but not exhausting all metamorphoses of spindle-form in coral reef fishes, the rays, especially sting-rays, should be included. In these the extreme flattening of the body largely resides in the wing-like, muscular pectoral fins. These three

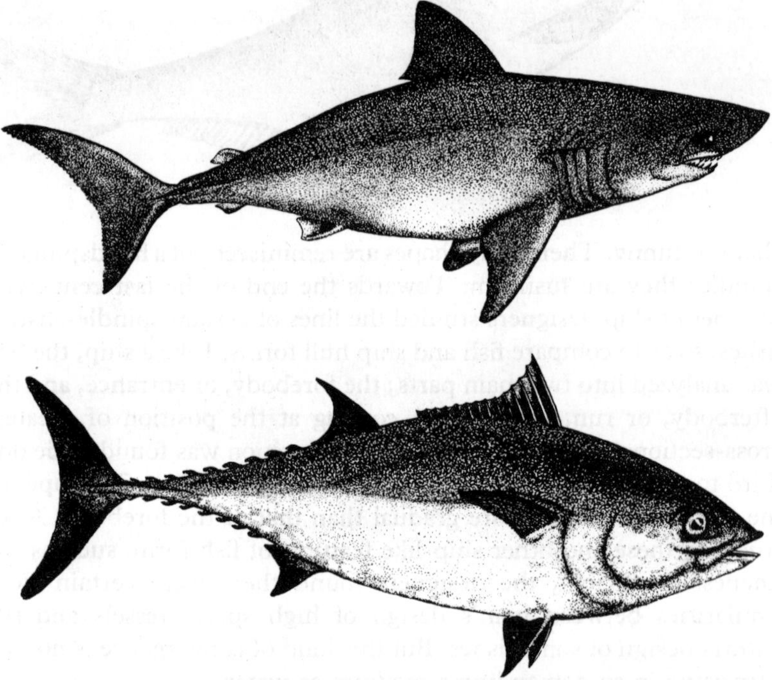

Figure 2. Convergences between two very distantly related fishes. Above: mackerel-shark (*Lamna nasus*). Below: blue-fin tunny, *Thunnus thynnus*. (After A. P. Andriashev.)

quite unlike kinds of 'flatfish' have very similar habits: they lie, as we have seen, quietly on the sand, their eyes scanning the water for prey. Here, surely, is evidence that the way of life is manifest in the forms of fishes.

A fish's swimming style is both a function of its shape and of its inner structural plan. When a spindle-shaped fish, such as a herring or mackerel, is moving at cruising speed, its head may be seen to nod from side to side, while the tail describes a much wider arc. But the movements of fishes are illusory or impossible to follow without a ciné camera, and here we owe much to the studies of Sir James Gray [4]. He has written thus:

> As observed by the human eye, the motions of various types of fish appear to vary considerably from one species to another. At one extreme is the eel, which, during motion, is characterized by distinct waves of curvature which pass alternately down each side of the body from head to tail. At the other extreme is the mackerel or trout which appears to progress by means of transverse strokes of the expanded caudal fin. An examination of successive instantaneous photographs shows, however, that the nature of these two types is essentially the same, for in all cases, waves of curvature pass along the body with increasing amplitude as the hind end of the fish is approached.

To complete this spectrum of fish waves (we shall consider their propulsive effects at a later stage) there are the tunny-like fishes relatives of the mackerel (figure 3). When a blue-fin or yellow-fin tunny or a marlin is swimming, the only part of the body that really bends is the slim tail stalk. In these fishes most of the vertebrae are

Figure 3. The 'spectrum' of body waves in swimming fishes. Left to right: tunny, mackerel, dog-fish and eel.

closely, but not rigidly, interlocked, and the body muscles are designed to transmit most of their pull to the tail stalk, so moving the hard tail fin from side to side with quick, powerful strokes. Travelling waves of curvature have been virtually eliminated.

The flexures of the fish body arise from the activity of the series of muscle blocks or myotomes that lie on either side of the backbone. These muscles are massive, which is one reason why fish make such good food. In a fairly active species, the goldfish, the weight of the muscle blocks is about two-fifths the weight of the fish. This proportion is increased to a half in the lively dace and to two-thirds in a lusty trout. In the tunny-like fishes nearly three-quarters of the weight is sheer body muscle. A thousand-pound blue-fin tunny will thus be driven by muscles weighing over 700 pounds. We can begin to see why the entire width of the North Atlantic is no barrier to this species.

When the skin is stripped from a fish's flanks, these muscles are revealed as a series of closely packed segments that pursue < shaped, or ≷ shaped courses across the body (figure 4). Each segment consists of thousands of minute muscle fibres, which run fore and aft between the tough connective tissue septa that separate one muscle segment from the next. In the jawed fishes each muscle segment is also divided into an upper and lower half by a horizontal

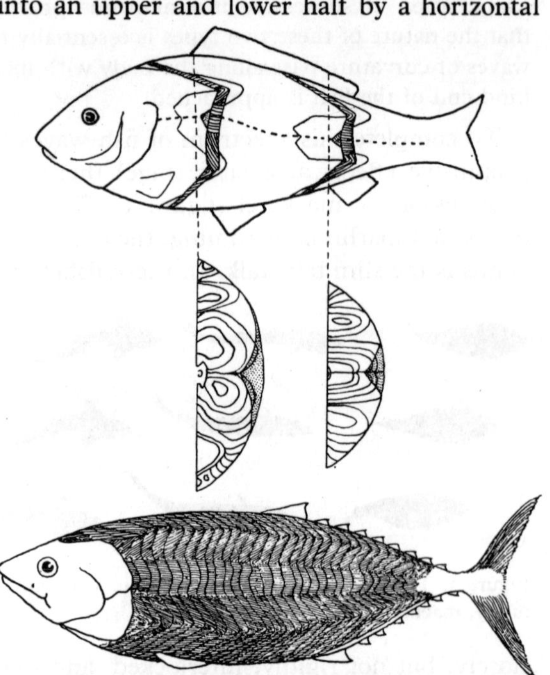

Figure 4. The muscle segments (myotomes). Above: a carp, showing the outer appearance (shaded) and inner conical extensions of two muscle segments. In cross sections of the fish the serial nesting of these conical parts appears as groups of rings. Below: the muscle segments of a skipjack (*Katsuwonus pelamis*). (After T. Kafuku.)

septum that extends down the middle line of the body. There is a close correspondence between the numbers of muscle segments and vertebrae, the two series being arranged so that the contractile muscle fibres pull across the joints between the vertebrae. In some eels there are well over a hundred vertebrae and muscle blocks: at the other extreme, the number of these elements in certain box-fishes (Ostraciontidae) is no more than fourteen.

In a shark or bony fish, the outer appearance of a muscle segment gives little clue to its entire three-dimensional form. As the segments pass inwards to their attachments on the backbone, they form four series of nested cones (figure 4). The more active the fish, the greater the development of these conical Chinese puzzles. In the mackerel sharks and tunnies in particular, the cones of the muscle segments are very pronounced. The main mechanical effect is that the contractions of any one muscle segment cover several vertebrae and are not parallel to the backbone, but oblique, both towards the head and the tail. The integral outcome is the gentle bending of the body in a concave wave that deepens as it travels from head to tail, so sweeping the tail fin to one side of the line of motion. As a series of muscles on one side of the backbone is contracting, the opposite series is relaxed, but slightly stretched, and it is along this series, of course, that waves of convexity are formed (figure 3). Then the rôles of the two sets of muscle blocks are reversed. The entire series of swimming movements is a rapid switching of waves of contraction and relaxation from one side of the body muscles to the other.

When startled, a lively kind of fish seems to do no more than deliver a few sudden sweeps of its tail to dart away from a hovering or resting position. Such a fish must be superbly adept at obtaining a purchase on its fluid surroundings. If the fish is now to swim at a steady speed, the thrust it exerts on the water must equal the resistance encountered along its path. During any interval of time, the motive energy of the fish will be equal to the product of this resistance and its swimming speed. Clearly, the lower the resistance the greater will be the speed for a given expense of energy. It is also fair to infer that the more cunning the streamlining, the lower will be the resistance and the more readily and quietly will the fish move through its environment. Here, then, are two important problems and they are by no means fully solved. How does a fish obtain its thrust? What is the nature of the resistance it encounters?

Despite their rather uniform but demanding surroundings, fishes are the most protean of all animals with a backbone. We have seen that a wide range of body forms is paralleled by an extensive spectrum of muscle wave forms, which would suggest that part of the secret of fish movement is to be sought in the elegant flexures of their bodies.

Eels are a good starting point. They have a small tail fin, yet they move through the water with seeming ease. As waves of flexure pass down the body, the rear part of each wave of convexity will exert an obliquely backward thrust against the water. Along the swimming path, the thrusts from the curves along the left side of the body must balance those from the right side. The line of motion will thus lie between these two equal and opposite sets of forces (figure 5).

Thinking of eel-like swimming, it would seem that the thrust developed by a spindle-shaped fish must come from its pliant after-body, mainly from the tail and its fin. When the tail sweeps from one side towards the fish's line of motion, the flexible caudal fin is set obliquely backward: so, to a lesser extent, is the surface of the tail itself. During any such half-stroke, any backwardly inclined surface must exert some thrust to the rear. In fact, the thrust coming from fin and body surfaces will depend directly on their area, the speed and span of their lateral movement, and on their angle of attack

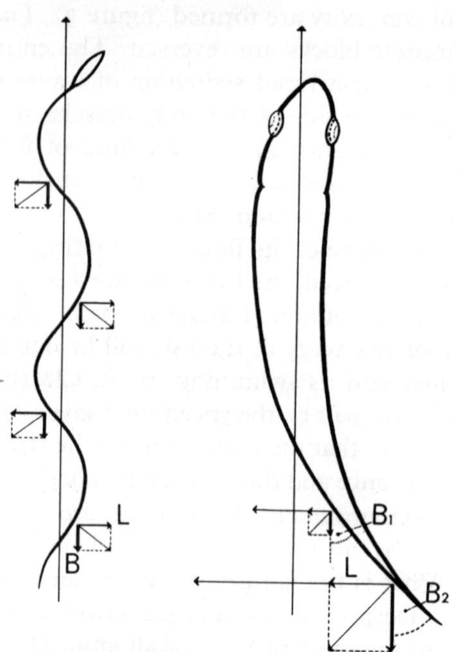

Figure 5. The propulsive forces exerted by an eel-shaped and a spindle-shaped fish. Undulations of the body, which travel down the fish and push diagonally backwards on the water, can be analyzed into lateral (L) and backward (B) components. The useful force exerted by the latter depends on: (a) the area of the moving surface, (b) its speed, (c) extent of its lateral swing and (d) its angle of attack. In a spindle-shaped fish, the tail fin, which provides most of the thrust, surpasses the more forward parts of the body in (a), (c) and (d).

(figure 5). Using these relationships, and knowing the precise movements of a swimming fish, one can estimate the proportional thrust developed by each relevant surface. In a deep-bodied fish like a bream (*Abramis*), for instance, about 45 per cent of the entire thrust comes from the tail fin. The corresponding figures for goldfish and dace are 65 per cent and 84 per cent [5]. In these three species, then, the more active and streamlined the fish, the more effective is the tail fin as a propellor.

This seems a good point to come to the other extreme of our 'spectrum of fish waves' (figure 3). The tunny-like fishes, as we saw, are driven by immense muscles that pull on a slim but strong tail

Figure 6. Fin patterns of some teleost fishes. Top to bottom: gobioid fish (*Acanthogobius*), rock-fish (*Sebastodes*), whitefish (*Coregonus*), dolphin-fish (*Coryphaena*) and marlin (*Makaira*). Note in particular the increase in the aspect ratio (fin-span^2: area) of the tail-fin in this series of fishes. The more active and speedy the fish the greater is this ratio.

stalk. There are no waves of flexure moving down the body, which must mean that virtually all the thrust is derived from the powerful beats of the inflexible, lunate tail fin. A tunny deprived of this fin would be as crippled as a destroyer without propellors.

Other fishes that move quickly in the turbulent, surface waters of the ocean – flying-fishes, rainbow-runners, amberjacks, wahoo, sword-fish, marlins and mako-sharks – also have stiff tail fins, all shaped something like a sickle. Part of the secret of speed must reside in such a fin-shape, which has a high aspect ratio [fin span^2: fin area (see also p. 21 and figure 6)]. These oceanic fishes have evidently 'discovered' the great thrust that is got when a swept-back tail fin is oscillated rapidly at small angles of attack. Such a fin is, we presume, the aquatic counterpart of a high aspect-ratio aerofoil, which when set at a small angle of attack generates a high lift force. Loss of flexibility and lessened angles of attack must be more than counterbalanced, so far as speed is concerned, by the area, shape and rapid strokes of the tail fin. But a flexible, fan-like tail fin, the area of which can be altered during its strokes, should give a fish a smoother control of its motion [5].

Many kinds of machines are now being shaped to travel through air and water with the least possible disturbance of the medium. The more the disturbance, or turbulence, the greater the resistance to movement. A perfectly streamlined body, the designers' dream, creates no disturbance as it moves, but even so, there must still be a friction between the medium and the surface of the body. The most artfully fashioned forms must always be subject to the viscous nature of water or air, and fishes are not exempt from this rule.

A mackerel has an elegant, muscular body with smoothly flowing lines (figure 66). It simply has to be better streamlined than the gravely moving lumpsucker, but how does it compare with the cigar-shaped form of a flying-fish (figure 13)? This is a difficult question, the main reason being that little is known of the flow of water around fishes in motion. Rhythmic flexures of a fish body are easier to film than the elusive, changing patterns of streamlines and turbulence that impinge on the skin.

Over fifty years ago, Professor W. S. Clemens, who had a special interest in the streamlining of mayfly nymphs, studied the pull, or drag, of moving water on variously shaped wax models, each of the same weight [6]. The models were streamed at the end of a line leading to a delicate spring balance, which thus gave an indication, in grammes, of the pull of the current. The more the pull, the more the resistance of the model.

One set of measurements is particularly interesting, and, at first sight, surprising. When a cone was streamed so that its apex was foremost, the pull was nearly twice that measured when the base, which had rounded edges, was facing the current. This disparity of drag must be a reflection of differing patterns of flow, which are represented in figure 7 and can be explained in the following way: whereas water will flow smoothly down a cone that is pointed into a current, this smoothness will end when the base is reached. The inertia of moving water is such that it cannot round sharp bends

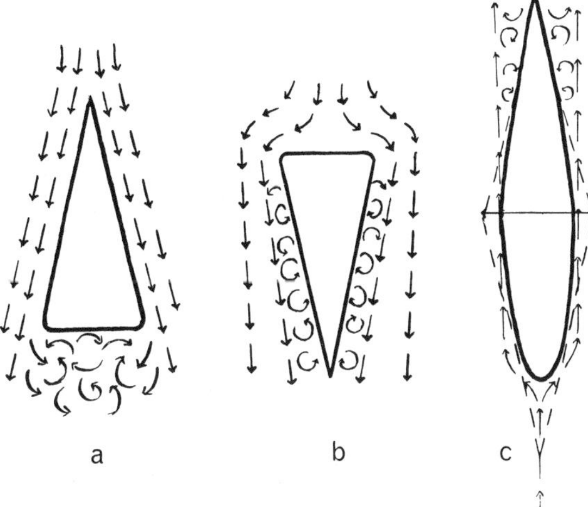

Figure 7. The eddies behind a cone in a stream (a) evidently exert more drag than the eddies along a cone that is streamed base foremost. (b) When the two cones are put together and shaped into a fusiform body, the turbulence and drag are still further reduced.

without becoming markedly disturbed. The space, as it were, behind the base of the cone will be filled in by a series of relatively large eddies, which will exert a drag to the rear. When the cone is reversed, the water will be forced to flow, without much disturbance, over the base, and after it has passed this surface, the lines of undisturbed flow will be fairly close to the tapering surface of the cone. The intervening space will be occupied by a turbulence of small eddies, which must exert less drag than the larger swirls behind the cone in the reversed position.

This interpretation is certainly over-simplified, but the experiments illustrate important aspects of streamlined form. The shape of the entering or upstream part of a body which bears the water flow up to the position of maximum cross section, is less critical than the remaining downstream section.*

A cone pointed into the current can be given a downstream section in the shape of another cone with a matching base. The more gradual the taper of the downstream cone, the more smoothly will water be led to this cone from the other. There will be the extra frictional resistance of water against the added cone, but this will be more than offset by the marked reduction of the wake and its drag. If the angular transition between the cones is now rounded off, to give the water a still smoother run, the whittled-down form will bear a fair likeness to the body of a fish (figure 7). It will recall the ship designers' study of fish form. Thinking also of muscle cones, composed of thousands of minute contractile cylinders, the muscle fibres, we may even remember Cézanne's insight: 'Tout dans la nature se modèle selon la sphère, le cône et le cylindre.'

Looking more closely at the front part of a moving body, it is easy to see why some tolerance can be given to its form. The water impinging on this part is forced to follow its lines, as may be seen by glancing along the bows of a ship. But just after the water has passed the plane of maximum cross section, there is a danger of the flow, especially if it is rapid, becoming separated from the body. If this happens, the divergent gap between the undisturbed flow and the surface of the body will be filled with swirls, all playing a part to increase the drag. Similarly, beyond the shoulders of a ship, the water hugging the hull hisses with turbulence. To keep the flow attached along the greatest possible surface, upstream and downstream sections should merge along gently curving lines, as indeed, they do in many fishes, dolphins and whales. But when they move at high speeds, some turbulence is bound to arise.

The flexible principle of upstream form also seems to apply to fishes. Considering only the more active species, which need good lines, there is a great difference between the blunt-headed flying-fishes and their relatives, the half-beaks, all with a lance-like extension of the lower jaw. There is a similar contrast between the bullet-headed tunnies and the sword-fish with its broad blade, or the marlins and their spikes (figure 8).

A considerable diversity of upstream form is also found in fishes

* In one experiment the base of the cone must be the upstream part, but this is represented by the entire taper of the cone in the reverse position, which is completely without a downstream run. Hence the large swirling wake and the larger drag.

of the horse-mackerel and amberjack family (Carangidae). In short, the tail fins of all these active fishes look more alike than their heads, which leads us to another aspect of the 'cussedness' of water.

As well as the resistance that is tied to body form, the oscillations of the tail induce a further kind of drag, one revealed in the wake. In his fine book, *This great and wide sea*, R. E. Coker [7] says that much of the problem of motion in water is bound up with the filling in of the space that is continually being vacated by a moving aquatic animal. He adds: 'Anyone who has felt the downward and backward 'drag' on the stern of a speed boat when the power is suddenly cut off can appreciate the fact that replacement of water behind is as much a feature of movement as displacement of water ahead.' But, as I have already written, very little is known of the flow of water around fishes in motion. Yet there does appear to be some connexion be-

Figure 8. Similar tail-fin shape but different forms of head. Above left: a flying-fish (*Cypsilurus*) and a half-beak (*Hemirhamphus marginatus*). Below left: a yellow-fin tunny and a sword-fish.

tween the shape of the tail fin and the drag induced by its beat. Dr D. R. Gero [8] has seen that the tips of this fin produce less turbulence than the centre portion. Thus, the more forked the fin, the less should its oscillations trouble the water. In other words, as the aspect ratio is increased the drag diminishes. Here, then, is good cause for speedy fishes to have forked, or better still, sickle-shaped tail fins. Looking back to the preceding paragraph, it also enhances our observation that the tail fins of some fishes look much more alike than their heads. Because water flows as it must, the greater hydrodynamic scope given to the head end has found expression in greater evolutionary experiment – in a flexibility that Nature must deny to a tail fin fashioned for speed.

A third source of impedance, the friction between the medium and the surface of a body in motion, has already been introduced. The water flowing over the skin of a swimming fish will not slip past with frictionless ease. Water has a certain stickiness or viscosity, a quality that tends to make it cling to the fish as a thin film, so producing a shearing stress or skin friction. There is the idea that the slimy skin of fishes enables them to slip through the water with greater facility. But Dr Gero took a plastic model of an airship, and having smeared it with mucous got from the skin of a bowfin (*Amia calva*), found that the film of slime made not the slightest difference to the rate at which the model fell through a column of water [8]. An old, and rather beguiling notion, would thus seem to be unfounded.

This more intimate, frictional tug of water on a fish must directly depend on its swimming speed and on the nature and extent of the wetted surface. The exposed area of a shark's skin is much increased and roughened by the thousands of tiny denticles that stud its surface (figure 16). Many bony fishes have well-formed, over-lapping scales, which although covered by the epidermis and a film of mucous, leave some outer imprint of their pattern. But we have little idea of the extra effect such regular roughening may have on skin friction. If this effect is quite small compared to the drag associated with the body form and the wake, which may well be true for fishes that swim at slow to moderate speeds, the hydrodynamic advantage to be got from a smooth skin is hardly worth having. But when a fish has a body built for swift motion, reduction of skin friction may be a vital factor. Certainly, the skin of the tunny-like fishes is beautifully smooth. In species like the frigate-mackerels, the skipjack and little tunny, the greater part of the body surface is scaleless, and even in the scaled shoulder region, or corselet, the scales are set well below the surface of the skin, a condition also found in the fully scaled bluefin tunnies and albacore. Adult sword-fishes have a perfectly smooth, scaleless skin. In the marlins (*Makaira*), the skin is somewhat roughened by small, denticle-like scales, but once more, most of the scales are concealed well below the epidermis.*

We have seen that the drag of water on a fish is likely to increase with its swimming speed. The precise nature of this interplay has yet to be elucidated, one simple difficulty being the lack of reliable data on the speeds attained. Quite recently, however, much has been learned, particularly from ciné films of fishes swimming in suitably designed 'fish-wheels'.

The idea of a fish-wheel is simple and ingenious [9]. The fish swims in the hollow, water-filled rim of a large Perspex wheel, which is turned at a speed equal and opposite to that of the fish. The animal

* In the skin of tunnies and marlins, oily tissues are sandwiched between layers of connective tissue. Such an arrangement may well act as a 'turbulence-damper'.

can thus be kept in one position with respect to two suitably placed ciné cameras, one giving a side view, the other a view from above. Accurate records will then be available of the fish's speed, the frequency and amplitude of its body and fin movements and of the precise way these movements are made.

These studies are being pursued at the Zoological Laboratory, Cambridge, where Richard Bainbridge is working with certain fresh-water fishes, such as dace, rainbow trout and goldfish. In general, he finds their swimming speeds to be directly related to the size of the fish and to the frequency and amplitude of the tail beats, the frequency being expressed as the number of beats per second and the amplitude as the side to side distance of sweep of the tail fin.

Some records may be cited [10]. A young dace measuring 9 centimetres (about $3\frac{1}{2}$ inches) swam at a speed of 50 centimetres per second (1·12 miles per hour) when its tail was beating ten times per second. At the same frequency of tail beat an older fish spanning 24 centimetres (about $9\frac{1}{2}$ inches) swam at over three times the speed of the younger fish. To reach this speed (170 centimetres per second or 3·8 miles per hour) the smaller fish had to swing its tail twenty-five times per second.

Such findings are to be expected, but the virtue of fish-wheel work is that precise relationships can be revealed. Above low values of tail frequency (five beats per second), Dr Bainbridge was able to equate the speed of swimming (V) to the length of the fish (L) and frequency (f) of the tail movements. The equation he got is:

$$V = \tfrac{1}{4}[L(3f-4)],$$

V being expressed in centimetres per second, L as the length of the fish in centimetres, and f the frequency of tail beats per second.

Thus, as a fish grows in size, it will move at a steadily increasing speed for a given frequency of tail beat, or, looked at in another way, to maintain a given speed, fewer beats will be required in a given time. The larger the fish, of course, the greater the body surface thrusting against the water and the wider its amplitude of sweep. In fact, the distance of sweep of the tail fin proves to be about one fifth the length of the fish.

The maximum observed speeds of fishes range from about a half to forty-four miles an hour. The lower figure was got from gobies (*Gobius minutus*) while the fish flashing through the water at 44 miles per hour was a blue-fin tunny. The second highest speed comes from a wahoo (*Acanthocybium*), a torpedo-shaped fish related to the tunnies which reaches a length of 6 feet or more. A hooked fish shot away with a hundred yards of fishing line at 41 miles per hour. These records, and a few others will be found in the following table.

TABLE I

Maximum observed speeds of fishes [11 and 12]

SPECIES	LENGTH	SPEED	
	(cm)	*cm/sec*	*m.p.h.*
Goby (*Gobius minutus*)	6–7	27	0·6
Stickleback (*Spinachia spinachia*)	10·0	72	1·75
Herring (*Clupea harengus*)	20·0	160	3·6
Sea trout (*Salmo trutta*)	20·0	240	5·4
	30·0	320	7·2
Rainbow trout (*Salmo gairdneri*)	15·0	160	3·6
Dace (*Leuciscus leuciscus*)	15·0	175	3·9
Goldfish (*Carassius auratus*)	15·0	145	3·2
Striped bass (*Roccus*)			12·0
Dolphin (*Coryphaena*)			20·0
Wahoo (*Acanthocybium*)			41·0
Blue-fin tunny (*Thunnus*)			44·0
Pike (*Esox lucius*)	16·5	210	4·8

Considering the most reliable figures, which also include the observed top speeds of fishes such as bleak, carp and barracuda, it appears that near or at their maximum speeds spindle-shaped fishes cover from nine to thirteen times their own length in a second. But a speed of ten lengths a second seems to be the most that can be sustained for a period of about a minute.*

This introduces a more meaningful aspect, the relation of speed to stamina. Considering any one fish, it is reasonable to expect that it will be most quick to tire at its maximum speed.

Comparison of one species with another is a little more complex and requires careful observation. For instance, sea trout swim faster than herring of the same size (table 1) but herring seem to have greater staying powers.

Close study has been given to speed and stamina in a second series of fish-wheel trials, again using goldfish, dace and rainbow trout. Taking fishes of various sizes, Dr Bainbridge [12] graphed the top

* Considering speed alone, this figure can hardly be applied to very large and active fishes. A 10-foot blue-fin tunny (a rare sized fish) swimming at ten times its own length in a second would be travelling at over 65 miles per hour! What of a 20-foot white shark?

speeds they could sustain for periods ranging from one to twenty seconds. Very high speeds were sustained for no more than one to five seconds. Between times of ten to twenty seconds there is much less difference in the sustainable, and much lower, speeds. In fact, each graph levels off towards the economical or cruising speed of the species concerned, a speed that can be maintained for relatively long periods. Evidently, and surprisingly, the cruising speed of young rainbow trout is less than that of medium-sized dace and goldfish.

The graphs also contain another surprise. For time intervals from five to twenty seconds, the speeds a goldfish can sustain are somewhat greater than the corresponding figures of the dace and appreciably more than those of the rainbow trout. The good performance of the goldfish is even more remarkable when it is remembered that the relative mass of muscle is least in this species. But if, as in airships, the drag due to body form is least at a fineness ratio (length of body to the mean of the maximum height and breadth) of 2·5, the goldfish comes closest to this desirable figure, one that is more nearly attained by blue-fin tunnies and skipjack. But, unlike the tunny, what the goldfish gains from body form it loses in muscular 'engine' power. In fact, when due allowance is made for both the relative amount of muscle and the fineness ratio, there seems little to choose between a dace, a goldfish and a rainbow trout. Clearly, there are other influences at work.* Are these hydrodynamical and/or physiological in nature?

A natural question, and also a complex one, concerns the relation between performance in a fish-wheel and activity in nature. Rainbow trout and dace or dart will often snap up their prey during a quick rush, an aptitude that seems to be reflected in the fish-wheel, where they swim faster than the goldfish for intervals of a second. But why should the goldfish be endowed with the best longer-term stamina? A dace may spend much of its life breasting currents and a rainbow trout is apt to leave its lively stream and make its way to the sea. However, closer thought can only point to a simple conclusion: we know more about the ways of a fish in a fish-wheel than we do in nature. Will all be clear when we have a more intimate knowledge of the lives of rainbow trout, dace and goldfish?

* There is some connexion between the staying powers of a fish and the amount of red muscle in its muscle-segments or myotomes. Fishes such as pike, which stalk their prey and seize it in a swift dart, have little red muscle compared to species that are more continuously active. Tunnies, for instance, have more red muscle than most other kinds of fishes. Red muscle fibres, which contain the oxygen-catching pigment, myoglobin, are smaller and hold more fatty material than the white fibres.

At all events, trial by fish-wheel leads to valuable insights, the first being that the ability of fishes to sustain high speeds is less than we were once inclined to suppose. In turn, this realization sharpens our appreciation of cruising speeds, which are calm ways of moving when motive power is held in reserve and muscular fuel is used no faster than it can be supplied.

Keeping enough power in reserve can mean life rather than death. While swift and unceasing motion may well be the lot of blue-fin tunnies and their kind, quick acceleration rather than sustained swiftness is likely to be the vital aptitude of most kinds of fishes. A quick, evasive dart may keep a fish from the jaws of a predator, but not always. Sometimes the swift rush of a shark, a pike or barracuda will end in a good meal. Here are some figures for both prey and predators. Starting from rest, species such as rainbow trout, pike, carp, rudd and dace (size range $5\frac{1}{2}$ to 9 inches) covered about 2 inches in one-twentieth of a second, which is an acceleration of over 40 yards per second per second [13].

Returning to cruising speeds, those with the time to 'stand (or float) and stare' know well enough that many of the movements of fishes are calm and unhurried. Fishermen of all kinds also know that there are daily and seasonal rhythms in their activities. The drifter-man not only waits for the seasonal arrival of herring in his waters, but his nets must be shot in readiness for the evening 'swim'. Anglers become more alert with the approach of times when fish are usually on the take. In a coral reef many kinds of fishes, such as various wrasses and parrot-fishes, are active by day and rest at night. Towards nightfall, moray-eels and squirrel-fishes emerge from their daytime coral shelters and begin to hunt for their food. At much the same time there is a surge of activity in the deep ocean, when many kinds of small luminous fishes leave the deep layers that are their living space by day and begin a strenuous climb to the plankton-richer surface waters. Here they feed by night, but before daybreak they have vanished into the depths of the sea.

This is simply to say that not all of the fishes are on the move for all of the time. In fact, for some, or even most of the day, a good many are doing nothing in particular. There is every gradation between the restlessness of a tunny and the stillness of a hawk-fish, the one driven by the 'needs' of its powerful muscles, the other perched on a twig of coral, waiting for something edible to pass within reach.

Chapter 3

Form and Motion (2)

Fin patterns

The kind of life led by a fish is expressed not only in the form and flexures of its body, but also in the pattern and motions of its fins. A nest-building stickleback shows this to perfection. Fins are quite unsuited for grasping: the fish must use his jaws to build the nest. Such an activity requires a long and varied series of regulated movements of the body and this precision would be impossible without the lively and adaptable motions of an array of flexible fins. To see a stickleback at work is to be made aware of fins at the height of their evolutionary development.

Fins are as old as fishes. The earliest of the jawless fishes, the pteraspids, which were swimming from 400 to 300 million years ago, had some kind of tail fin. Some of their relatives, the cephalaspids, were better equipped. Besides a well-formed tail fin, there were one or two dorsal fins along the back; an anal fin on the underside of the tail; and, emerging from the shoulder region, was a pair of paddle-like pectoral fins.

Jaws can grasp and bite, but these ancient, feeble-mouthed fishes must have been restricted to feeding on small organisms and organic wastes. Slow, exploratory movements are best suited to such ways of making a living. Indeed, except for the anaspids, which are believed to have fed on plankton, the jawless fishes do not have the look, or the fin pattern, of active fishes and most of them were heavily armoured.

The evolution of jaws must have been the most far-reaching single event in the history of fishes. Means of capturing larger and faster moving prey had to be matched by bodies shaped and muscled for quicker motion. Movement towards the prey and an accurate snap of the jaws demand balance and control, qualities that can only be attained by the development of a suitable pattern of fins.

One major group of jawed fishes, the placoderms, was short-lived in geological time. These need not concern us, except by way of noting that some of them possessed a second set of paired fins, the pelvics, with an attachment to the underside of the abdominal region. The first known representatives of the other two major groups, the cartilaginous and bony fishes, look as though they mean business. The body is spindle-shaped and bears a full complement of well formed fins. Considering first the median members, there are one or two dorsal fins, a single anal fin, and an asymmetrical caudal fin, formed round the upturned end of the tail. The lower, fin-ray supported part of the tail fin thus comes to be much larger than the corresponding upper web. There are two sets of paired fins. The pectorals emerge low down from the shoulder region, while the pelvics are set well back from the pectorals on the abdominal underparts.

Many of the modern cartilaginous fishes, including most kinds of sharks and dog-fishes, still have such a body shape and fin pattern (figure 2). Clearly, these features have stood the test of time. But they have only persisted, in their entirety, in one small group of bony fishes; the sturgeons, spoonbill and paddle-fish (figure 83). The change – and it has significant functional implications – is in the tail fin, which in nearly all other modern bony fishes is symmetrical in form, and in some species is reduced or even absent. But this modification excepted, many of the more primitive teleost fishes (figure 84 and plates 4 and 5), such as tarpon, ten-pounders, herrings, bone-fish, milk-fish, osteoglossids, salmon, characins and carp-like fishes (cyprinoids), have much the same fin pattern of the earliest, ray-finned bony fishes. The living coelacanth *Latimeria* has a fin array (figure 81) that is closely similar to its Upper Devonian and Carboniferous relatives, but these already possessed a symmetrical, three-lobed tail fin. The modern lung-fishes also have a symmetrical *tail* (figure 82), but there is no trace of a caudal fin in the continuous median fin that fringes the downstream part of the body.

Returning to the teleost fishes, the fin patterns of the more advanced forms have diverged considerably from the more primitive herring-like condition. One striking change is in the position of the pelvic fins. In the spiny-finned fishes (Percomorphi) and others, notably the cod-like fishes (Anacanthini), these fins have moved close to the pectoral pair (figures 74 and 86, plate 24). They may be set just behind the pectorals, or just below, or they may even be in advance. At the same time, the pectoral fins have moved up the sides of the shoulders to a lateral position. Like the evolution of a symmetrical tail fin, this new setting of the pelvic fins has a far-reaching functional significance. Indeed, Professor J. E. Harris [14] has argued that the two changes are closely connected. Much of what follows in this section is due to his work and to the earlier studies of Dr C. M. Breder [15].

As a shark or a dog-fish swims, the side to side sweep of the tail fin is not only helping to drive the fish forward, but is tending also to raise the tail, thus depressing the head. This lift force comes from the large lower flap of the asymmetrical caudal fin (figure 9). But a shark or a dogfish is perfectly capable of level 'flight'. The lift due to the tail fin is, in fact, being counterbalanced by that coming from the large pectoral fins. Like the wings of an aircraft, these fins have a

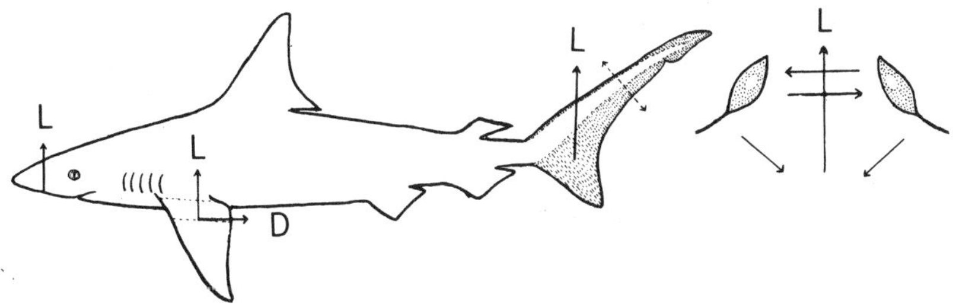

Figure 9. In moving from side to side the flexible lower lobe of a shark's tail-fin exerts a lift (L) as well as a propulsive force. Lift is also obtained from the down-warped pectoral fins and the underside of the snout. The pectoral fins also produce a drag (D).

gentle downward inclination looking backwards from the leading edges, and, just as wings give a lift, so do these fins. During level swimming, the lift coming from the pectorals must be exactly offsetting the body-tilting, head-depressing effect of the tail fin. More precisely, the pectoral and caudal lift forces are counterbalanced about the centre of gravity of the fish. The integral effect is a steady upward force acting through this centre or point of balance [16].

There is need for this lift. If a shark or dog-fish stops swimming, it will begin to sink. Being without a buoyant, gas-filled swimbladder, these fishes and their relatives are heavier than sea water. The combined effect of paravane-like pectoral fins and an asymmetrical tail fin will enable the fish to maintain its level in the water. For species that range over the deep ocean, the whale-shark, basking-shark, mackerel-sharks, thresher-shark, blue-shark, white-tip shark and so on, this facility for keeping height in the water is particularly important (the shallow-sea dog-fishes and ground sharks can rest on the sea floor). In the open ocean the best feeding for these fishes, whether plankton for a whale-shark or skipjack for a mako-shark, is in the surface layers. This is their living space and here they must stay.

Earlier in this chapter we saw that the evolution of jaws must have been the most important single event in the history of fishes. Here the implication will be that the corresponding turning point in the evolution of the bony fishes was the acquisition of a swimbladder. Considering first the tassel-finned fishes, the once abundant Dipnoi have handed on a lung-like swimbladder to their few modern representatives, the Australian, African and South American lungfishes. A calcified swimbladder is one of the most distinctive features of fossil coelacanths, but this organ is reduced in the living *Latimeria*. Among the ray-finned fishes (Actinopterygii), with the exception of the sturgeons, the few survivors of the once dominant palaeoniscoid and holostean groups – the bichirs, reed-fishes of the first group, and the garpikes and bowfin of the second group – also have a lung-like swimbladder. Such a swimbladder is found in a few primitive teleosts, but in the overwhelming majority it acts as a buoyancy device, making the fish weightless in water.

This weightless, just buoyant condition will also be attained, or nearly attained, by a fish using its swimbladder as a lung. In other words, there being little or no tendency of the body to sink, there is no necessity for fins designed to produce a lift. Paravane-like pectorals and an asymmetrical caudal fin are unthinkable on a neutrally buoyant bony fish. Both kinds of fins are free, as Professor Harris remarks [14], to evolve in other directions. We can begin to see why a symmetrical tail fin, or tail, producing a simple forward thrust, is common to nearly all the modern bony fishes.

The change from asymmetry to symmetry was not the work of a geological moment. It must have been relatively quick, if indeed it occurred, in the coelacanths, for the earliest known Upper Devonian species have a symmetrical, three-lobed tail fin. The change was also accomplished in certain lung-fishes during the 60 million years or so of the Devonian period. But in the complex of evolutionary radiations that formed the ray-finned fishes, the passage from the rather stiff-looking, asymmetrical tail fin of the earliest palaeoniscoid fishes to the elegantly symmetrical member that is found in the very earliest of the teleosts occupied more than 150 million years. We must remember, however, that most of the ray-finned fishes, which had free run of the waters until the teleosts arrived, possessed heavily armoured scales and a dense, inner skeleton besides. With such weight to carry they may well have needed some uplift from the tail and pectoral fins.

The development of versatile, nicely controllable fins is just as

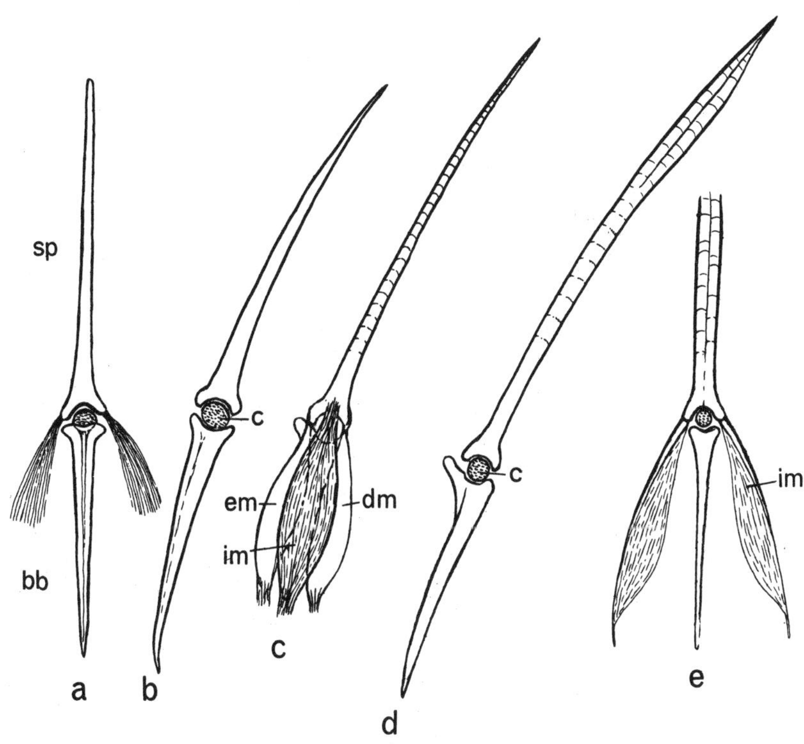

Figure 10. Fin-rays and supporting basal bones (bb) of a teleost fish. Front (a) and side (b) views of a spiny ray (sp) are shown on the left. Next (c and d) are seen a simple segmented ray and a branched segmented ray. The extreme right (e) shows part of a segmented ray in front view. Note the double nature of the segments. The base of each ray forms a ball and socket joint with a rounded cartilage (c). Inclinator muscles (im) move the rays from side to side, while elevator muscles (em) and depressor muscles (dm) raise and lower the rays.

significant a feature of teleost evolution as the change in tail-fin form. In the earlier ray-finned fishes, the rays supporting the fins are very numerous and heavily built. These fins look stiff, as indeed they are in the sturgeons. But in the teleosts the rays are not crowded together and each can be moved by a special set of muscles (figure 10).

A median fin can be spread or collapsed or undulated; movements that are also possible in the paired fins, which have a further capacity for forward and backward motion. If tail-fin symmetry became possible with the acquisition of a swimbladder, so did fin flexibility. When a fish becomes weightless in water it is at once presented with the freedom to manoeuvre in all directions, as man with an aqualung has recently discovered. But this freedom cannot be exploited without the lively and adaptable motions of an array of flexible fins. Water is never quite still: often there is a complex of currents and swirling, ever-changing motions. If, for instance, a fish is hovering over its territory, or lurking near its refuge in a coral head, all manner of precise compensatory movements of the fins are needed for its station-keeping. A hovering fish must also breathe, but the forward motion imparted by the out-flow of water from the gill chambers can be countered by the rhythmic, backing movements of the pectoral fins. Quick strokes of the pectoral fins or undulations down the dorsal and anal fins are the usual means of locomotion of many kinds of teleosts. One pectoral may be thrust out to act as a pivot, about which a fish may make a sharp turn. Lastly, many fishes use the pectoral fins as brakes, which brings us back to earlier remarks on the position of the pelvic fins. Yet before resuming this topic, we should note that none of the above fin movements is to be found in the heavier-than-water, lift-needing sharks and dogfishes.

As Dr Breder has written: 'The fins of sharks are very limited in their movements practically their whole function being that of keels and rudders. On this account, sharks are unable to make an abrupt stop, sweeping to one side of an obstacle instead, for they are unable to use the pectorals or any other members as "brakes"' [15]. What would have been the course of vertebrate evolution had the cartilaginous fishes acquired a swimbladder?

Use of the pectoral fins as brakes is strikingly displayed by many teleosts, particularly those belonging to the spiny-finned groups, such as freshwater perch and sun-fishes, sea-bass, groupers, snappers, amberjacks, mackerel and their allies, in which these fins are set high up on the sides of the shoulder region. When one of these fishes needs to pull up, the pectoral fins are simply thrust out and held on either side of the body, in which position they will obviously exert a sudden and powerful drag on the animal's motion. But the pelvic fins, which are inserted directly below, or a little ahead of, or behind, the pectorals, also play an essential part. Briefly, this is as follows.

After the pelvic fins of a sunfish (*Lepomis auritus*) were removed, Professor Harris [16, 17] found that whenever the fish used its pectoral fins as brakes, it not only began to lose speed, but also soared upwards in the water. The outspread pectorals were exerting both a drag and a lift, the latter being due to the fact that the plane of the pectorals is inclined slightly forward from the vertical axis of the fish. The function of the pelvic fins is to counteract this lift.

They are spread and moved outwards from the body and then, almost simultaneously, appropriate muscles raise the hindmost fin-rays, so making the trailing edge of the fin higher than the stiff, spine-supported leading edge. Having this inclination, which is contrary to that of a shark's pectorals, the pelvic fins of the sunfish will produce both a drag and a downward force, the latter counteracting the lift coming from the pectorals.

The positions of these fins are no less significant than their inclinations. As the pectorals are placed high on the shoulders, the combined effect of their drag and lift forces can be made to pass through the point of balance (the centre of gravity) of the fish, so eliminating pitching tendencies. The action of the pelvic fins would also make the fish pitch were they not placed below the pectorals and thus close to the centre of gravity. Here their pitching effect will be small and readily offset by slight adjustments of the pectoral fins. The entire four-fin braking action is thus not only very effective, but proceeds without loss of stability. There could be no better instance of the 'evolutionary freedom' available to fishes with a swimbladder. Symmetry of tail-fin form, flexibility of fin motion, and now paired fins acting as brakes; all these potentialities and others have been realized in fishes that are weightless in water.

Watching the elegant and easy motions of fishes, we can readily forget that steady, balanced movements are quintessential to their existence; if the motion of a fish is unsteady, a flash of silver from its flanks or disturbance of the water may easily attract a predator. Underwater swimmers, who must sometimes think of their own lives, are better aware of this fact. Here, for instance, is sound advice from two experienced underwater naturalists, Carleton Ray and Elgin Ciampi [18].

> Many fishes feed mainly by sight, and those dangerous to man are usually the faster-moving fishes, such as barracuda. They usually hunt down their prey with quick, darting strikes. Any splashing about the surface, especially in water with poor visibility, may resemble a school of

frightened fish, or a wounded fish, causing barracudas or other fishes to strike blindly.

Bright, shiny objects, such as rings or buckles look like the bright side of some fish and attract predators. When fishes are hurt they often show their light undersides.

Smooth, regular movements will attract very little attention under water, but jerky, splashy movements are known to scare most fishes away and to attract such larger predators as barracudas and sharks.

But as well as keeping quiet about its presence, steady motion is essential when a fish is taking its food. A sea-horse aiming its pipette-like snout at a hopping copepod crustacean needs perfect control. A shark must strike quickly and accurately if it is to secure a squid or a mackerel.

The paired fins, as we have seen, can control pitching. Any tendency of the body to roll or yaw, that is to veer from the line of motion, must also be checked. As seen in profile, the larger the expanse of the median fins (dorsal, anal and caudal) and the deeper the body, the less will be the inclination to roll. Rolling can also be counteracted by compensatory movements of the paired fins. Concerning yawing, Harris writes: 'Primitive sharks, with long bodies, can be stabilized by a comparatively small posterior median fin area. The shorter bony fish, and particularly those with deep bodies like coral-reef forms, require very much larger median fins to provide this automatic stability.' [14]

Fitness of fin pattern to mode of life is also evident in other types of slim-bodied predatory fishes. In gar-pikes (figure 83), pikes (figure 84), the deep-sea melanostomiatid and malacosteid fishes, needle-fishes, sticklebacks (figure 57) and barracuda, the dorsal and anal fins are opposite one another and inserted on the downstream part of the body, close, or fairly close, to the tail fin. As feathers make an arrow fly true, so these fins keep a darting fish from veering off paths that have also been judged by eyes.

Motion without body flexure

When a pike is stalking a minnow or any other seizable prey, it glides quietly through the water, moved by steady undulations of its dorsal and pectoral fins. If it needs to quicken its pace, the other fins begin waving as well. It is not until the prey is within striking distance and squarely in view of both eyes that sudden flexures of the body impel the pike forward in a quick raptorial dart.

In a surprising diversity of fishes, flexures of the body are not the main means of locomotion. These fishes are driven entirely or almost entirely by intrinsic movements of the fins. In a good many species the caudal fin is quite well developed, but oscillations of the tail are employed only when they are thoroughly alarmed.

At one extreme, the tail and its fin have become virtually useless as propellors. This is plainly evident in most kinds of rays, whose sole means of locomotion are centred in their wing-like pectoral fins (plate 4). Considering the entire ray order (Batoidea) and the first two major groups – saw-fishes and guitar-fishes – the trunk and tail are not only shark-like in form, but are flexed and sculled in a shark-like way. Undulations of the pectoral fins are at most an auxiliary means of locomotion. Sculling motions of the tail, carried out in a rather lazy fashion, also propel the electric rays, but in the skates, sting-rays, eagle-rays, cow-nosed rays and devil-rays, the tail is reduced to a simple trailing appendage (plate 3). Within the last five ray families not only are the pectoral fins very large and packed with serial muscles, which move corresponding radial cartilages, but there is a connexion between the mode of life and the way the pectoral fins are moved.

Taking first the more sedentary way of life, skates and sting-rays spend much of the time on the sea floor, either searching or lying in wait for their prey, which consists largely of molluscs, crustaceans and fishes. When these rays are on the move, undulations follow one another down the pectoral fins, the height of the waves increasing from the front to the middle of the disc, thereafter decreasing. Like the waves that move down the flanks of spindle-shaped fishes, the backwards facing parts of these travelling pectoral waves push against the water, so providing a succession of thrusts (see plate 3).

The eagle-rays, cow-nosed-rays and devil-rays lead a more active, free-swimming existence. When they are hungry the first two kinds move slowly over the sea floor, searching for molluscs and crustaceans, but the devil-rays use their satanic horns to funnel and guide small fishes and plankton into a wide and capacious mouth, where the plankton is screened by an elaborate sieve of platelets, set across the entrances to the gill slits. All these rays are capable of moving swiftly through the sea, impelled by a graceful, heron-like flapping of an elegantly-shaped pair of pectoral fins. If these species came from bottom-dwelling ancestors, not only have pectoral waves been abandoned for a more powerful way of swimming, but, correlated with the assumption of free-water flight, the eyes, like those of sharks, are set on the sides of the head – not on the top, as in sting-rays and skates.

Apart from the eels, many bony fishes have a very reduced tail fin. To select but one habitat, the deep-sea floor, here live chimaeras, halosaurs, notacanths and rat-tails, all with a long tapering tail ending without any trace of a caudal fin. Undulations down their long dorsal or anal fins or quick strokes of the pectorals may well be useful means of locomotion, but further thrust can also be provided by waves of flexure receding down their bodies.

In most kinds of tube-mouths (Solenichthyes), however, the structural plan has become so curiously transformed that body waves are either precluded or virtually useless as means of locomotion. The exceptions are the cornet-fishes (*Fistularia*), which can certainly scull with their tails. So can the trumpet-fishes (*Aulostomus*) (plate 21), but for most of the time . . .

They seem to float through the water with no body motion whatsoever. Actually the nearly transparent dorsal and anal fins are the active propulsion units, waving rapidly to and fro. If the fish has to move quickly, it does so by a strong lateral motion of the tail as do other fishes, but the trumpet-fish would rather not so exert itself. It far prefers to move on its slow course, head up or head down, backwards or forwards, surveying all with its large eyes [18].

Undulations down the dorsal and anal fins are even more the main means of propulsion in the snipe-fishes (*Macrorhamphosus*) and the bellows-fish (*Centriscops*), the bodies of which are invested by a skin that is armoured and stiffened by rough scales and bony scutes. Driven by these fins, the bellows-fish has the power to dart suddenly and quickly in any direction and it has also been seen to swim backwards with an easy graceful motion.

The functional design of the shrimp-fishes (Centriscidae) is even stranger (figure 55). There is a strait-jacket of translucent, immovable plates around a strongly compressed body, which ends not in the caudal fin but in the spinous dorsal member. This curious fin pattern is related to a down-curving of the rear part of the backbone, the tail thus being directed downwards and backwards from what seems to be the underside of the after body. The short tail and its fin separate the anal and soft dorsal fins (figure 55), the whole forming a bizarre but perfectly functional propellor unit. Bodily flexure plays no part in a shrimp-fish's life, much of which appears to be spent hovering or swimming in a head down position. To see a school of shrimp-fishes on the move, all standing on their heads, each being propelled by a fussy synchronous flickering of the soft dorsal and anal fins, is one way of realizing how remarkably adaptable is the teleost sort of bony fish.

All these tube-mouths have a relatively small tail fin: in the pipe-fishes and sea-horses (Syngnathidae) this member is either much re-

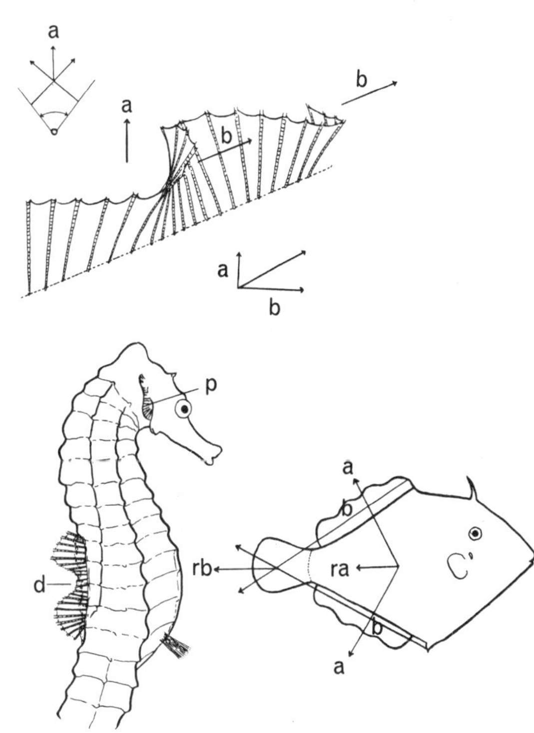

Figure 11. Above: Showing how waves down a teleost fin produce both a backward thrust (b) and an upward thrust (a). How the latter is produced is shown in the diagram on the left. The resultant force of (a) and (b) is shown in the diagram on the right. Below left: a sea-horse moving by undulations along the pectoral (p) and dorsal (d) fins. Below right: a file-fish (*Monacanthus*), showing how waves down the opposed dorsal and anal fins produce an upthrust (a) and a backthrust (b). The resultants of both (ra and rb) provide a propulsive force. (After Breder and Edgerton [19].)

duced or absent. From just behind the head to the tip of the tail, the body of a pipe-fish is encircled by a series of jointed bony rings. Such an armature is also found along the prehensile tails of sea-horses, but the more rigid trunk region, which merges with the arching, mobile head, is encased in a series of interlacing bony plates. Pipe-fishes, such as *Nerophis*, are able to twine their tails around the stems of marine plants and sea-horses are well known for the ways in which their coiling, mobile tails can fasten on to similar growths. When freely on the move, however – and pipe-fishes have been seen from bathyscaphes in the mid-waters of the deep ocean – they are driven by rapid undulations of the dorsal fin, aided by similar motions of the pectorals.

High speed camera studies of sea-horses have revealed that the individual rays of the dorsal and pectoral fins oscillate from side to side as quickly as seventy times in one second [19]. A fin in motion is thus a series of fin-ray metronomes, each slightly off beat with the next, all combining to form flickering sequences of waves that push against the sea (figure 11). While moving the fish at no more than very modest speeds, the direction of fin thrust is variable. If, for instance, the undulations of the fins are small in amplitude, the swing of the fin-rays being short, the thrust produced tends to be parallel to the base of the fin. When the arc of fin-ray swing is wide, the push is away from the fin base.

Sea-horses (plate 40) and pipe-fishes (figure 52) are usually found among seaweeds, eel-grass and turtle-grass. While moving – and being concealed – in these marine jungles, neat and varied adjustment of the swimming path is clearly more appropriate than speed. Equally, precision of manoeuvre is also needed whenever one of these fishes is following the slow but devious path of a small, planktonic food organism. The prey will not be sucked into the mouth unless the aim of the pipette-like snout is both timely and accurate.

The diversity of living spaces in coral reefs and atolls is matched by a diversity of the sizes, forms and colour patterns of coral fishes (plates 20 and 21 and figures 75 and 76). Their ways of swimming are no less diverse. Eels, squirrel-fishes, red mullets, groupers, snappers, lethrinids and many others are propelled by body flexures. Even more kinds of coral fishes depend on the pectoral, dorsal and anal fins for their motive power. Sculling motions of the tail and its fin are only employed when these fishes are really hard pressed.

Wrasses, parrot-fishes, butterfly-fishes and surgeon-fishes are driven by synchronous rowing strokes of their pectoral fins. The tail fin is simply used as a rudder and stabilizer. In an entire order of teleost fishes (Plectognathi), movement tends to stem from the activities of the opposed dorsal and anal fins. The Balistidae or trigger-fishes (figure 12) are propelled by a simultaneous side to side flapping of these fins, but in the file-fishes (Monacanthidae) the driving force, as in sea-horses and pipe-fishes, comes from waves passing along series of rays (figure 11). Side to side sculling actions of the dorsal and anal fins, aided by pectoral movements, is the preferred way of swimming of the trunk-fishes (Ostraciontidae), puffer-fishes (Tetraodontidae) and porcupine-fishes (Diodontidae) (figure 12). In the ocean-going sun-fishes (Molidae), propulsion is centred so exclusively in the synchronized side to side paddling motions of the high dorsal and anal fins that the body muscles have been virtually lost. The muscles moving the fins are correspondingly enlarged. Just behind these fin propellors, and fringing the truncated, rear part of the body, comes the curiously modified caudal fin (figure 12).

Apart from the ocean sun-fishes, the ways of swimming of the plectognaths can almost be discerned in their design. Trunk and tail are not built for flexural modes of locomotion, curvature of the spinal column being restricted by a small complement of vertebrae (fourteen to twenty). The muscle segments are also rather weakly developed, while the tail fin has a low aspect ratio. More specifically, all but the tail of a trunk-fish is encased in a firm bony box. Although not so confined, the deep bodies of trigger-fishes and file-fishes are stiff and invested with a leathery skin bearing bony scales or small spines. The porcupine-fishes and most of the puffer-fishes have a globe- or blimp-shaped body covered by a thick skin bristling with large or small spines.

All of these plectognaths and the other coral fishes considered above have a small or rather small mouth and special feeding habits calling for neatness of manoeuvre. And from time to time, or for the entire span of their adult life, they must thread their way among the thickets and labyrinths of coral. Like sea-horses and pipe-fishes, their means of movement are suited to their surroundings and ways of making a living (see also p. 326).

The plectognaths are not the only order of marine teleosts to have abandoned body flexure as a main means of propulsion. The John Dories (*Zeus*) (plate 16) and the boar-fish (*Capros aper*) (plate 41) make great use of undulations down the soft dorsal and anal fins, swimming rather like a file-fish. Tacking through the water, a John Dory will furtively stalk a smaller fish, which is snapped up by a sudden thrust of its very protrusible jaws. Other members of the John Dory order (Zeomorphi) also have opposed soft dorsal and anal fins composed of numerous rays, together with a reduced caudal fin of low aspect ratio and a rather stiff body, characters which strongly suggest that they swim like *Zeus* or *Capros* or a file-fish.

To end but by no means exhaust this review of marine fishes with other than flexural forms of locomotion, we may turn to bottom-living fishes. Sculpins and bullheads (Cottidae) and certain scorpion-fishes (Scorpaenidae) have the habit of skulking on the sea floor, but on the approach of a small fish or other suitable prey they dart and pounce, aided by quick strokes of their pectoral fins; these are also

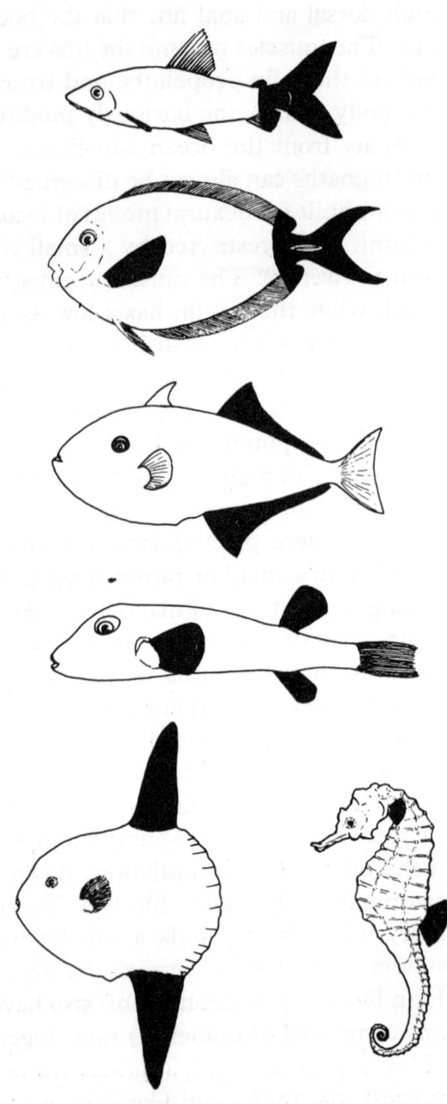

Figure 12. In a great many fishes, such as the characin *Leporinus* (top), propulsion is centred largely in the tail and tail fin, but in numerous species the tail and its fin are not the usual means of locomotion. Surgeon-fishes (second from top) cruise by rowing with their pectoral fins (black), but use the tail region if necessary. *Xanthichthys* (third from top) and other trigger-fishes swim by a side-to-side flapping of their dorsal and anal fins, while in *Sphaeroides* (fourth from top) and other puffer-fishes, these two fins and the pectorals are used as propellors. In the ocean sunfish (bottom left) the caudal region is strangely reduced and modified, motion being centred in the sculling motions of the opposed dorsal and anal fins. Sea-horses (bottom right), which have prehensile tails and a very small tail fin, swim by means of undulations along their pectoral and dorsal fins.

used as props when these fishes are resting. Many kinds of gobies use their pectorals as sculls, the tail fin simply serving as a stabilizer and rudder.

Concerning freshwater fishes, strong simultaneous beats of the large pectoral fins are the usual means of propulsion of the perch-like darters of North American rivers, the kind of motion produced matching their common name. When it is moving quietly around its habitat, the bowfin, another North American fish, is propelled by undulations of its long dorsal fin. If speed is required, the body muscles are brought into play.

Unlike the bowfin, long-bodied electric fishes depend entirely on the motions of a long median fin. Keeping the long axis of the body straight and rigid – odd behaviour for eel-like forms – the knife-fishes (Gymnotidae) and the related electric-eel (*Electrophorus electricus*) (figure 39) which live in tropical South America, are moved by undulations of a long anal fin. In *Gymnarchus niloticus* from tropical Africa, the long dorsal fin provides the propulsion (figure 41).

In the most diverse group of electric fishes, the Mormyridae, which are related to *Gymnarchus* and also live in tropical Africa, there are two main modes of swimming. Most species move by sculling motions of the tail stalk and fin, but certain species, like *Mormyrops attenuatus* and *Isichthys henryi*, swim by synchronous undulations of the opposed dorsal and anal fins. The striking and significant feature is that all electric fishes swim in a stiff-bodied way, so as to keep their electric organs in line with the long axis of the body [20]. As we shall see in Chapter 10, this posture of the body is essential for precise functioning of the electric field, a physical possession that is very much a part of their lives.

Flying-fishes

The living space of the most highly evolved flying-fishes (Exocoetidae) is in the clear blue surface waters of the tropical and subtropical ocean. When disturbed by a ship or harassed by their enemies, they leave the sea for the air, to glide rapidly and briefly over their element. It would be pleasing to think that their flight is a form of play, but how can we be sure of this? We do at least know that dolphin-fishes (*Coryphaena*) have such a lust for flying-fishes that they will often hurl themselves out of the water in pursuit of their gliding prey. With surer intent, a *Coryphaena hippurus* will follow a flying-fish by swimming swiftly beneath it, ready to seize it when it re-enters the sea. Sea birds, such as terns and noddies, also account for a great many flying-fishes.

The firm, cigar-shaped bodies of flying-fishes range in length from about 6 to 18 inches. The large, wing-like pectorals, ribbed with strong but flexible fin-rays, emerge high up on the shoulders, level with the large round eyes. In certain flying-fishes, such as *Cypsilurus*, *Danichthys* and *Prognichthys*, the pelvic fins, which are inserted well back along the abdomen, are also large and wing-like (figure 13). The tail fin is not only firmly knit and sickle-shaped, but has an extended lower lobe.

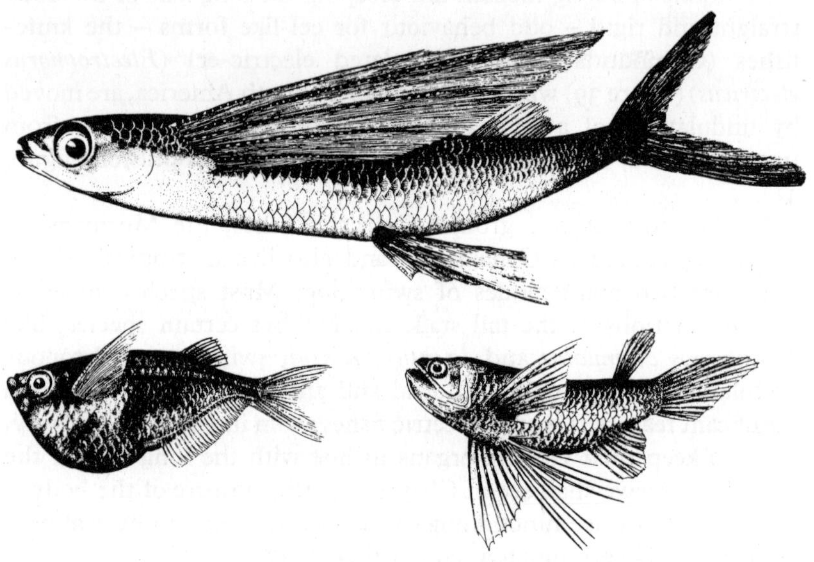

Figure 13. Three flying-fishes. Above: an oceanic flying-fish (*Cypsilurus lineatus*). (After A. F. Bruun.) Bottom left: freshwater hatchet-fish (*Carnegiella*). Bottom right: African butterfly-fish (*Pantodon buchholizi*).

Before breaking surface, a flying-fish is moving at high speed, impelled by rapid and vigorous strokes of tail and fin. After the flurry of emergence, it may be airborne in a second or less. High speed photography is clearly necessary if precise appreciation of the take-off is to be gained, and the following description is based on pictures of *Cypsilurus* species.

The fish darts out of the water at a swimming speed of about 15 to 20 miles per hour and opens its pectoral fins, the pelvic members being closed. The long axis of the body is inclined at an acute angle of about 15° to the horizontal, the head being raised, the tip of the lower and longer lobe of the tail fin trailing in the sea. The tail is now beating at about fifty times per second, the tip sculling from side to side, the fish rapidly gathering momentum and moving into the wind. When air speed is gained, the pelvic fins are unfolded, so providing a lift force that brings the body more nearly to the horizontal plane and raises the tip of the tail fin out of the water. With both pairs of wings now fully spread, the fish glides swiftly over the sea. In a few seconds air speed is lost and the fish, which may be gliding alone or in company with others, drops into the ocean. But this is not bound to happen. By closing its pelvic fins, the fish can lower the tip of its tail fin into the water, regain speed and take off for a second glide, which may be followed by a third, or fourth or even a fifth. Hundreds of observations, made by Dr Leonard Schultz and others in the Pacific, show that a single glide, lasting from one to ten seconds, is the usual mode of flight. Glides lasting from four to seventeen seconds and consisting of two take-offs were not uncommon. One fish resorted to five take-offs and was in the air for twenty seconds. The distance covered may be anything from twenty to several hundred yards, the flight being along a straight line or a curve.

While there is no way of knowing how flight was evolved in the Exocoetidae, study of their relatives, the gar-fishes (Belonidae) and the half-beaks (Hemirhamphidae) provides some inkling. According to C. M. Breder, certain gar-fishes and half-beaks

> . . . are expert at leaping and skittering over the surface of the sea, their bodies are held at an angular elevation of about 30° or even more in an extremely rigid manner, and their submerged tail is vibrated rapidly. This is usually alternated by short leaps from which they alight tail first for further activity, or may sometimes plunge head first for submersion [15].

Give a half-beak a pair of large pectoral fins and it may glide rather than leap after the take-off. Is it altogether surprising that such provision actually obtains in an Indo-Pacific species, *Euleptorhamphus viridis* ? This flying half-beak, in which the pectoral wings are about half the length of the body, has been seen to fly for a distance of about 50 yards, a flight requiring two take-offs. An Atlantic species (*Euleptorhamphus velox*) with shorter wings is said to take to the air, only to fall over on its side and then 'skip' like a flat stone.

One other group of marine fishes, the flying-gurnards (Dactylopteridae), which are equipped with a pair of very long, wing-like pectorals, are also said to be able to glide. Whatever their powers of flight, they are certainly less than those of flying-fishes, if only that they are not so active and their wings are so delicately built.

Bird-like flight, sustained by strokes of pectoral 'wings' appears to be confined to the small, 1 to 2½-inch freshwater hatchet-fishes (Gasteropelecinae) of Panama and tropical South America, and the butterfly-fish (*Pantodon*), of West African rivers (figure 13). Seen from above with their pectorals unfolded, the hatchet-fishes look very like the marine flying-fishes but their bodies are actually deep and compressed. Much of the body depth is fashioned about the large fan-shaped coracoid bones of the pectoral girdle, which are fused together to form attachments for the powerful muscles that pull the 'wings' downward and backward. Two opposing sets of muscles return the fins in readiness for the next propulsive down-beat.

Hatchet-fishes live near the surface, feeding on insects that alight or fall on the water. In all probability they fly to escape from their enemies. Naturalists have seen them skimming over the surface, the rounded ventral keel or the tail fin being immersed, the pectorals beating against the water. After take-off runs of 40 feet or more, they flew above the water for 5 or 10 feet. But in one species (*Carnegiella vesca*), as Dr Stanley Weitzman found, a take-off is not necessary. Here are his observations: 'Five specimens had been transferred from an aquarium to a half-full, wide mouth, gallon jar. As I bent over to observe these fish from above, one of them leaped from the water and hit my wrist at a point about 2½ feet directly above the water's surface. I distinctly heard a faint buzz as the fish passed a few inches from my ear. Undoubtedly the pectoral fins were in motion and probably aided the fish in its flight, although much of its initial velocity was undoubtedly gained by the leap from the water. A 2½-foot vertical jump is very high for a 1¼-inch fish, and since the fish hit my wrist with such a force that it would have gone higher, I can only conclude that the pectorals materially aided the fish in its leap. Exactly how flight is accomplished and how much true flight occurs with the aid of the pectoral fins is still unknown. The use of high speed photography might aid in determining these matters. Although it seems that these fish use their pectorals as an aid in maintaining velocity in the air they undoubtedly do not possess true self-directional flight as do insects and birds' [21].

Like this hatchet-fish, the butterfly-fish (*Pantodon buchholzi*) of West African fresh waters has marked powers of leaping out of an aquarium. This small fish, which is related to the Osteoglossidae

(p. 365), has large, fan-like pectoral fins, which may be moved up and down but not folded back against the body. The pectoral muscles, particularly those that produce the down beat, are very large, the entire arrangement suggesting that *Pantodon* is capable of true, bird-like flight [22]. But proper study of its aerial activities in nature has yet to be made.

Gliding flight in fishes is thus confined to the ocean, flapping flight to fresh waters; associations that are not likely to be co-incidental. If we knew more of the interplay of air and sea surface, gliding over the ocean, which is also a habit of flying-squids, might well be seen as a most fitting use of natural forces.

Fins as limbs

There is good structural evidence that the ancestor of the most ancient amphibians was an early kind of osteolepid (tassel-finned) fish. The first amphibians not only inherited fish jaws and some form of air-breathing swimbladder, but paired fins that would be turned into limbs. The steps in the last transmutation are unknown, but it is likely that fins had to be used as limbs out of sheer necessity. Study of Devonian sedimentary rocks indicates that some of the tassel-finned fishes lived in lakes and pools that were liable to dry up. In such arid times, certain of these fishes may well have escaped death by urging themselves over the land with the aid of their paired fins, so, perhaps, finding a larger more permanent body of water.

Having a muscular, paddle-like base, braced by a firm inner skeleton and an outer covering of scales, each of the paired fins of these early tassel-finned fishes had a certain fitness for use as a limb on land. In modern times fully formed fins of this kind are only found in the Australian lung-fish (*Neoceratodus forsteri*) and the coelacanth (*Latimeria chalumnae*), but neither fish is known to move over dry land. The lung-fish can certainly use its paired fins for resting on the bottom of a river, and the same may well be true of the coelacanth in the sea. African lung-fishes (*Protopterus*) will even use their paired fins, which are reduced to narrow appendages, for walking in a rather sprawling fashion along the bottom.

Use of the paired fins as limbs, whether underwater or on land, is by no means a unique feature of the tassel-finned type of fin. Skates (Rajidae) will move along the sea floor by kicking back with their pelvic fins, so obtaining purchase from the substratum. Numerous teleost fishes also use their pelvic fins as legs, the angler-fishes

(*Lophius*) moving along the bottom by means just like those of skates. Lizard-fishes (Synodontidae) and gurnards (Triglidae) are able to creep along the sea floor on their pelvic fins, movement in the latter fishes being aided by the lower, finger-like rays of the pectorals. Certain blennies, as Carleton Ray and Elgin Ciampi found, use their pelvic fins '. . . like feet in walking or even like hands in manipulating objects. The authors have had blennies fearlessly crawl over their bodies using these ventral fins.' [18]

Of all these fishes, including the tassel-finned groups, the paired fins look most limb-like in the angler-fishes, well called the Pediculati (Latin *pediculus* = little foot). The pectoral fins have the rays inserted

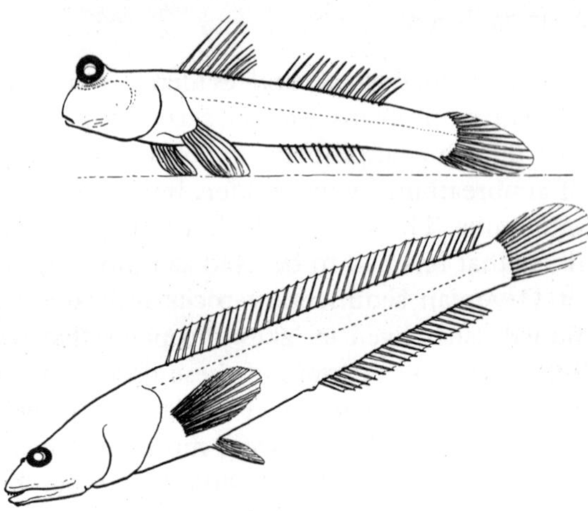

Figure 14. Two fishes that move on land. Above: African mud-skipper (*Periophthalmus papilio*). Below: Asiatic snake-head (*Ophicephalus striatus*). (Drawn from specimens in the Museum of Comparative Zoology, Harvard University.)

on lobes that may be long and narrow, the whole fin resembling an arm with a many-fingered hand. In the frog-fishes (Antennariidae) the paired fins are not only used as limbs, but come closest to being grasping organs.

Sargassum-fishes (*Histrio histrio*) (plate 14) clamber and crawl among floating jungles of sargassum weed, using their fins as hooks and legs. Here is part of an account by W. K. Gregory and G. M. Conrad of their manoeuvres: '. . . one could see the large paw-like pectoral fins being turned downward so that they could extend beneath the huge throat and clasp the fronds of seaweed, along which the fish walked by alternately moving his pectoral and pelvic "feet". At other times the pectoral paddle, by virtue of its elbow-like joint was turned upward in seeking contact with the fronds that were somewhat above the horizontal line of the fish.' [23] But a sargassum-fish can also swim with surprising speed, as I once found when trying to catch one in a dip-net.

A number of freshwater and marine teleost fishes are well able to move over land. A fish in water has little or no weight to carry, but once it 'walks' on land the out-of-water weight must be supported as soon as its body is raised off the ground. Perhaps it is not surprising that but one group of fishes, the mud-skippers (Periophthalminae) are the only really adept fin walkers. And it is hardly coincidental that most of the fishes able to stay on dry land for lengthy periods, such as the cuchia (*Amphipnous*), snake-heads (Ophicephalidae) (figure 14), climbing-perch (*Anabas*) and certain catfishes (*Clarias* and *Saccobranchus*) are able to breathe air (pp. 91–98).

Serpentine fishes, such as freshwater eels (*Anguilla*) and the Asiatic cuchia move on land, as in water, by the waves of curvature that pass down their bodies. Indian catfish are said to progress on land '. . . by lateral strokes of the tail, aided by undulatory movements of the long anal fin and rowing movements of the pectoral fins'. The Indian climbing-perch (*Anabas scandens*) has sharp spines on the lower parts of the gill covers, which are spread out alternately and fixed on to the ground whilst a push is given by the pectoral fins and the tail. Dr B. K. Das describes snake-heads as rowing on land with the pectorals, the head being raised. By such means these Indian fishes will move from pool to pool at night, and at times, no doubt, some find a new habitat when the old one has dried up or has become untenable.

Apart from the climbing perch, these fishes have thus adapted swimming movements for motion on land. Their smaller counterparts in the intertidal regions do not creep but hop, propelled by a flick of the tail. Mud-skippers will use this way of motion if seeking the safety of their burrows or the water. The tail is first flexed to one side, the stiff lower rays of the caudal fin being pressed into the mud. Then the tail is suddenly and powerfully straightened and the fish flies forwards, aided, it seems, by a lift from the pelvic fins. Most likely, the salariine blennies of tropical regions use similar means of leaping. Professor J. L. B. Smith has written that: 'Many leave the water freely, and hop about the rocks and reefs for quite long periods. On tropical reefs as one walks between the pools they jump ahead like a cloud of grasshoppers. . . .' [24] When the tide is out, the sheep's head molly miller (*Bathygobius soporator*), a goby that lives from the West Indies north to Cape Hatteras, leaps unerringly out of one pool to the next. Even if unable to see their objective, these fishes will face in the right direction with the tail curved to one side. 'Then', writes Lester Aronson, 'with a sudden snap, which was too rapid to be seen clearly, they would shoot through the surface of the water and fly through the air to the adjacent pool.' [25]

The mud-skippers, also called walking gobies (figure 14), are found over parts of the Indian and West Pacific Oceans and along the coasts of tropical West Africa. At low tide they spend much of their time moving over mud flats and mangrove swamps, feeding on insects, small crabs, molluscs and worms. Little appears to escape the notice of their independently rolling eyes, which project from a sort of optical turret.

As already stated, they 'walk' as well as hop. Each pectoral fin, which is hinged to a specially strengthened girdle, has a projecting muscular base. The lowermost pectoral and caudal rays and all the pelvic fin rays are also particularly strong, able, like the pectoral girdle, to withstand the strains and stresses of moving on land.

The ways of walking of the West African species, *Periophthalmus koelreuteri*, have been carefully studied by V. A. Harris [26]. When a fish is on the move, the strong, lower pectoral rays are bunched together to form a strut, the two fins acting like arms and crutches to impel the body forwards, the weight being taken by the pelvic fins, which act as legs but contribute little to the thrust. Using their paired fins in this way, these gobies can also climb rocks and the prop roots of mangrove trees. Climbing is particularly well displayed by those species that have retained the suctorial pelvic fins that are typical of most gobies (in the West African species, the pelvic fins have lost this kind of structure).

To end these two chapters, we may recall that each species of fish, and there are about twenty thousand, leads its own unique kind of life. This life appears in its form, fin patterns and movements. We have seen also that water imposes severe restrictions on a moving body. But the designs of fishes transcend this impedance: though living in a medium that is eight hundred times as dense as air, fishes easily rival birds in the wealth of their shapes, appendages and colour patterns.

Chapter 4

The Covering and Framework of form

THE life of a fish is bound to its form. One aspect of this may be seen through its growing life, which does not end, like that of birds and most mammals, on the attainment of sexual maturity. But through all the many changes of growth, an adult fish keeps much the same shape. Now the form of a moving fish stirs certain hydro-

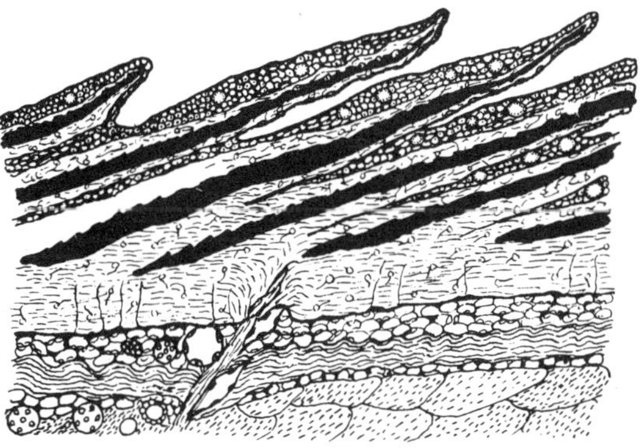

Figure 15. A section through the skin of a teleost fish, showing the overlapping scales (black). The scales lie in the dermis and are covered by epidermis. (After van Oosten [27].)

dynamic patterns, which express the individuality of its kind. An ill-shaped fish will be handicapped compared to normal individuals when it tries to catch food, avoid enemies, move from place to place or mate. Schooling species will not accept a mis-shapen fish with abnormal ways of swimming.

The form of a fish is not only bounded by the outer cells of the skin, the epidermis, but this very form and its retention largely reside in a thicker and fibrous under layer of skin, the dermis (figure 15). The scales are embedded or concealed in the dermis and in development they are wholly or largely derived from this layer. But before reviewing the dermal skeleton, which can be a living kind of plate armour, some attention must be given to the epidermis.

This outer layer, which consists of layers of cells held together by viscous material, is the surface of contact between a fish and its living space. The epidermis contains nerve endings that are sensitive to touch and it gives rise in its embryonic state to the taste buds and the lateral line system, a special set of aquatic sense organs found only in fishes and certain amphibians. Light organs (p. 173) are also formed in the epidermis, but the cells holding pigments, known as chromatophores, are scattered in the dermis. Poison glands, which usually discharge their venom by way of a spine, are another product of the epidermis [27].

The epidermis contains mucous cells (figure 15), which secrete the slippery, slimy substance well known to fishermen, fishmongers and cooks. The film of mucous protects the skin against bacterial and fungal infection and it seems to play a part in regulating the permeability of the skin to water and salts. The idea that mucous is a lubricating slime, enabling a fish to slip more easily through the water, does not appear to be true (p. 22). But as a swimming teleost flexes its body, the over-riding play between the scales on a curved part of the body may well be oiled by the mucous film, which will also protect the delicate epidermis from abrasion.

The two modern groups of jawed fishes are readily distinguished by the nature of their outer and inner skeletons. The cartilaginous fishes or Chondrichthyes – sharks, rays and chimaeras – have a gristly inner skeleton, whereas most bony fishes fully deserve their name. Moreover, the outer skeleton of sharks, rays and chimeraras is unlike its counterpart in bony fishes in that the scales have the structure of teeth, and are rightly called denticles. Teleost scales are typically thin bony plates, but the bichirs (*Polypterus*), the reed-fish (*Calamoichthys*) and the garpikes (*Lepisosteus*) have kept the ganoid type of scale that was once so characteristic of the early ray-finned fishes, the palaeoniscoids and holosteans. The teleosts have lost the hard, glassy substance known as ganoin that capped the scales of their holostean ancestors.

Turning to the tassel-finned bony fishes, the scales of the living lung-fishes also lack the hard dentine-like layer or cosmine and the underlying vascular bone that covered the scales of their early fossil relatives. The scales of the ancient coelacanths were quite thin, the cosmine layer being restricted to fine ridges or tubercles. *Latimeria*, an arch conservative, has kept this kind of scale.

Each denticle of a cartilaginous fish consists of an expanded calcified base, held firmly by the fibres of the dermis, and a backwardly curving spine or tubercle that emerges from the skin (see figure 16). It is this outer part that is like a tooth, for it is composed of dentine and capped with enamel. In most sharks the denticles are very small and they invest the entire body in a series of criss-crossing rows (figure 16), this whole lattice-work of scales and the thick fibrous dermis forming a flexible armour. The pattern of scales is seen to pleasing effect when shark skin is prepared as shagreen for ornamental purposes.

Saw-fishes (Pristidae) and guitar-fishes (Rhinobatidae), rays that swim like sharks, have a similar complete covering of denticles, but

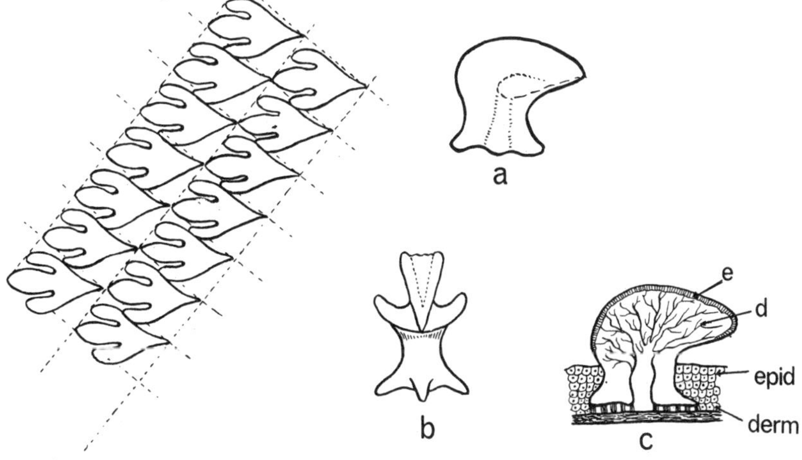

Figure 16. Left: A small part of the lattice-work of scales on a spiny dogfish (*Squalus cubensis*). A single denticle is seen from the side (a); from the front (b); and in longitudinal section (c). e, enamel; d, dentine; epid, epidermis; derm, dermis.

the species that move by undulations or beats of muscular, wing-like, pectoral fins have either a naked skin or one with a very incomplete armature (plate 3). In the skates, the denticles are scattered sparsely, sometimes in patches, over the upper surface of the head and pectorals, but they are usually more numerous over the tail. Sting-rays (Dasyatidae) and eagle-rays (Myliobatidae) either have very few denticles, or none at all, while the devil-fishes (Mobulidae) have a naked skin.

The interplay of factors in the moulding of functional design is largely concealed in the past, but it is hard to resist the idea that the reduction or loss of armature was connected with the evolution of pectoral propulsion in the rays. Whether these fins are undulated, as in skates and sting-rays, or flapped, as in eagle-rays and devil-rays (plate 3), the partial or total loss of heavy denticles would not only lighten the load, but improve the flexibility of the wings. If the moveable, rayed parts were thickly studded with denticles, a good deal of muscular effort would be wasted in flexing the armature. The response of skin and scales might well be too sluggish for the execution of rapid movements.

As if to make up for the lack of protection over the most prominent and fleshy parts of the body, many skates are armed, particularly on the back and tail, with large thorn-like denticles. Sting-rays and eagle-rays, with even less or no armour at all, carry a serrated, dagger-like spine (perhaps a modified denticle) at the base of the tail, and the spine is grooved to receive the venom of a poison gland.

To return to sharks, their armour is visibly flexible enough for their ways of moving, though some of the larger species do appear to swim in a rather stiff, if elegant, fashion. Yet the flexures of a swimming shark are not so curved as the undulations that flow down the pectoral fins of a skate. And we have already seen that the guitar-fishes and saw-fishes, which are rays with shark-like ways of swimming, have a complete armature of denticles. If there is a linkage between the mode of locomotion and degree of armature, this is what we might expect.

The electric rays (Torpedinidae) swim by lashing a stumpy, shark-like tail and they have a completely naked skin. However, the greater part of the body is disc-like, the pectoral fins being relatively large. Yet space that might be given to pectoral muscles is taken up by a large pair of electric organs, which are themselves modified (hypobranchial) muscle fibres (figure 40). At all events, an electric ray can deter with a powerful shock, as I am well aware, having stepped on one in the Red Sea. It hardly needs protection in the form of scales.

The scale patterns of bony-fishes must also conform to their ways of swimming. To take one extreme first: eels, which have a naked skin or very small scales, are able to bend their bodies into series of elegant waves, and this is the least of their flexural capacities. If they were covered with large, overlapping scales, these movements could be hampered. But the loss of scales may be more related to their 'muscling-in' habits (p. 86), and it would certainly be wrong to conclude that eel-shaped fishes must necessarily be scaleless. The West African reed-fish (*Calamoichthys calabaricus*) looks like an eel, and is covered with a heavy armour of fairly large ganoid scales. It swims in an eel-like way and it can bend its body into the shape of a U, which means, of course, that the scale pattern is perfectly adaptable to flexure.

The wave-like flexures produced by the muscle segments of a swimming fish cover a wide spectrum, ranging from the flowing undulations of eels to the scarcely perceptible bending of a tunny's body, driven by rapid beats of a sickle-shaped tail-fin. The skin must yield to these flexures and the degree of response must be largely resident in the fibre patterns of the tough dermis.

Some of the fibres are gathered in sheaves that extend from the outer to the inner dermis, but the main tracts of fibres hug the body surface, pursuing spiral courses. While the layering of these tracts varies from one kind of fish to the next, there is one over-riding arrangement: some of the fibres follow left-handed spirals; others right-handed spirals. The two sets of fibres thus form minute lattice-works over the entire outer surface of the muscle segments (the skin covering the head will not be considered here). Even apart from the other fibres, these intersecting tracts may be packed into a thick layer, right-handed tracts sometimes alternating with left-handed ones. In fact, the skins of some fishes can be tanned to make a fine-grained leather.

More precisely, these fibre tracts are disposed around the body in left- and right-handed geodesic spirals, a geodesic, like a great circle course over the earth, being the shortest distance between any two points on a curved surface. If the skin is stripped from a fish and laid out flat, the criss-crossing fibres will tend to fall into straight lines. Indeed, over the nearly flat underside of a sole or plaice, or any other flatfish, the fibres come close to pursuing straight paths across the body.

The criss-crossing angle between the fibres, measured at corresponding points, varies from one kind of fish to another, but the salient feature is that in deep-bodied fishes, these angles are smaller than those in slimmer species (figure 17). As a fish swims, the skin with its microscopic trellis-works of fibres will thus harmoniously and pliantly conform to the muscle waves that curl down the body. A fish in its skin is rather like a sailor asleep in his hammock. The meshes of the hammock will expand and relax with all the movements of his sleeping form. But just as the stretch of a hammock is limited, so is the interplay of fibres in the skin of a living fish. The skin is pliable, but there is a limit to its 'give', which must differ from one form of fish to the next.

Scales are creations of the developing dermis, and it is hardly surprising that they too are disposed in rows that spiral round the body in left- and right-handed directions, following the geodesic courses of the fibres. Looking closely at a part of the skin, the scales of sharks

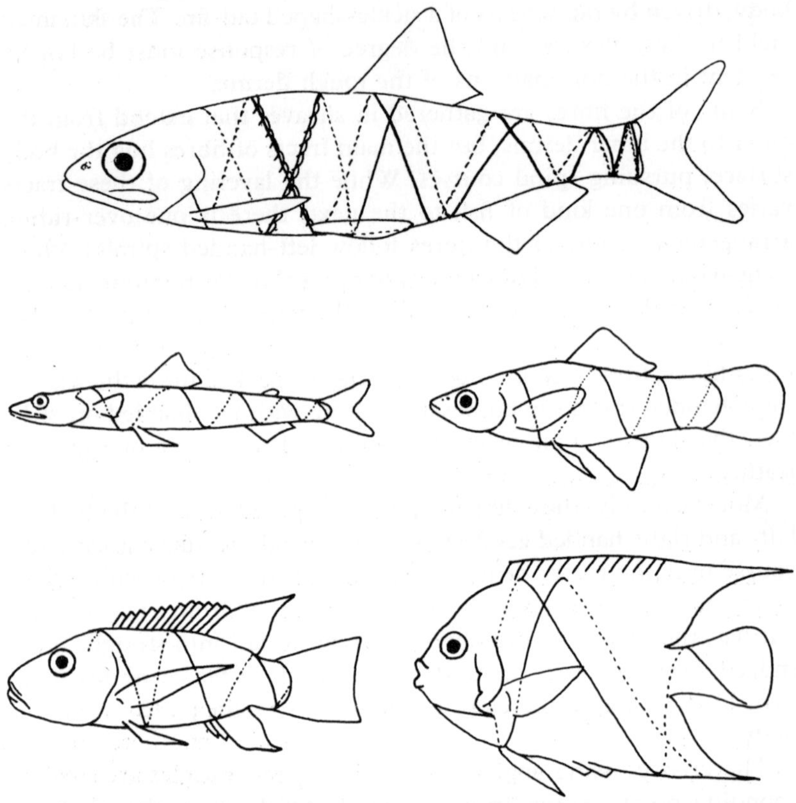

Figure 17. Geodesic scale-rows in fishes. Above: tarpon, *Megalops atlanticus*. Middle left: lizard-fish (*Synodus foetens*). Middle right: top-minnow (*Platypoecilus maculatus*). Bottom left: cichlid (*Tilapia macrocephala*). Bottom right: angel-fish (*Angelichthys isabelita*). (After Breder [28].)

and bony fishes fit into the meshes of a diamond lattice-work, as may be seen by turning to figure 16. By comparison, the peg-and-socket scale parquetry of the bichirs, reed-fish and gar-pikes is a little more complicated. Some impression of the entire spirals of scales may be got from figure 17. In the tarpon, for instance, one row will be seen to wind round the fish in a continuous course from head to tail. An intersecting row swaddles the fish in the reverse direction [28].

To visualize the entire nexus of scaly geodesics is well nigh impossible, but the effort to do so may lead to the impression that the scale patterns of fishes are not only pleasing but mechanically apt. The whole moving pattern of stresses in the skin of a swimming fish, an investment over intergrading curved surfaces, is thus harmoniously linked to pliable structural patterns of skin fibres and scales. But the skin, as we saw, has a limited flexibility and so controls the shape and hydrodynamic performance of each particular species. By themselves the overlaps between the scales of bony fishes must set a limit to a flexure. On the concave side of a curving part of the body the overlaps will increase, but there will come a point when the scales will be pressed closely together and so resist further change. On the convex side the limit will largely reside in the fibrous microlattices of the dermis.

The size of the scales is another factor. To consider only spindle-shaped fishes with fairly large scales, such as the herring and carp and their relatives, there is a gradual reduction in the size of the scales from head to tail fin. The most supple part of the body, the tail, is thus invested with smaller scales than the less mobile trunk. Flexural mobility of the fish body also depends on the structure and number of vertebrae in the backbone (p. 58).

There are myriads of curved surfaces in plants and animals, and it is pleasing to realize that geodesic spirals are widespread in nature. Thinking only of fibres, those in the skin of various kinds of worms and in the body wall of sea anemones are disposed like the dermal fibres and scales of fishes. Cellulose fibrils that strengthen the cell walls of plants may also spiral round the cell. Many more instances could be given but perhaps part of one of Pascal's *Pensées* will now be apt: 'Nature diversifies and imitates.'

The inner skeleton

The inner framework of a spindle-shaped fish – and most species have this form – looks like an arrow. The skull corresponds to the arrow-head, the backbone to the shaft and the tail fin to the feathers. The tail 'feathers' help to make a darting fish fly true, but their main use is in propulsion, and in acting as a rudder for changing course. They form the terminal driving surface of a fish and they articulate with a jointed shaft that is flexed by powerful muscle segments – a bow spring built round the arrow. The aim and intent of the 'archer' reside in the central nervous system of the fish, which receives intelligence of the outer world through the sense organs.

The inner skeleton also comprises the pectoral and pelvic girdles, with which the paired fins articulate, and the basal supports of the median fins. The fins themselves also have inner supporting rays. Lastly, there are various kinds of ribs attached to the backbone. The framework of the gill arches is considered as part of the head skeleton.

This outline applies only in part to the jawless lampreys and hagfishes, which have a simpler but special kind of skeleton, formed, except for the 'backbone' of cartilage, or gristle. This 'backbone' or notochord, is an unjointed flexible rod with a central core of fluid-filled cells, which is invested by a thick fibrous sheath and an outermost elastic coat. The median fin fringing the tail contains a rank of radial cartilages, but there are no paired fins. In the lampreys much of the head skeleton consists of an elaborate basket-work supporting the gill pouches and cartilages bracing the muscular, tooth-studded tongue. The part of the skull containing the brain is for the most part a simple trough, roofed over with connective tissues. Hagfishes have large jaw cartilages and a latticed cartilaginous framework for the nasal tube and olfactory capsule, but the gill pouches lack a branchial basket (see p. 82 for the respiration of lampreys and hagfishes).

Sharks, rays and chimaeras, the cartilaginous fishes (Chondrichthyes) also have a gristly skeleton, but in most respects it is more elaborate. Cartilage is a hard, jelly-like stuff formed of certain kinds of proteins and toughened by fibres. It is much more tensile than bone, and is also much less resistant to the complementary forces of compression. But where extra strength is needed, it is reinforced by small plates of apatite (see p. 61), which fit together rather like the pieces of a simple jigsaw puzzle. This calcification is usually confined to the outer parts of the cartilages, although the vertebrae, as we shall see, are notable exceptions.

Each vertebra of a jawed fish has a main body, the centrum, straddled by a neutral arch, through which runs the spinal cord. In the tail region there are also underlying haemal arches holding the caudal artery and vein. The centra are formed around the notochord, which is the first part of the backbone to develop. The notochord is, indeed, the first functional 'backbone' of all fishes with a free-swimming larval stage (figure 64). As the young fish grows, the notochord is invaded by cartilage-building cells and broken up into a series of cylindrical centra, which develop across the fibrous divisions between the muscle segment. At the same time the arches are formed and in the bony fishes the cartilage is later replaced by bone.

Except for the gar-pike, the vertebral centra of jawed fishes are deeply concave at each end. They are connected by short ligaments which are attached to the rims of the concavities. The space between one centrum and the next is filled with the unconstricted remains of the notochord, rudiments of which may also be found in the body of the centra. In certain sharks, the sturgeons, lung-fishes and various deep-sea teleosts, the whole or part of the notochord remains unconstricted.

The intervertebral filling, which has the consistency of a watery jelly, plays an important part during the swimming of a fish, when the backbone is being flexed by the strong contractions of the attached muscle segments. Here is an apt description of the jelly in action, written over one hundred and fifty years ago by Everard Home: 'The fluid contained in the cavity being incompressible, preserves a proper interval between the vertebrae to allow of the play of the lateral elastic ligaments and forms a ball round which the concave surfaces of the vertebrae are moved, and readily adapts itself to every change which takes place in the form of the (intervertebral) cavity.' [29]

But these fluid ball-and-socket joints will not entirely buffer the vertebrae against strains. On the concave side of a curving section of the backbone, the outer and lateral parts of the centra will be those most affected by the transmitted forces of compression. On the convex side, the ligaments will take up part of the tensile stress, but part will be conveyed to the centra, and again it will be the lateral sectors, particularly the outermost parts, that will be subject to most pull. A moment later, as a swimming undulation passes down the body, the convex part of the backbone will have become concave and the stresses will be reversed. The left and right halves of the centra must then be built to withstand alternating forces of compression and tension.

The gristly vertebrae of sharks are aptly calcified to withstand these forces, and the more active the shark, the greater the reinforcement. Some of the vertebrae of a mackerel shark (*Lamna nasus*), a very active species, are drawn in figure 18. Looking at the outer part of the vertebrae, calcified struts appear to run fore and aft along the lateral parts of the centra. But each 'strut' is actually the base of a wedge of calcification that narrows as it passes into the body of the centrum. These wedges have a radial disposition in cross section (figure 18), each reinforcement being nicely placed to withstand the stresses of tension and compression, which are mainly along fore and aft lines and decrease on moving inwards to the mid points of the centra. The wedges are also thickest and most closely set in the lateral sectors of the centra, the parts subject to most stress. It is little wonder that this radial pattern of reinforcement has been repeated again and again in all manner of sharks and rays (figure 18).

The vertebrae of sharks are also well designed in another way. The flexibility of any section of the backbone will depend on the limits of

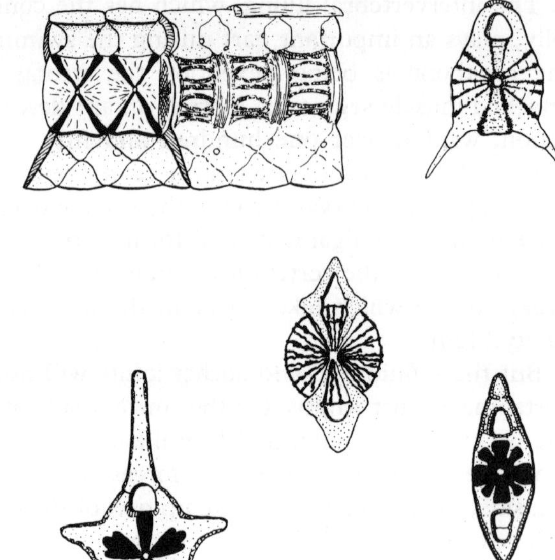

Figure 18. Above: Calcified regions (shown black) in the vertebrae of a mackerel-shark (*Lamna*). A cross section through a vertebra is seen on the right. Below: left to right, cross-section through the vertebrae of a skate (*Raja*), a white shark (*Carcharodon*) and a guitar-fish (*Rhinobatus*). (After Goodrich [30].)

interplay between the vertebrae. When a shark or a dogfish is swimming, the most mobile region of the body is that part between the pelvic and caudal fins. It is just here that the number of vertebrae are doubled, there being two to each muscle segment; this is known as diplospondyly. Each of these vertebrae is thus half the length that one might expect, the whole doubled series endowing this motor region with extra flexibility.

The tail fin of a shark differs from that of most living bony fishes in being markedly asymmetrical. The larger lobe below the upturned part of the backbone is supported by basal cartilages and fringed with numerous bristle-like rays (ceratotrichia) made of a horny material. The rays are the main supports of the small upper lobe of the tail fin (figure 19). The dorsal and anal fins are also braced by a basal skeleton of cartilages which merge with a more flexible fringing web containing horny rays (figure 19).

Again, the paired pectoral and pelvic fins of sharks have a basal framework of parallel cartilages and a distal stiffening of fine horny rays (figure 19). But these rays are small or absent in the greatly en-

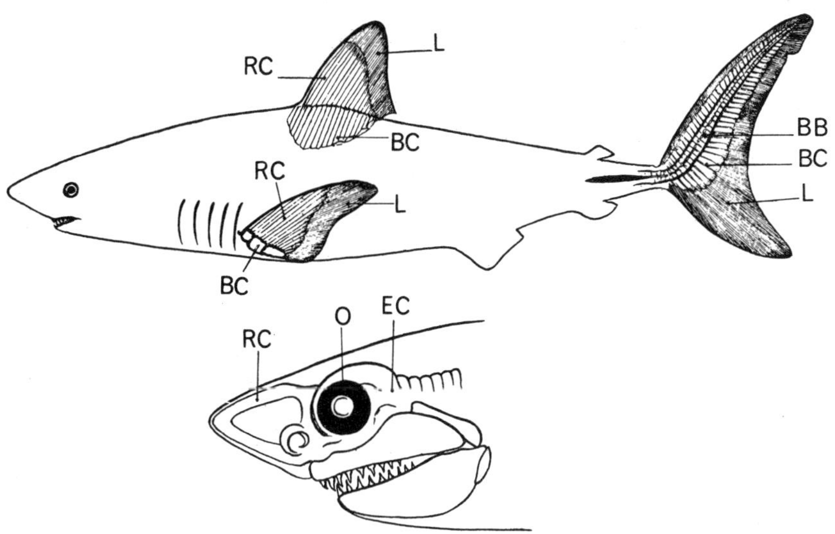

Figure 19. The fins and skull of a mako-shark (*Isurus*). The fins are supported by basal cartilages (bc), radial cartilages (rc) and thin, horny fin-rays called lepidotrichia (l). Note the upturned backbone (bb) that forms the axis of the caudal fin and the large lower lobe of this fin. In the skull are seen the rostral cartilages (rc) that support the snout, the orbit (o) and the ear-capsule (ec).

larged pectoral fins of the rays, which are built around long jointed rays of cartilage called radials that are attached to large basal cartilages running fore and aft from their articulations to the pectoral girdle (figure 20). This girdle differs from that of a shark in completely encircling the trunk and in being attached to the upper side of a strengthened, anterior section of the backbone formed of fused vertebrae. As the entire motive life of most kinds of rays is centred about their pectoral wings, a firmly held girdle and stout basal cartilages are decidedly apt. And it is significant that the cartilaginous rays are strengthened by an inner, calcified core. It is these rays that are set in motion by the pull of the powerful pectoral muscles.

The skull of a fish, like that of any vertebrate, is a housing for the brain and the sense organs concerned with smell, sight and hearing [30]. In sharks and rays, the large paired organs of smell are en-

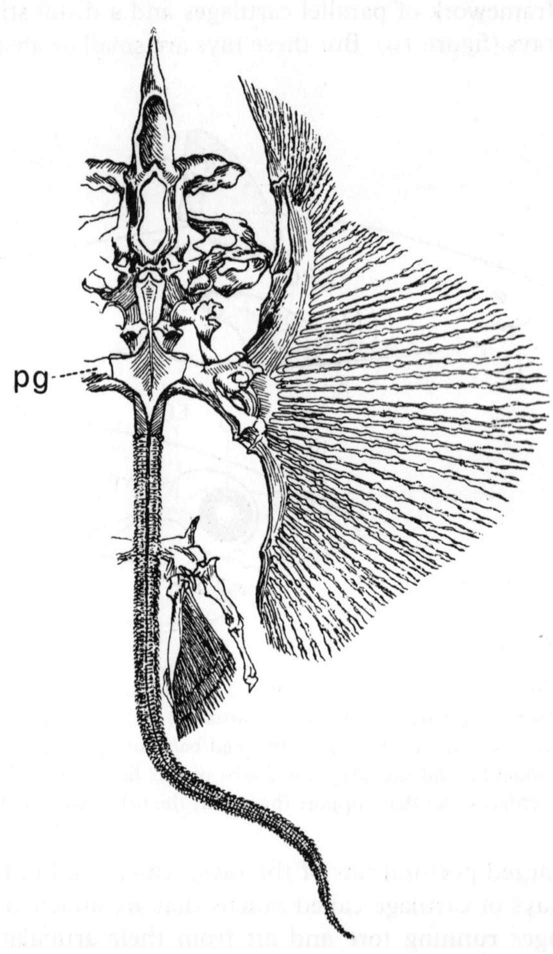

Figure 20. The skeleton of a skate (*Raja*). Note the strong attachment to the backbone of the pectoral girdle (pg) which supports the large wing-like pectoral fins.

closed in roomy capsules on each forward side of the gristly brain-box or neurocranium. The invisible parts of the eye-balls slip into bowl-shaped orbits at the sides of this cranium, which also encapsules the ears at either side of its posterior section (figure 19).

To the lower part of each auditory capsule is attached a cartilage or hyomandibular, the two forming a suspension for the tooth-bearing V-shaped upper jaw known as the palatopterygoid cartilage, which is also linked to the cranium by ligaments. In a few sharks the upper jaw articulates directly with the cranium. The lower jaw, Meckels' cartilage, articulates with the upper jaw. Both jaws are strongly reinforced by an outer calcified layer. Sharks with a powerful, shearing bite and rays that crush hard-shelled food have particularly strong jaws.

The fibrous framework of bone is hardened by dense deposits of apatite, a complex mixture of calcium phosphates and carbonates. Spicules of this mineral fit together in a way that makes bone very resistant to tension and compression. Indeed, bone can have a strength to density ratio that is more than half the comparable figure for hard steel. Bony fishes have this kind of skeleton, but the space available precludes any review of the many elaborations of bony architecture in the tassel-finned and ray-finned fishes.

The vertebrate masters of water, the teleost fishes, have greatly improved on the bony structure of the ancient ray-finned fishes, the palaeoniscoid and holostean groups. This ancient kind of bone texture is still found in some of the living relicts of these groups; the bichirs, reed-fish and gar-pikes. A skull or girdle bone of one of these fishes is made of very compact bone and it is quite thick. A teleost bone of comparable size is much lighter in weight, yet is very strong. The strength is largely focussed in radially disposed struts, regions where bony substance is concentrated to support the intervening thinner bone (figure 21). The vertebrae of teleosts also have a lighter open structure, and very often the centra are braced by fore and aft struts, like the calcifications in the vertebrae of sharks and rays.

Having this light but strong kind of bony architecture, and having lost the dense ganoin that armoured the scales of their ancestors, the teleosts have shed cumbersome weight. For the two kinds of tissues that contribute most to the weight of a fish are bone and muscle and both are conspicuously heavier than water. Now the teleost fishes have not only disposed of much skeletal ballast: they have also acquired a truly hydrostatic swimbladder, a gas-filled fibrous sac to make them weightless in water (see p. 67). The two changes may well have evolved in tandem, but at all events a light but strong framework for powerful muscles – the motors of a perfectly buoyant fish – must be part of the teleosts' freedom of the waters, part of the secret of their extraordinary success.

Fin flexibility and tail fin symmetry are also part of this secret,

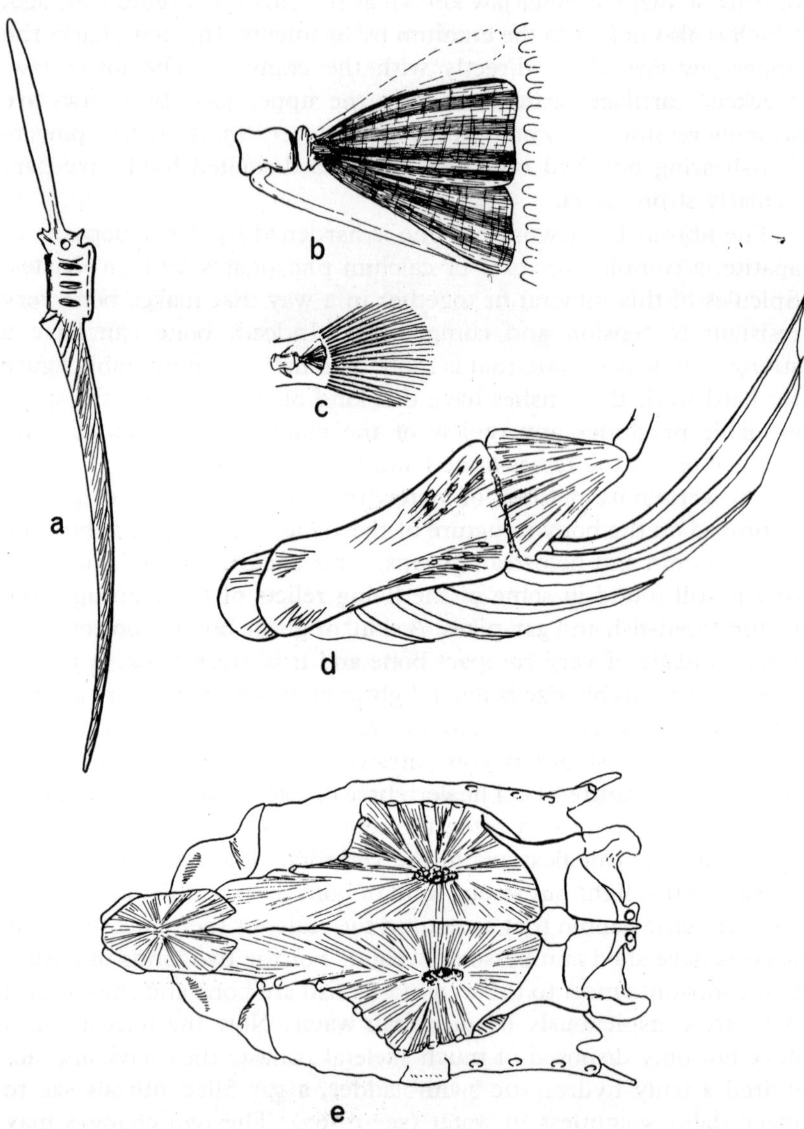

Figure 21. The light, radially strutted bones of teleost fishes. a–d Bones of a flounder *Lepidopsetta*. (a) the first caudal vertebra; (b) bones at the base of the tail fin, which is shown in (c); (d) the hyoid arch and attached branchiostegal rays; (e) top view of the skull of the salmon (*Salmo salar*), showing the radial strutting of the frontal bones.

two further refinements that are most fitting in a just buoyant fish. The fin rays of the teleost fishes are not crowded together and heavily armoured, as were those of most of the ancient ray-finned fishes. Each fin ray is moved by its own set of muscles, the whole series endowing each fin with lively and versatile motions. The muscular arrangement in a median fin is shown in figure 10, which also shows that each fin ray has its own supporting radial element. Here again the fin structure of teleosts has advanced beyond the condition found in most of the fossil ray-finned fishes in which the dorsal and anal fin-rays easily outnumbered the series of radial bones. Such lack of correspondence still exists in the sturgeons and their relatives, survivors of an ancient palaeoniscoid group. And no one can say that a sturgeon's dorsal and anal fins are capable of lively and versatile motions.

A symmetrical tail fin, which is so marked and pleasing a feature of the teleosts, was also rare in the fossil ray-finned fishes. But during the larval stage, the tail fin of a teleost passes through an asymmetrical phase (figures 66 and 67), when the lobe below the upturned end of the notochord is larger than the upper lobe that encircles the notochordal tip. The caudal fin rays develop in the lower lobe of the larval tail fin, but the finished member bears no outward trace of its asymmetrical beginning. Inwardly, however, the remnants of the upturned end of the notochord may be enclosed in a bony urostyle, a process attached to the last vertebral centrum.

Like the horny, bristle-like fin-rays of the sharks and rays, the main fin-rays of bony fishes are disposed along both sides of the fin, but for most of their length the corresponding halves fit closely together (figure 10) to form a series of bony rods.

These fin-rays also differ from those of cartilaginous fishes in being formed of a series of joints and in usually being branched at the tip. In the spiny-finned teleosts, however, some of these soft rays have lost their joints and double structure and now appear as solid bony rods with a sharp-pointed tip. Further account of these spines will be found on pp. 199–200.

The visible parts of the fins of ray-finned fishes are entirely supported by rays, which articulate with basal bones or radials that are housed under the skin of the trunk and tail (figure 22). At most these radials may project a little way into the base of the paired fins. The curious arm-like pectoral fins of angler-fishes and the muscular 'walking' pectorals of mud-skippers are conspicuous exceptions among the teleosts. In these fishes the radial bones brace the muscular, proximal parts of the pectorals and are essential to their visible mobility. Each of the paired fins of the bichirs and reed-fish has a muscular scale-covered lobe with an inner radial skeleton, though the reed-fish has no pelvic fin. But in no ray-finned fish is the basal architecture of the paired fins like that of the ancient tassel-finned fishes, still found in *Latimeria* and the Australian lung-fish. Each pectoral or pelvic fin has a muscular, scale covered lobe fringed by jointed rays and the inner radial skeleton of the lobe is so disposed

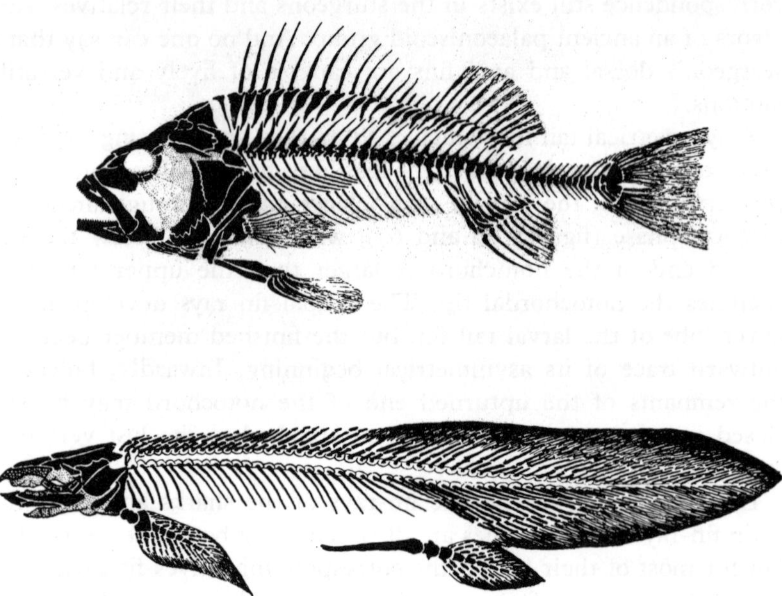

Figure 22. Skeleton of a spiny-finned teleost fish (above) and an Australian lung-fish (*Neoceratodus*).

that a single basal radial articulates with the pectoral or pelvic girdles. The entire radial skeleton usually has a central jointed axis with lateral extensions (figures 22 and 81). In many of the fossil tassel-fins and *Latimeria* the second dorsal and the anal fin also have a concentrated radial skeleton to support a muscular and visible part of the fin, though the radials are not disposed in a basal row as they are in the ray-finned fishes. These median fins, like the paired members, have a paddle-like appearance. When the Australian lung-fish is swimming quickly, which it does by undulating its body, the paired fins are held against the sides. But gentler movements may be accomplished by paddling or waving motions of the pectoral fins. Perhaps *Latimeria* uses its pectorals in the same way: it certainly has the power to rotate them about their articulations with the girdle. The longer based fins of the teleosts are more suited to undulatory motions. Even when the pectoral fins are being used as oars, they are undulated at the same time.

The skull of a tassel-finned or a ray-finned fish is a most elaborate piece of bony architecture, but the ground-plan is comparable with

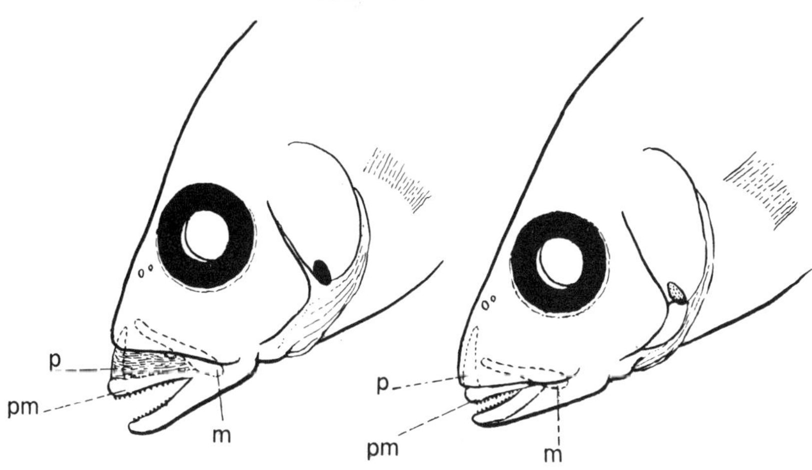

Figure 23. The protrusible jaws of a grunt (*Haemulon*). Each premaxilla (pm) has a pedicel (p) which slides against the forward part of the skull. The protrusion of the premaxillae is braced, guided and impelled by the maxillae (m). (Drawn from a specimen in the Museum of Comparative Zoology, Harvard University.)

the skull of a shark or ray. The brain-box or neurocranium is built of separate closely interlocking bones and is braced by a beam-like bone, the parasphenoid, that extends from the auditory to the nasal regions and forms the roof of the palate. In most bony fishes the lateral bones of the palate, the palatines and pterygoids, link with suspensory bones such as the hyomandibular and others that are attached to the two auditory capsules. But with certain rare exceptions, these side-palate bones, which may bear teeth, do not form the upper biting edge of the mouth as does the corresponding structure, the palatopterygoid cartilage, of the sharks and rays (figure 19).

This function is assumed by the premaxillary and maxillary bones. In the more primitive bony fishes both these bones bear teeth but in the more advanced teleosts the premaxillaries bar the maxillaries from participating in the gape. Most of these advanced species have premaxillaries that can be shot forward from their housing over the nasal part of the skull. The maxillaries brace and guide this movement in a way that can best be seen in figure 23 [31].

The acquisition of protrusile jaws must have played a great part in the overwhelming evolutionary success of the teleost fishes. Well over a half of the twenty thousand odd living species, notably the spiny-finned kinds, have such jaws. When protruded, they are particularly effective in grabbing or sucking food from the bottom and in securing prey from crevices of rock or coral. However, the sudden forward snatch of a John Dory's jaws is equally effective in taking fishes swimming at mid-water levels. As soon as the prey is seized the jaws are quickly withdrawn. But apt use of an extensible grap calls for precision of manoeuvre, which brings us once more to realize the significance of neatly controllable fins moving a neutrally buoyant body. To weightlessness in water, light but strong bony architecture, versatile fin movements and symmetry of tail-fin form, we can now add jack-in-the-box jaws, one more part of the successful history of teleost fishes.

Chapter 5

'Poissons sans Poids'

'And the fish suspending themselves so curiously below there – and the beautiful curious liquid.'
WALT WHITMAN, Assimilation

MAN with an aqua-lung enjoys a freedom that bony fishes have had for many millions of years – weightlessness in water. He has become 'l'homme sans poids'; able to hover at a particular depth with little or no effort. By carrying a buoyant breathing device, which counters the pull of gravity, a diver gains a greater freedom for underwater manoeuvre. He can begin to appreciate the nicely controlled motions of a just buoyant fish.

Like modern lung-fishes, many of the ancient tassel-fins may well have used their swimbladder as an aqua-lung, which could be recharged by swallowing air at the surface. The survivors of the earliest ray-fins: bichirs, reed-fish, gar-pikes, bowfin (but not sturgeons), also use the swimbladder as a lung, and so do a few primitive members of the last evolved and most diverse group of ray-fins, the Teleostii. But in most teleosts the primary activity of the swimbladder is to keep the weight of its possessor at the vanishing point. It is a hydrostatic organ. In some teleosts, however, this organ has acquired two other functions. By developing connexions with the ears it may act as a kind of hydrophone (p. 142), or it may be a sound-maker (p. 163). Here we shall simply be concerned with the hydrostatic use of the swimbladder in teleost fishes.

The swimbladder is a gas-filled, usually ellipsoidal, sac that is housed in the upper part of the body cavity below the backbone and kidneys (figure 24). It appears early in development as an outgrowth from the roof of the foregut, and in many of the more primitive, soft-rayed teleosts* the connexion with the foregut is retained as a pneumatic duct in the adult fish (figure 25). Such fishes have an open swimbladder. But in at least two thirds of the teleosts the swimbladder is completely closed (figure 25). The pneumatic duct is only found in the larvae of these fishes, who use it to convey swallowed air to the cavity of the developing swimbladder. Having thus inaugurated the buoyant function of the swimbladder, the duct then soon degenerates. In the jewelled cichlid (*Hemichromis bimaculatus*), however, and probably in other fishes with a closed swimbladder, the

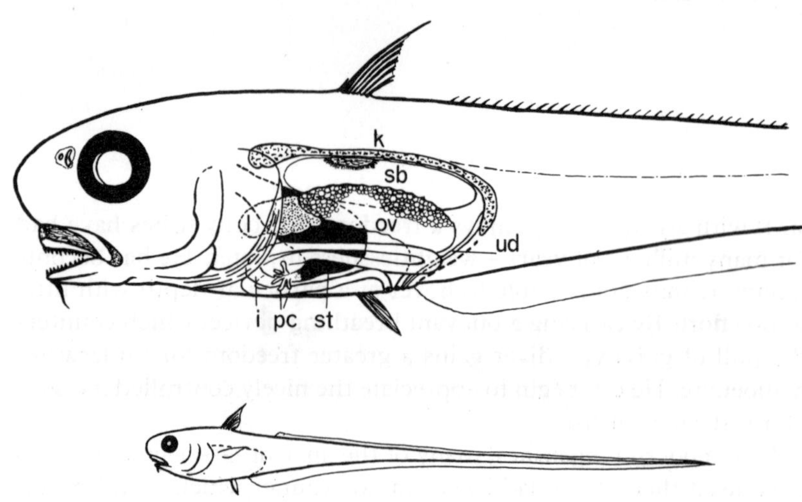

Figure 24. The inner organs of a rat-tail (*Trachonurus*), one of the deep-sea, macrourid fishes. (k) kidney; (ud) urinary duct; (sb) swimbladder; (ov) ovary; (l) liver; (st) stomach; (pc) pyloric caeca; (i) intestine. (Drawn from a specimen in the Museum of Comparative Zoology, Harvard University.)

larval fish never has a true pneumatic duct and gas is formed by large cells within the sac. Fishes with a closed swimbladder include a number of primitive deep-sea fishes such as stomiatoids and salmonoids, which are related to the herrings, and nearly all the more advanced kinds of teleosts.† But in a number of orders, such as the

* E.g. herrings, salmon-like fishes, osteoglossoids, notopteroids, mormyroids (Isospondyli); pikes (Haplomi); carp-like fishes, characins and catfishes (Ostariophysi); and eels (Apodes).

† Lantern-fishes (Iniomi); halosaurs and notacanths (Heteromi); flying-fishes, half-beaks, gar-fishes and skippers (Synentognathi); trout-perches (Salmopercae); toothed-carps (Microcyprini); sea-horses, pipe-fishes, etc. (Solenichthyes); sticklebacks (Thoracostei); cod-like fishes and rat-tailed fishes (Anacanthini); opah, oar-fishes, etc. (Allotriognathi);

flatfishes (Heterosomata) and the cling-fishes (Xenopterygii), the swimbladder is absent, an aspect that will be considered later in this chapter.

Freshwater teleost fishes have more capacious swimbladders than the marine species, which is to be expected. Sea water, with a density of about 1·026, gives more support to an underwater body than freshwater, which has a density of 1·0. A freshwater fish thus requires a larger internal float. Knowing the density of fish tissues, it is easy to calculate the proportion of the body volume that must be given to the swimbladder if the fish is to be weightless in water, that is neutrally buoyant. Let us first consider a freshwater fish. If this fish displaces 100 millilitres of water, the volume (v) occupied by the tissues can be got from the equation:

$W = \rho v$, where W is the out-of-water weight of the fish in grammes and ρ is the density of its tissues (1·076).

Thus, 100×1 (the density of freshwater) $= 1{\cdot}076\ V$.

$V = 93$ millilitres.

The remaining volume of the fish ($100 - 93 = 7$ millilitres) will thus be occupied by the swimbladder.

Similarly, the weight of a marine fish displacing 100 millilitres of sea water will be $100 \times 1{\cdot}026$ grammes.

Thus, $102{\cdot}6 = 1{\cdot}076 V$.

$V = 95{\cdot}0$ (approx). The volume of the swimbladder is $(100 - 95)$ 5 millilitres.

If the swimbladder completely eliminates the weight of a fish in water, we should expect the capacity of this organ to be about 7 per cent of the body volume in freshwater fishes and 5 per cent in marine ones. Requisite measurements of these two volumes in a series of freshwater and marine teleosts show that these percentages are matched, or nearly matched, in living fishes. That the swimbladder is doing its unseen work as a hydrostatic organ can be readily observed. A pike, a carp, a characin, a sword-tail, a sea-horse, a stickleback, a cod, a John Dory, a perch, a trigger-fish, or any fish with a properly capacious swimbladder, can suspend itself – 'so curiously below there' – with little or no effort. Fishes without a swimbladder, such as sharks and rays, mackerel, blennies and flatfishes are heavier than sea water. They soon begin to sink if they relax their efforts [32].

John Dories, etc. (Zeomorphi); squirrel-fishes, etc. (Berycomorphi); perch-like fishes (Percomorphi); trigger-fishes, file-fishes, etc. (Plectognathi); toadfishes (Haplodoci) and spiny-eels (Opisthomi).

A fish with a swimbladder is thus relieved of a continual wastage of energy. If deprived of this float, it would have to sustain an underwater weight equivalent to 5 or 7 per cent of its weight in air. While hovering, for instance, it would have to tread water with a force equal to its submerged weight. When we realize that even an active pelagic fish rarely exerts a propulsive force of more than 25 to 50 per cent of its weight in air for more than a very brief period, the force such a fish needs merely to keep its level seems quite considerable [33].

The heaviest tissues in a fish are bone and muscle, but the only

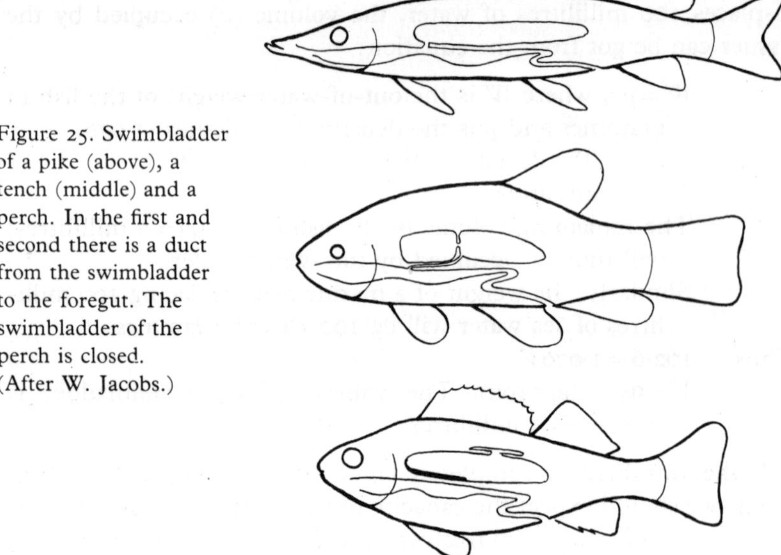

Figure 25. Swimbladder of a pike (above), a tench (middle) and a perch. In the first and second there is a duct from the swimbladder to the foregut. The swimbladder of the perch is closed. (After W. Jacobs.)

one that is lighter than water is fat. The body fluids of marine teleosts, having a lower salt content, are a little lighter than sea water. If a marine fish without a swimbladder is to be weightless in water, about a third of the weight of its tissues must consist of fat. Perhaps very oily fishes, such as the castor-oil fish (*Ruvettus pretiosus*), come close to containing this quantity of fat. At all events, certain deep-sea sharks, such as *Centroscymnus coelolepis*, contain as much as 90 per cent of the hydrocarbon, squalene, in their livers, which may make up a quarter of the total bulk of the shark. Moreover, squalene is so light – its specific gravity of 0·86 may be set against the 0·93 of cod liver oil – that these sharks must be neutrally buoyant in sea water.

Some of the fishes that swim in the mid-waters of the deep ocean either have a reduced swimbladder, which may be invested with fat, or none at all. But compared to the equally numerous species that have a highly developed swimbladder, these fishes possess poorly ossified skeletons and spare muscular systems. Measurements of the buoyancy of two such species (*Gonostoma elongatum* and *Xenodermichthys copei*) showed that they were only slightly heavier than sea water, an outcome that is almost entirely due to the paring down of bone and muscle substance [33]. In deep-sea angler-fishes and gulper-eels these two tissues are even more reduced. How all such fishes with a light chassis and low power engines manage to catch their prey will be considered in Chapter 17. But to anticipate, we may note that the angler-fishes are well named. Why hunt for food when you can fish with a luminous bait? (see p. 305).

To recapitulate, a fish will be weightless in water if the capacity of its swimbladder is about 5 or 7 per cent of its body volume, depending on whether it is a marine or freshwater species. Ideally, the swimbladder must be kept inflated at this proper volume and at a pressure equal to that of the surrounding water, which increases by one atmosphere, or 14·7 pounds per square inch, for a descent of 10 metres. A fish swimming very close to the surface will simply be subject to the pressure of the atmosphere, but if it dives to a depth of 10 metres, the pressure will increase to two atmospheres. At this depth the swimbladder will be compressed to half the volume it had at the surface. If the fish returns to the surface, the sac will regain its original volume, for the pressure will then be decreased by a half.

More generally, the swimbladder of a diving fish will be compressed by the increase in pressure, when the fish will become heavier than water. To regain buoyancy it must inflate the sac to the proper volume. If it is moving towards the surface the swimbladder will expand as the pressure decreases. The fish must now deflate the sac. To keep control of its movements it must keep control of its buoyancy.

Deflating the swimbladder

When a fish with an open swimbladder is ascending, gas expelled through the pneumatic duct issues as bubbles through the mouth or gill cavities. Fishes containing a closed swimbladder do not have this means of rapid deflation. Instead, a special expanse of the swimbladder wall houses a dense network of blood capillaries, which lie close to the contained gases, mainly nitrogen, oxygen and carbon

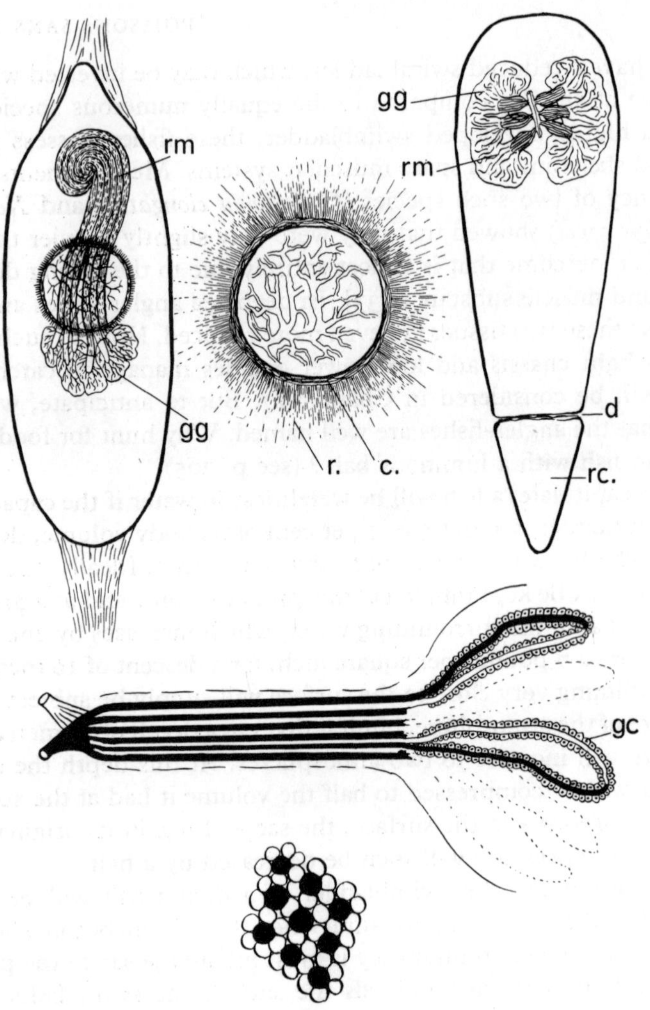

Figure 26. The swimbladder (top left) of an abyssal rat-tail (*Lionurus carapinus*), and a three-spined stickleback. In the rat-tail note the six very long *retia mirabilia* (rm), each of which supplies a lobe of the gas-gland (gg) and the circular 'oval'. The stickleback has six much shorter retia (rm) to feed the gas-gland (gg). And, whereas gas can be lost from the swimbladder of a rat-tail through the oval, which can be opened and closed by radial (r) and circular (c) muscles, the stickleback deflates its swimbladder through a rear chamber (rc), which is separated from the forward, gas-producing chamber by a diaphragm (d). Below: diagram of a *rete mirabile*, showing the close and parallel association of arterial capillaries (white) and venous capillaries (black). Gas gland cells (gc). Under this diagram are shown a few of the capillaries, as they can be seen in a cross section of the *rete*. Each capillary is about one hundredth of a millimetre in diameter.

dioxide. Any of these gases will diffuse into the network so long as its pressure within the swimbladder is greater than that in the capillary blood (carbon dioxide, being more soluble in water than oxygen, diffuses about twenty-five times as fast). Besides this difference of gas pressures, the value of which will depend on the depth, the rate of deflation of a closed swimbladder is proportional to the area and intricacy of the capillary network and the pace of its circulation.

In many fishes, such as sea-horses, pipe-fishes, sticklebacks, John Dories, various perch-like species, gurnards and toadfishes, gas absorption takes place in a rear chamber of the swimbladder, which is separated by a diaphragm from the front, gas-producing chamber (figure 26). Lantern-fishes, many cod-like fishes and rat-tailed fishes, numerous perch-like fishes and trigger-fishes have what is called an 'oval', a special thin-walled, capillary area that is usually found on the roof of the swimbladder (figure 26). Whether chamber or 'oval', these resorbent parts are so beautifully contrived that they merit further attention.

Considering first the 'oval', this is provided with circular and radial muscles (figure 26). When the radial muscles contract, the circular members being relaxed, the 'oval' is stretched out, so exposing its capillaries to the swimbladder gases. When sufficient gas has been lost the oval can be pinched off from the main swimbladder cavity by the constriction of the circular muscles. As we shall see later, the opening and closing of the oval are under nervous control.

The corresponding movements of the posterior swimbladder chamber have been nicely followed in the wrasses (Labridae). On contraction of a layer of muscle fibres, which are confined to the walls of the gas-secreting chamber, the diaphragm is pulled forward close to the gas gland, so enlarging the rear chamber. The capillary network, now being spread and dilated, can do its work, for the thin cell layer that covers it is permeable to gases. When deflation is finished, the muscles relax and the diaphragm slides back to the rear end of the sac, which is now lined by an epithelium that is impermeable to gases. The gas gland, which is now fully expanded and ready, if need be, to do its work, is simply a transmuted part of this lining. Again, these movements are regulated by nerves [34].

Inflating the swimbladder

Fishes with an open swimbladder recharge it by swallowing air at the surface, which is forced down the pneumatic duct into the sac.

In roach, tench, carp, bream, minnows and other fishes of this order (Ostariophysi), the pneumatic duct develops a swelling known as the pneumatic bulb, one function of which is to pump air into the swimbladder. Complete inflation may take an hour or two, after which the gas tension is probably in excess of the surface atmospheric pressure. In carp-like fishes, at least, the swimbladder gases are certainly maintained at a pressure slightly higher than that of the surrounding water.

Some fishes with an open swimbladder, such as whitefish (*Coregonus*), pike, carp, goldfish and eels are able to secrete gases into the swimbladder cavity. Goldfish take from five to seven days to refill an empty swimbladder, but freshwater eels require no more than twelve to twenty-four hours. In all such fishes the inner lining of the sac is modified into gland cells (the gas gland), which receive blood through a special capillary system known as a rete mirabile. Gland and retia are intimately concerned in gas production (p. 75). These two structures, as we might expect, are highly developed in eels, which need never swim to the top to recharge the swimbladder.

This gas-producing complex is also found in all fishes with a closed swimbladder. The rete mirabile, which means 'wonderful net' is not well named. It is certainly wonderful, but the capillaries do not form a network. Any rete mirabile, which would be better called a fascis mirabilis or 'wonderful bundle', consists of regular and intimate associations of arterial and venous capillaries, which follow parallel courses and carry blood to and from the gas gland (figure 26). The capillaries are so neatly packed together that a cross section of a rete looks like a mosaic or parquetry based on a pattern of hexagons (figure 26). Each capillary is about one hundredth of a millimetre in diameter, and in a large rete there can be many thousands. The quintessential feature is the close and very large surface of contact between arterial and venous capillaries.

Such contact is vital, both to build the gas pressures needed to recharge the swimbladder and to keep gases within the sac. The internal gas pressure is necessarily the same as that exerted by the water surrounding the fish, whose swimbladder should stay at a certain volume if the fish is to be properly buoyant. To restore the volume of a compressed sac, the retia and gas glands must generate pressures, mainly of oxygen and nitrogen, that are individually greater than the combined pressure of gases within the swimbladder. But the combined pressure of gases dissolved in water is equal at most to one atmosphere. Moreover, the partial pressure of oxygen, which forms most of the swimbladder gas in deep-sea fishes, is no more than one fifth of an atmosphere. Starting with this pressure, a deep-sea fish living, say, at a depth of 2,000 metres, must build a pressure of 200 atmospheres of oxygen to secrete fresh gas into its swimbladder. To maintain this pressure, the leakage of gas must be kept to a minimum.

Like the gills, which provide nearly all of the swimbladder gases, the retia mirabilia are designed, so to say, on the counter-current principle. The gills are efficient catchers of oxygen because blood and water flow in opposite directions across myriads of lamellae (p. 89). Counter-current flow in the retia, as we saw, involves two sets of parallel, contiguous capillaries. One ingoing arterial set carries blood to the gas gland; the flow in the outgoing venous set is away from the gland.

To appreciate the rôle of the retia in preventing the loss of gas, consider the deep-sea fish from a depth of 2,000 metres. The pressure of oxygen in the blood entering the arterial capillaries is about one-fifth of an atmosphere; the pressure at the beginning of the outgoing venous capillaries will be close to 200 atmospheres, the pressure of gas within the swimbladder. If there was no contact between the ingoing and the outgoing blood, the swimbladder would soon lose its oxygen. Through counter-current contact, gases can diffuse from the outgoing blood – which holds the greater concentrations of gas – to the ingoing blood. In fact, gas exchange between the two sets of capillaries is so thorough that very little gas is lost [35].

While this mechanism explains the maintenance of pressure, it will not account for the building of gas pressures. Each of these must be generated, as we saw, at a pressure greater than that within the swimbladder. But if gases can be added to those saved and transported back to the swimbladder, pressures could be built by counter-current multiplication. With each successive circulation of the blood, more and more gas could be concentrated at the gas gland end of the arterial capillaries. In this way, the pressure of any gas would eventually become greater than the combined pressure of gases within the swimbladder.

It now seems that one function of the gas gland is to produce a substance that releases these additional gases from the outgoing blood in the retia. The substance may well be lactic acid, which on entering the outgoing capillaries will have two effects on the blood. It will reduce the solubility of all gases and release some of the oxygen bound to haemoglobin. The pressure of each gas in the outgoing blood will thus become greater than that in the ingoing blood. As long as this extra pressure is maintained, gas will diffuse from outgoing to ingoing capillaries and so be carried towards the gas gland, where it will be concentrated and multiplied in the manner already described. Oxygen, nitrogen, and even the rare gases such as argon, are evidently secreted in this way [36].

The plan of the gas-secreting complex varies a great deal among teleost fishes. The number of retia, for instance, ranges from one to hundreds. In most deep-sea fishes there are from one to eight, but there is a more significant aspect. The deeper the living space, the greater is the length of the retial capillaries. The longest, measuring 25 millimetres or more, are found in the swimbladders of macrourid and brotulid fishes that live at abyssal depths of 2,000 metres and beyond (figure 26). The greater the depth, and the consequent pressure of gas within the swimbladder, the more difficult it is to prevent the loss of gas. Hence the longer capillaries, which provide extra surface, *inter alia*, for the exchange of gases. At the same time, the longer the length of these minute blood vessels, the greater will be the gas multiplying factor in the retia. This remarkable aspect is surely a fitting *coda* to this brief account of biological engineering at its best.

By these two contrary processes, gas elimination and gas secretion, the swimbladder tends to be kept at a just buoyant volume. Both processes are under nervous control and the end is weightlessness in water. Whatever movements the fish may make, whether up or down, this nervous control continues to act towards this desirable end. Now the swimbladder, like the gut from which it is derived, is innervated by parts of the autonomic or involuntary nervous system (in fact, both the inflatory and deflatory parts of the swimbladder get their nerve supply from branches of the vagus nerves). When a fish becomes over-buoyant, excitation of appropriate nerves causes the deflatory part to expand and begin its work. At the same time, the activity of other nerves leads to contractions of the gas secreting tissues, which are thus put out of action [34]. The rôles of the two parts are reversed when the fish becomes heavier than water.

But how does a fish 'know' whether it is over or under buoyant? One seemingly obvious way is through the response of sensory nerve endings to the stretching or slackening of the swimbladder wall. Recent critical tests with minnows (*Phoxinus*) were particularly revealing. The fish responded to a small decrease in pressure by a quick descent, accomplished by upward beats of the pectoral fins. After a small increase in pressure, contrary movements of the pectorals carried the minnow upward. These reactions are evidently initiated through nerve-endings in the swimbladder wall. By registering rapid changes of pressure, these endings must send signals to the brain, which can bring into play the nerves leading to the inflatory or deflatory part of the swimbladder [37].

Living space and swimbladder

Fishes with an open swimbladder, nearly all of which belong to the carp, characin and catfish order (Ostariophysi), abound in fresh waters; but there are relatively few in the ocean, the main kinds belonging to the herring family (Clupeidae) and the eel order (Apodes). That the ocean is the headquarters of teleosts with a closed swimbladder may be seen by turning to the list of such fishes on p. 68. Only one group, the trout-perches, is entirely confined to fresh waters. However, most of the toothed-carps and certain of the perch-like fishes, notably the sunfishes (Centrarchidae), freshwater perches (Percidae) and all but a few of the many cichlid fishes, spend their entire lives in freshwater habitats.

The earliest teleost fishes are likely to have had an open swimbladder, which is found in the most primitive living species, whether freshwater or marine. During their gradual occupation of the ocean – when they were also evolving into the existing types of teleosts – most of the early marine fishes may well have been forced, as it were, to acquire a closed swimbladder; for marine habitats, being so much deeper than those in fresh waters, pose special hydrostatic problems.

Perhaps it is not surprising that nearly all the freshwater species that can secrete gases do this very slowly. Freshwater fishes with an open swimbladder, few of which live at a depth greater than 10 metres, are within easy reach of the surface, where, by swallowing air, they can recharge this organ to a just buoyant volume. But a great many marine teleosts have depth ranges that extend well below a level of 10 metres. Consider a fish with an open swimbladder that lives at 50 metres, which is less than a moderate depth in shelf waters. The climb to the surface might not be strenuous, but having inflated the swimbladder to the proper volume, the fish would be faced with a rise in pressure from one to six atmospheres as it returned to its 50 metre habitat. At the end of the dive, the swimbladder would be compressed to one sixth of the volume it had at the surface. To inflate the sac to a just buoyant volume, the fish would have to produce the requisite amount of gas at a pressure of six atmospheres. It would then need well developed retia mirabilia and gas glands. The only way such a fish could obviate the need for gas secretion would be to pressurize the swimbladder at the surface, which would require a great deal of time and a powerful pneumatic pump. The retention of an open swimbladder and air-gulping habits would thus not appear to make much biological sense in deeper marine environments.

This 'sense', which is fitness of structure to function, has been attained in two ways. In most groups of marine fishes the pneumatic duct has been lost and superseded by a gas-absorbing capillary network, the now closed swimbladder being provided with well developed retia mirabilia and gas glands. The second way is seen in the eels, which develop these gas-producing tissues, but keep the pneumatic duct. As we have seen, eels need never visit the surface to replenish the swimbladder, and, indeed, they rarely seem to use the pneumatic duct as a means of deflation. Instead, this duct has become enlarged and provided with a capillary network and it acts much like the 'oval' or resorbent chamber of a closed swimbladder. To all intents and purposes, then, eels live and behave as though they had a closed swimbladder.

This still leaves the herring-like fishes (Clupeidae), which are the only diverse group of marine teleosts with an open swimbladder. This appears to lack a gas-producing system. But during part of their lives, some of the clupeids, such as herring, sprat and pilchard, move up and down in the sea each day. During their ascent the emission of gas through the pneumatic duct would seem to be a quick way of keeping pace with the diminishing pressure, but curiously enough it is in these three species, and a few others, that the swimbladder has a second exit near the anus (figure 37). In the herring and sprat, at least, deflation of the swimbladder is by way of this second opening to the exterior. The pneumatic duct is simply used to fill the swimbladder, which may well be inflated to give the fish neutral buoyancy at an intermediate level in its vertical range.

Fishes with a fully developed swimbladder are able, as we saw, to stay poised in mid-water with little or no effort. To keep station they need only back water, usually with the pectoral fins, so as to offset the propulsive force of water being pumped out of the gill chambers. But there are many teleosts that spend a great part of their lives resting on the bottom. When such a fish takes off it must keep moving to stay at a given water level. Having no swimbladder, or a reduced one, it is heavier than water. Under-buoyant, bottom dwelling fishes like this* are particularly diverse between the tide marks and at deeper levels over the continental shelf. The life of these fishes is centred at the interface between sea and land. Some, like the bullheads, scorpion-fishes, lizard-fishes and flat-heads, lurk on the sea floor in wait for their prey. Gobies, sea-snails, lump-suckers and cling-fishes have pelvic fins that form part or the whole of a suction disc for adhering to weeds, rocks, stones and shells. Held in this way, they may resist or elude the tug of tide and current. Flatfishes emerge from sand or mud to glide slowly over the sea floor in search of their food, resting from time to time on the tips of their dorsal and anal fin rays. The habit of all these fishes is to lurk, cling, burrow, creep, glide or dart. Like Antaeus, son of Poseidon, they derive their strength from contact with mother Earth. A swimbladder is of no use to such fishes and it has accordingly been lost, although it may still be found as a larval organ in the pelagic young of certain flatfishes, bullheads, gobies and others.

Almost without exception, the fishes that move above these Antaean species, fishes that swim freely and easily in mid-water space, have a capacious swimbladder.† In cooler waters there are numerous kinds of herrings, together with cod-fishes and hake. Except for some of the herrings, most of these shelf fishes live within a limited depth range. The lower limit of depth may be the sea floor, which some of these fishes use as a resting or 'sleeping' place. In one significant respect the upper limit is more critical. This I first appreciated when fishing in the clear waters of a Red Sea coral reef; trigger-fishes, butterfly-fishes and wrasses nearly always made for my bait and by reeling it in I could entice them to follow it, but only to a limited extent. They would rise after the bait for a short distance and then lose interest. Having a closed swimbladder, they are without means of rapid deflation. Indeed, experiments have shown that such fishes cannot tolerate more than a 25 per cent increase in the volume of the swimbladder. Rapid movements that take them

* E.g. gobies (most species), blennies, weevers, bullheads (Cottidae), sea-poachers (Agonidae), lizard-fishes (Synodontidae), sea-snails (Liparidae), flat-heads (Platycephalidae), lump-suckers (Cyclopteridae), certain scorpion-fishes (Scorpaenidae), cling-fishes (Gobiesocidae), flatfishes (Heterosomata) and bat-fishes (Ogcocephalidae).

† Among the best known tropical representatives of these mid-water swimmers are flying-fishes (neritic species), half-beaks, garfish, snappers (Lutianidae), groupers, sea-bass, etc. (Serranidae), squirrel-fishes (Holocentridae), damsel-fishes, butterfly-fishes, wrasses, parrot-fishes, grey mullet, barracuda, red mullet, surgeon-fishes, trigger-fishes and file-fishes.

above the ceiling of their living space must thus be decidedly restricted.

A closed swimbladder is found in the smaller fishes, such as skippers and flying-fishes, that swim in the surface waters of the open ocean, but it may be reduced or absent in dolphin-fishes (*Coryphaena*) and certain tunnies. Evidently, the loss of a hydrostatic organ is of no consequence to these powerful and restless fishes. Tunnies, at least, are full of buoyant oils and the greater the store of these food reserves, the less the energy needed to counter a tunny's tendency to sink. At all events, tunnies have energy enough to swim at their accustomed levels in the turbulent upper ocean and to pursue the most wide-ranging migrations.

Moving down to the mid-waters of the deep ocean, about half the species of luminous fishes that swim between depths of 200 and 1,000 metres have a capacious swimbladder, one that is well endowed with gas-producing tissues. Yet the species without a swimbladder may still be close to proper buoyancy simply because they have lightly ossified skeletons and sparely developed muscles. This is certainly true of the deeper-living (1,000 to 3,000 metres) mid-water species such as ceratioid angler-fishes and gulper-eels, none of which has a swimbladder. It looks as though this organ is unable to function at great pressures, but at least half of the abyssal fishes that live on or near the deep-sea floor, at depths of 2,000 metres and beyond, have a capacious swimbladder with extremely long retia mirabilia. This resilient adaptability, and indeed the entire range of swimbladder development in deep-sea fishes, seems to be intimately related not, surprisingly, to pressures but to the economy of life in all but the deepest parts of the ocean [38]. Further consideration of this aspect of deep-sea biology is left to Chapter 17.

Chapter 6

The Capture of Oxygen

OXYGEN is more readily available to men than to most kinds of fishes. For air-breathing animals, certain fishes included, 'air represents an ocean of oxygen compared to sea water' (or to fresh water) [35]. The natural fact is that air contains thirty to forty times as much oxygen as can dissolve in an equivalent volume of water. More precisely, a litre of air contains 210 cubic centimetres of oxygen, whereas the same volume of well-aerated water holds no more than 10 cubic centimetres of this vital gas.

Living organisms need oxygen for the release of energy, much of which comes from the oxidations of sugars and fats. After suitable transformations, this energy becomes available for all the elaboration that is life. Energy is not only required for the activities of the body – for the contractions of muscles, the secretions of glands and the conduction of nerves – but also for the building of biochemical compounds essential for growth, maintenance of living structure and reproduction. Much else could be added.

Fishes obtain oxygen by moving water, a medium eight hundred times as dense as air, over their gills. The oxygen diffuses from the water into the blood that flows in the gills, where it is chemically caught by the haemoglobin of millions of red cells and then circulated as oxyhaemoglobin to all parts of the body. This pigment not only carries oxygen but enhances the rate of transfer of this gas from water to blood.

In keeping with this biochemical fitness of the blood, the gills of fishes are superbly designed to intercept and capture the rather small quantities of oxygen that come their way. But to do their work, they must obviously receive an adequate flow of water. How is this achieved?

Considering the dense respiratory fluid they must move, it need be no surprise that the gill pumps of fishes are highly developed. In lampreys and hagfishes the gills are essentially a series of paired pouches with openings to the pharynx and the exterior. The gill filaments, the red, blood-rich parts of a gill that absorb oxygen, are attached to the walls within the pouches, which in lampreys are supported by a sinuous latticework of cartilage. All lampreys have seven pairs of gill pouches, each of which has a separate opening to the water (figure 79). Except for the genus *Myxine*, in which the ducts from each series of gill pouches lead to a common external opening (figure 79), this individual exhalent system is also found in the hagfishes (Myxinidae), which have five to fourteen pairs of pouches.

The active breathing movement in lampreys is due to the strong contractions of muscles in the outer walls of the gill pouches, which on being squeezed expel water to the exterior with considerable force. As soon as these muscles relax and the gill pouches, aided by the recoil of the elastic cartilages, begin to expand, water is drawn in to the pouches, so irrigating the gill filaments [39]. Such ways of breathing fit well with the habits of lampreys. When a parasitic form is feeding, it may spend a long time attached to its prey, clamped by the suctional powers of its mouth disc. Held by this disc, migrating lampreys will inch their way up rock faces and vertical walls. To build their nests, lampreys (usually the males) employ their suckers to remove stones from the gravel, which may mean much hard work. During all such activities, irrigation of the gills can proceed and be adjusted to suit the animals' needs. If the respiratory stream had to come through the mouth, this would obviously not be possible.

Expulsion of water from the gill system is also the active breathing movement of hagfishes. Rhythmic contractions of muscles in the walls of the gullet, gill pouches and gill ducts aid the velum, a special water-pumping part of the foregut, to force water through the external openings of the gill pouches (figure 79). When these muscles relax, the velar cavity and pouches expand, so sucking in a fresh supply of water, which enters the single nostril at the tip of the snout, flows down the nasal passage and enters the foregut. The intake of water is thus quite unlike that of lampreys, in which the nasal passage ends blindly in a hypophysial sac. Just how a hagfish manages to breathe when it is feeding, say on a dead fish, is not clear, for although water does not enter the mouth, a mass of food in the buccal and velar cavities might seriously hinder the circulation. One suggestion is that a feeding hagfish may be able to support an oxygen debt, accumulating the waste produces of metabolism, such as lactic acid, until oxygen is available for their oxidation [40]. It is also possible that some oxygen may be absorbed throu the skin.

In most jawed fishes the respiratory current enters the mouth, and after flowing over the gills, is expelled backward through appropriate external openings, but in the chimaeras (Holocephali) and lungfishes (Dipnoi), water passes from the nostrils into the mouth. The vascular, oxygen-catching filaments are developed on the sides of a close-set series of pharyngeal gill slits, which are separated by narrow gill arches with an inner, jointed skeleton. Each arch thus bears a forward and a backward set of filaments, the entire structure forming a complete gill, or 'holobranch', the whole series overlapping like the slips of a louver (see figure 27). Nearly all the jawed fishes have five pairs of gill slits, the exceptions being the shark genera *Heptranchias*

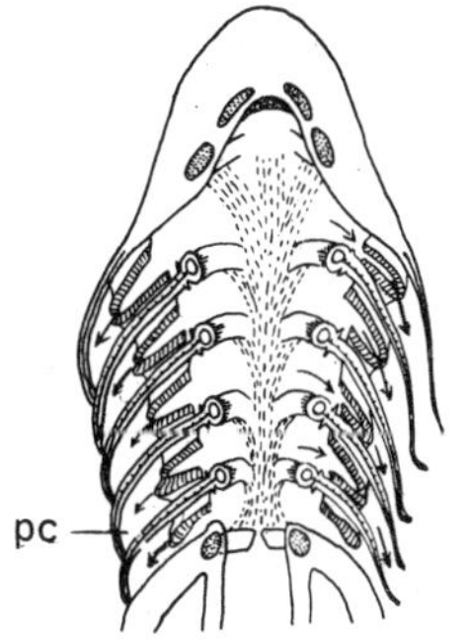

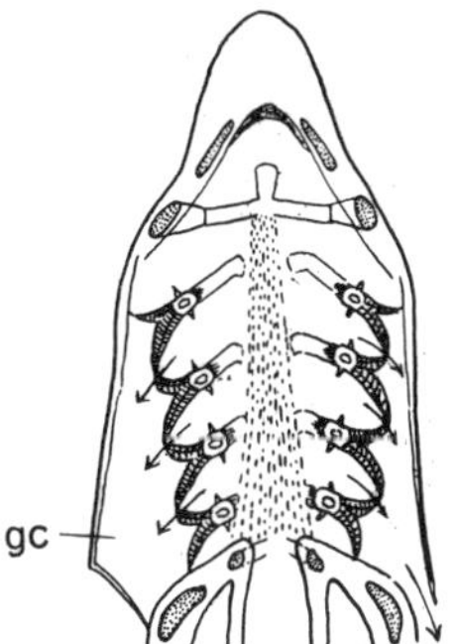

Figure 27. The gill system of a shark (left) and a teleost fish (right). The gill filaments are cross hatched. On the left of each figure the parabranchial chambers (pc) of the shark and the gill chamber (gc) of the teleost are shown in an expanded phase, when water is sucked from the mouth cavity (shown by arrows). On the right of each figure is shown the expulsion of water from the gill system.

(seven pairs), *Chlamydoselache*, *Hexanchus* and *Pliotrema* (six pairs) and the chimaeras (four pairs).

In the sharks and rays each gill slit has a separate, valvular opening to the exterior, whereas the gills of the bony fishes and chimaeras are covered by an operculum with an exhalant opening at the rear (figure 27). Lastly, between the first gill slit and the ear capsules is a pair of reduced slits known as the spiracles. In the chimaeras and nearly all the bony fishes they are obliterated during development, but in most rays and some sharks and dogfishes they are the inlets for some or all of the respiratory current.

Water is drawn into the opening gape of a jawed fish by the expansion of the buccal cavity. The periodic nature of this intake – one inspiration being separated from the next by closure of the mouth – would hardly lead one to suppose that water moves through the gills in a virtually continuous stream. But the recent work of Dr G. M. Hughes and Dr G. Shelton [41] has confirmed that this is both a fact and a vital respiratory adaptation. Clearly, the more continuous and steady the flow, the more thoroughly will the gills be able to remove the oxygen dissolved in water. Remembering that water contains relatively little oxygen and that many jawed fishes have high oxygen requirements, this is surely a great advance on the 'tidal' method of gill ventilation in lampreys and hagfishes.

Two living pumps are at work when a cartilaginous or a bony fish is breathing. Contraction of the mouth cavity, together with that of the gill pouches in sharks and rays, acts as a pressure pump to force water through the gills. The other kind of pump generates suction to draw water from the mouth cavity into compartments on the far side of the gills. In bony fishes and chimaeras this pump acts through the expansion of the opercular cavity between the gills and gill covers, the latter being pulled outwards by appropriate muscles; but in sharks and rays each gill slit has its own suction pump, which is fully expanded when the louvered gill flaps are rounded and turned back in the closed position (figure 27).

Looking more closely at part of the sequence of events in a bony fish, as the pressure pump gives way to the suction pump, the mouth is opening and the opercular gill cavity contracting, so expelling water to the exterior. It is at the end of this brief period, when the gill cavity is almost fully contracted and sealed, that a small amount of water may actually be sucked from this cavity into the mouth. But for the greater part of the breathing cycle, or the whole of it in some species, the pressure of water in the mouth cavity exceeds that in the gill chambers.

The kind of life led by a fish is aptly expressed in the form and function of the two pumps. Study of active fishes showed that either the power of the buccal pressure pump was about equal to that of the gill chamber suction pump, as in the herring (*Clupea harengus*) and rainbow-trout (*Salmo gairdneri*), or that the former was the more effective, as in the horse-mackerel (*Trachurus trachurus*). But when certain kinds of active fishes are swimming they may simply keep the mouth open and the gill chambers expanded, so funnelling a continuous stream of water through the gill filaments. Skipjack (*Katsuwonus pelamis*) and other scomboid fishes are known to make use of their motion to irrigate the gills in this way. Indeed, if mackerel are

1. Camouflage. Top: A sharply broken colour pattern on a young *Plectorhynchus chaetodonoides*. Middle left: Sargassum-fish (see also Plate 14). Middle right: A butterfly-fish (*Chaetodon auriga*) showing a stripe through the eye and an eye-spot on the dorsal fin. Below: An Indo-Pacific wrasse (*Coris angulata*) with two large eye-spots.

3. Rays in motion. Top left: Five little devil-rays (*Mobula*) and, top right: two spotted eagle-rays (*Aetobatus*). Almost touching the lowest devil-ray is a cow-nosed ray (*Rhinoptera*). These kinds of ray swim by flapping their large and muscular pectoral fins, but skates, such as the thorn-back *Raja clavata* (bottom right), swim by undulations that move down their pectoral fins. Guitar fishes, *Rhinobatus* (bottom left), which have rather small pectoral fins, swim by sculling motions of the tail.

2. Above: Young pike trying to swallow three-spined sticklebacks. After Hoogland, Morris and Tinbergen [108]. Below: Two fishes with venomous spines. A stone-fish (*Synanceja*), on the left and a toad-fish (*Halophryne*).

4. Freshwater fishes of the great order Ostariophysi; From top to bottom: A South American characin (*Hemiodus semitaeniatus*), a vegetarian; A pike-like, predatory characin, *Luciocharax insculptus*, from Colombia; A flying hatchet-fish *Gasteropelecus maculatus*, W. Colombia to Panama; A predatory characin, *Pygocentrus piraya*; A characin, *Distichodus antonii*, from the Congo.

5. Freshwater fishes of the great order Ostariophysi; from top to bottom; A South American mailed cat-fish, *Hoplosternum thoracatum*; a mailed (loricariid) cat-fish, *Farlowella gracilis*, from South America; an Asiatic cyprinid, *Aristichthys nobilis*, which feeds on detritus; An African cat-fish, *Synodontis nigrita*, and to its left an Asiatic loach *Misgurnus anguillicaudatus*; the Asiatic grass-carp, *Ctenopharyngodon idellus*, which feeds on vegetation.

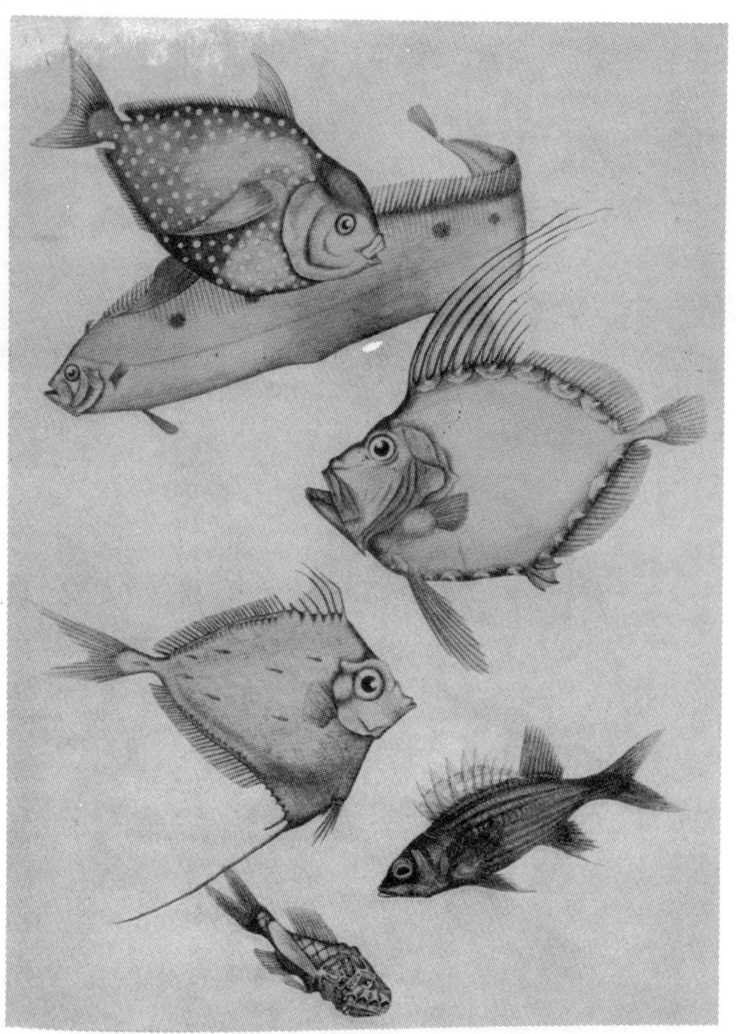

6. Fishes of the orders Allotriognathi (1 and 2), Zeomorphi (3 and 4) and Berycomorphi (5 and 6). From top to bottom:
(1) the opah (*Lampris luna*);
(2) the ribbon-fish, *Trachypterus arcticus*;
(3) the American John Dory (*Zenopsis ocellata*);
(4) *Xenolepidichthys americanus*;
(5) *Melamphaes mizolepis*, a deep-sea berycoid;
(6) A squirrel-fish, *Holocentrus marianas*.

7. Fishes of the sunless mid-waters of the ocean. The long serpentine fish extending from top to bottom is a gulper-eel (*Saccopharynx*). The other fishes, from top to bottom, are a female angler-fish (*Neocratias spinifer*) with attached parasitic male; a female angler-fish (*Linophryne arborifer*) with a luminous chin barbel as well as a luminous lure; a bristle mouth (*Cyclothone pallida*); a female angler-fish (*Photocorynus spiniceps*) and another female angler-fish (*Lasiognathus saccostoma*).

8. Midwater fishes of the twilight zone (200–1,000 metres). Above: two hatchet-fishes (*Argyropelecus lychnus*); middle: a star-eater (*Astronesthes similis*), and bottom: a deep-sea salmonoid (*Opisthoproctus grimaldii*).

9. Midwater fishes of the twilight zone. Above *Chauliodus* angling with its long second dorsal ray and attracting two deep-sea mysid shrimps (*Gnathophausia*). The lower picture is of a searsid fish (*Barbantia curvifrons*) with a luminous spark-producing shoulder organ.

10. Top: a mako-shark (*Isurus*) pursuing squid. This shark is swimming over the back of a whale-shark (*Rhincodon typus*). Below the mako is a hammerhead shark (*Sphyrna zygaena*). Bottom left a tiger-shark (*Stegostoma tigrinum*) and bottom right a sandbar shark (*Carcharhinus milberti*).

12. Plectognath fishes. The two bottom pictures are of a Gulf of Mexico puffer-fish (*Sphaeroides nephalus*), before and after inflation. The top left figure is of an Indo-Pacific trigger-fish (*Balistapus undulatus*). Below this is an inflated porcupine-fish (*Diodon*). The two top right pictures are of an Indo-Pacific box fish (*Ostracion lentiginosus*). The upper is a front view of an immature fish, the lower of a male fish.

11. Three cleaning relationships. Top left: a butterfly-fish (*Chaetodon nigrirostris*) cleans a red mullet (*Pseudupeneus dentatus*). Below: two neon gobies (*Elecatinus oceanops*) tending a Nassau grouper (*Epinephelus striatus*), both based on illustrations by Rudolph Freund in an article by Conrad Limbaugh in the *Scientific American*, August 1961. Top right: a rainbow wrasse (*Labroides phthirophagus*) cleans a moray-eel (*Gymnothorax eurostus*).

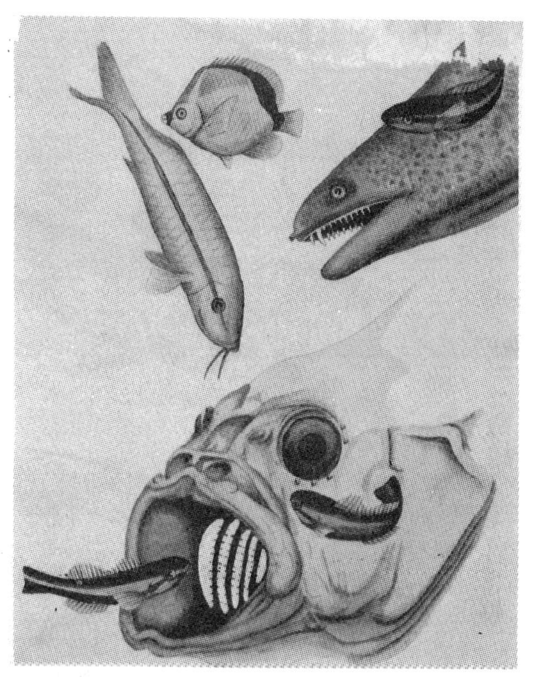

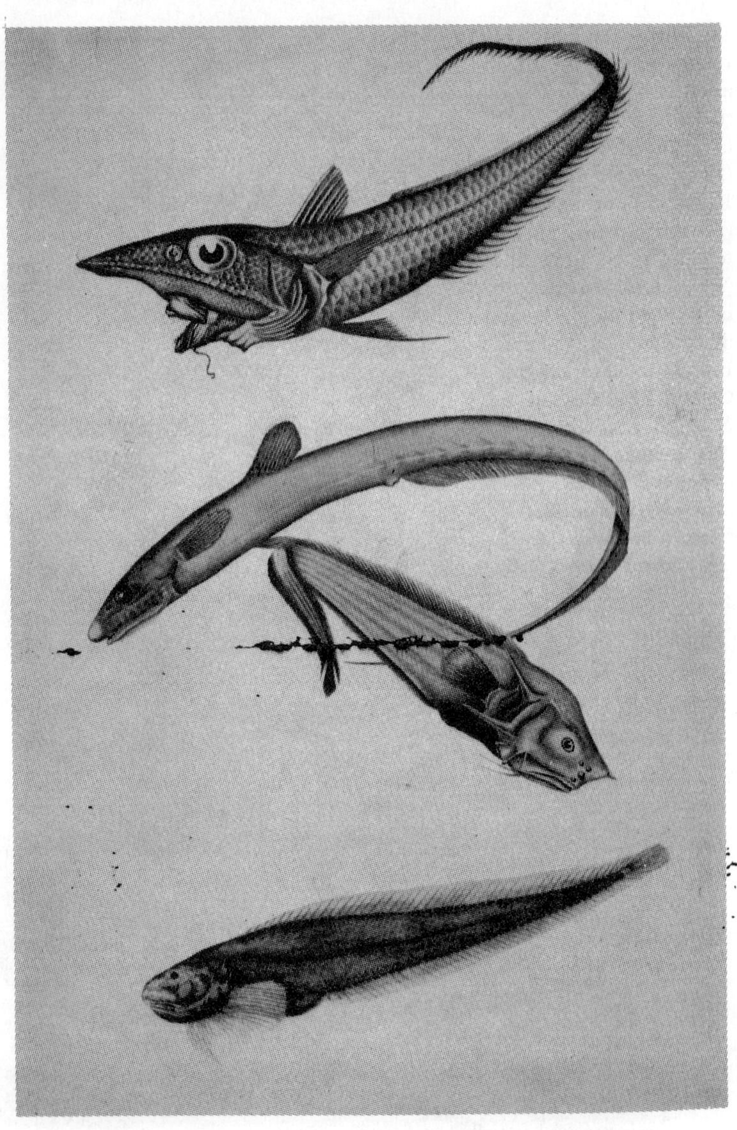

13. Benthic deep-sea fishes. Top: a rat-tail (*Coelorhyncnus flabellispinis*) from slope waters of the Indo-Pacific; top middle: a halosaur, bottom middle: a brotulid (*Acanthonus*), and bottom: a sea-snail (*Careproctus amblystomopsis*) taken at a depth of 7230 metres in the Kurile-Kamchatka Trench.

prevented from swimming freely, they will die from lack of oxygen. Yellow-tails (*Seriola*) and horse-mackerel keep their mouth and gill chambers open during relatively long pauses between successive inspirations. Sharks swim with open gape and flaring gill flaps, breathing movements being used only when they are idling, or resting on the bottom in the case of those which are benthic in habit.

Fishes that skulk on the sea floor or river bed for long periods have differently accented ways of irrigating their gills. Weevers (Trachinidae), dragonets (Callionymidae), bullheads, scorpion-fishes, flatfishes and angler-fishes are some of the better known teleosts with such a mode of life, a lurking life punctuated by periods of activity, as when they feed, court, breed and so on.

Compared to the active species, they have more capacious gill chambers, most of the increased volume being due to a greater development of the membranes that act as outlet valves. These branchiostegal membranes are extensions of skin from the edges of the gill covers, below which they are wide and ribbed with special rays from the hyoid arches (figure 21). As the gill chambers expand, appropriate muscles pull these rays into the sides of the fish, so keeping these cavities closed. When the fish starts to exhale, other muscles move the rays outwards and open the 'exhaust' valves, through which water is forced by the bellows action of the gill covers.

Large capacity gill chambers should make good suction pumps, which is what Dr G. M. Hughes found in a recent series of experiments. Measurements of the pressures in the buccal and gill cavities of the father-lasher (*Cottus bubalis*), the dragonet (*Callionymus lyra*) and two flat-fishes, the plaice (*Pleuronectes platessa*) and the lemon-sole (*Microstomus kitt*), showed that the opercular suction pumps are much the more effective means of irrigating the gills. As Dr Hughes remarks:

> The adaptive value of an increased development of the suction pumps in benthic forms is fairly clear when it is remembered that most of the time they live in water which is almost stationary. Selection has favoured the evolution of a mechanism which is well adapted to drawing a current across the gills during a relatively long part of the respiratory cycle. Such mechanisms are admirably adapted to ensure a steady flow across the gills without creating any disturbance of the muddy or sandy bottom [42].

Comparison of the dogfish (*Scyliorhinus caniculus*), which is a fairly active species, with the bottom-dwelling thornback-ray (*Raia clavata*) also revealed the greater importance of the parabranchial suction pumps in the ray. Moreover, as most kinds of rays spend much of their time resting on the bottom, perhaps half covered with sand or mud, the spiracles become the only inlets of the respiratory stream. When a bottom-dwelling ray is swimming or its head is raised off the substratum, some of this stream can enter the mouth. In the active devil-fishes (Mobulidae), however, the spiracles are very reduced and the entire respiratory intake is through the mouth.

There are exceptions to most biological rules, and in the present context there are the eels, most of which live on the bottom, burrowing in the substratum or hiding in holes and crevices of coral and rock. In eels the gill cover bones are considerably reduced, so limiting the bellows action and volume of each gill cavity, most of which is bounded by the gill membranes and their supporting branchiostegal rays. Indeed, the buccal cavity is especially large, the mouth and gills being comparatively far apart. As an eel breathes, it may be seen to force water towards the gills by a kind of swallowing action. It is thus not surprising that pressure records from the conger eel (*Conger conger*) revealed that the power of the buccal pump is greatly superior to that of the gill cavity suction pumps.

But why should eels be the exceptions? Dr William Gosline [43] holds that the special features of eel organization may be structural expressions of their marked habits of wedging themselves through small holes in the coral or rock. Hence the serpentine form, the heavy and usually scaleless skin, the absence of pelvic fins, the strong, arrow-head-like skull, and, as a respiratory adaptation, the reduction of the bony gill covers, which would be a hindrance and liable to damage as an eel forced its way into a crevice or under a rock. A sinuous, tough-skinned body, headed by a firm brain box, strong jaws and a flexible gill region seems the best organization for fishes with such marked powers of 'muscling-in'.

The frequency of the breathing movements of fishes, like those of other vertebrates, is related to body size; the pulse of the pumps of small species being quicker than that of larger ones. When idling, the breathing movements of sticklebacks (*Gasterosteus aculeatus*) and minnows number one hundred and fifty in a minute, whereas in a resting dogfish there are thirty to sixty cycles per minute. In a tench (*Tinca tinca*) the corresponding number is sixty. But habit and habitat must also be considered; for even when resting, the more active, free-swimming species breathe quicker than those that lurk on the bottom. In a herring 12 inches long, the rate is one hundred and twenty per minute and in a 15-inch rainbow trout ninety per minute. The figures for the bottom-dwelling and smaller bullhead and dragonet are forty per minute and twelve to twenty per minute respectively. When motionless on the bottom, the gill pumps of the plaice, thornback ray, conger eel and freshwater eel are turning over at a rate of thirty per minute.

In any fish, the quicker the respiratory pulse, the greater the flow of water over the gills in a given time. But the depth of respiration, the volume of water irrigating the gills during one breathing cycle, is also important. When the rainbow trout experiences a shortage of oxygen, the volume of the respiratory current may be increased to four times the normal amount, an increase that is largely due to a greater depth of breathing. In the freshwater eel, a similarly induced increase in the ventilation volume is about equally shared by appropriate changes in the frequency and the depth of breathing [44].

Water that has passed through the gills of a rainbow trout or an eel contains no more than one-fifth of its original content of oxygen. Capture of four-fifths of the rather small quantities of oxygen that flow their way must clearly mean that the gills are highly fitted for their life-keeping activities. Appreciation of how this comes to be involves a closer look at the form and function of the gill filaments, the long, blood-rich, finger-like processes that radiate in close array from the gill arches (figure 28).

The vascular, oxygen-catching parts of each filament are tightly packed series of small thin plates, usually called gill lamellae. These delicate extensions of gill tissue, which may be semicircular to leaf-like in form, represent an enormous surface over which water can flow. Some impression of this may be obtained from figure 28, which shows the arrangement of filaments and lamellae in a flying-fish. Clearly the more extensive the gills, the greater must be the surface through which oxygen can diffuse into the blood flowing in the lamellae. It is thus reasonable to expect that the more active fishes will have a relatively larger gill surface. The mackerel (*Scomber scombrus*) and the menhaden (*Brevoortia tyrannus*) have over 1,000 square millimetres of gill surface for every gram of their weight. Comparable figures for less active species, such as the puffer-fish (*Spheroides maculatus*), the toad-fish (*Opsanus tau*), the summer-flounder (*Paralichthys dentatus*) and the sea-robin (*Prionotus carolinus*) are from one-third to one-fifth of the above values [45].

The blood in each lamella, which flows in a fine network of capillaries, is brought very close to the gill surface, the intervening tissue being but a thin outer covering of delicate pavement cells. The paths of oxygen molecules diffusing from water to blood are thus made as short and rapid as is biologically possible.

There is an even more significant aspect of the blood circulation in the gill lamellae. The blood that is to absorb oxygen is pumped from

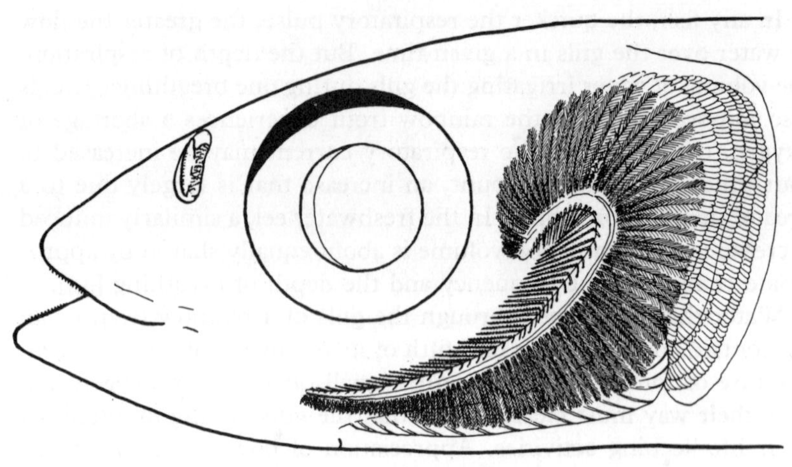

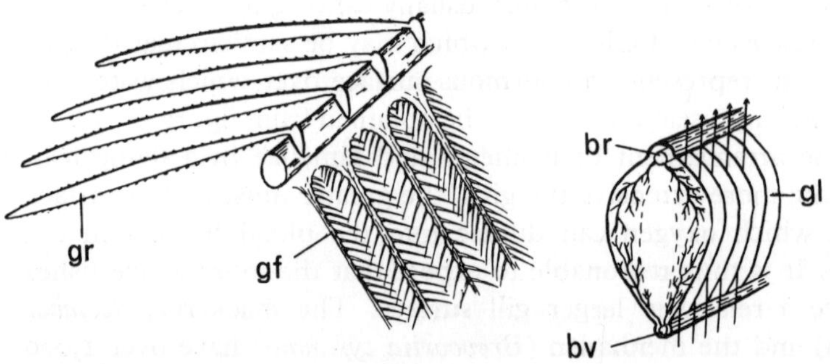

Figure 28. The head and large, elaborate gills of a flying-fish (*Oxyporhamphus micropterus*). In the upper figure, the first gill arch is seen in position. Part of this arch is seen (below left), showing four gill filaments (gf) and four gill rakers (gr). Note the numerous gill lamellae on each gill filament. Below right: a few of the gill lamellae (gl) enlarged showing the counter-current circulation of the blood (br to bv) and water (shown by arrows). (Drawn from a specimen in the Museum of Comparative Zoology, Harvard University.)

the heart to the gills by way of the ventral aorta, which sends a single vessel to each gill arch. After entering a gill filament, the blood is led to each lamella by a vessel that runs down one side of the filament. The blood then courses through the lamellae and is collected by an artery on the other side (figure 28). The vitally significant feature is that blood flows across the lamellae in the opposite direction to the water that streams between them (figure 28) [44]. This counter-current interplay of blood and water is a biological parallel to industrial heat exchangers, which employ a similar circulation to save calories by using the hot exhaust gases from a furnace to pre-heat the air for the combustion processes [35]. Limited oxygen supplies are thus used, like valuable calories, to the best advantage. In the gills, for instance, the blood about to leave the lamellae is losing its capacity to take up oxygen; but this is countered by its virtual contact with water having the highest natural content of oxygen and the lowest amount of carbon dioxide, certain concentrations of which can turn oxygen out of the blood. The complete effect is a very efficient interchange of oxygen and waste carbon dioxide between water and blood. Neat and striking proof of this was recently obtained by the Dutch biologists E. H. Hazelhoff and H. H. Evenhuis, who were able to reverse the flow of water through the gills of tench. With water and blood now flowing in the same direction, the amount of oxygen absorbed by the gills was only one-fifth of that extracted under normal conditions [46].

Naturally, the effectiveness of the contrary flow of water and blood depends on the thorough percolation of the gills by the respiratory stream. In lampreys and hagfishes, where the filaments are born on the walls of the gill pouches, and in sharks and rays, where each gill arch is enclosed in a pocket, the water is forced to flow through the filaments on its way to the exterior (see figure 27). The gill design of teleost fishes is very different in that the filaments are attached only at their bases to the gill arches, not fused along most of their length to pouch or gill arch. But during the respiration of a teleost the filaments are curved away from the arches by special internal muscles. The tips of one set of filaments are thus brought into close contact with those of the next adjacent set, the overall effect being the opposition of a zigzag wall of filaments to the water about to flow through the gills (figure 27). The respiratory stream is thus forced to percolate through the thousands of tiny meshes that are formed by the lamellae (figure 28).

Some of the main features of this section can now be drawn together. In trying to appreciate how fishes breathe in a dense medium that contains much less oxygen than air, we have reviewed the form and function of the gill pumps and looked closely at the fine structure and blood circulation of the gills. We have seen that these pumps are large and elaborate structures and they are moved by powerful muscles, the whole complex being able to keep the dense respiratory stream in healthy circulation through the gills. To make the best possible use of a rather sparse flow of oxygen molecules, the medium is pumped continuously through gills with a very large surface area and in which the blood flows very close to the water. The chances of oxygen molecules meeting those of blood haemoglobin are thus greatly increased, but the most cunning feature is surely the myriads of small counter-currents of blood and water in the gills, most fitting refinements of biological engineering. Here, in particular, a certain amount of detail has been unavoidable, but without it we could not have got so close to that part of life that is fish.

In conclusion, the review of these aspects of breathing leads to an important insight. Living in water, a medium much less suitable for breathing than air, fishes have to work hard to satisfy their needs for oxygen. It is largely because of this physical unsuitability of water that active rates of oxygen capture are only a few multiples of the rates when fish are resting. When we are really active we can take up twenty times as much oxygen in a given time than when we are sitting down. In insects this increase can be a hundred-fold or more. Having such contrasts in mind, there is surely much point to Dr F. E. J. Fry's conclusion [45] that the maximum growth of fishes is likely to be limited more by the oxygen-catching capacities of their gills than by the surface area of alimentary tract that is available for the absorption of digested foodstuffs. At the same time, let us not forget that fishes are the most diverse of all the vertebrate animals. They are the masters of their environment.

Considering the action of the gill pumps in fishes of differing activities and habitats, and realizing that the more active the fish the greater the surface area of gill tissue, it will be clear that oxygen needs vary from one species to another. When idling, a mackerel (*Scomber scombrus*) requires nearly three-quarters of a cubic centimetre of oxygen per hour for every gram of its weight. Under similar conditions, the puffer fish (*Spheroides maculatus*) needs less than one-tenth of this quantity of oxygen. The South American lungfish (*Lepidosiren paradoxus*), an air-breather, needs even less than the pufferfish. Looked at from another aspect, a freshwater eel is able to satisfy its oxygen requirements when each litre of water contains one cubic centimetre of oxygen. Three times this amount is inadequate for the normal needs of a rainbow trout [45]. Lastly, and surprisingly, crucian carp are able to pass the winter in waters that are devoid of oxygen. At higher temperatures (16°C) they tolerate such conditions for no more than a few hours [47].

When a fish is faced with a lack of oxygen, its breathing, as we have seen, becomes deeper and quicker. The respiratory pulse is, in fact, controlled by a nerve centre or centres in the hind end of the brain, the medulla oblongata. This centre is sensitive to a lack of oxygen, or an accumulation of carbon dioxide, and it responds by quickening the flow of impulses down the nerves that control the muscles of the gill pumps. But there are environments, particularly in tropical swamps and in the depths of the ocean, where the oxygen content of the water is barely detectable by chemical analysis. Concerning the numerous kinds of small deep-sea teleosts, such as lantern-fish and hatchet-fish, which can be caught in the oxygen minimum layers of the ocean, we have no idea. When we realize that over great parts of the Equatorial Indian Ocean and the Eastern Pacific Ocean between depths of about 100 and 1,000 metres, the oxygen content of the sea is less than a quarter of a cubic centimetre per litre, we can only wonder how countless thousands of lively lantern-fishes manage to get enough oxygen to breathe and to fill their swimbladders. But in swampy fresh waters the interface between water and air – between a dearth and an abundance of oxygen – can readily be reached. Knowing the protean adaptability of living organisms we might even expect that many kinds of freshwater fishes would have air-breathing organs. To these we may now turn, with the rider that there is at least one kind of air breathing fish in the sea, the tarpon. Besides this, two Indo-Pacific dogfishes (*Chiloscyllium indicum* and *C. griseum*), which can live a long time out of water, have been seen in an aquarium to gulp air through the mouth or inhale through one of the spiracles.

Air breathing fishes

If it were not for the fact that an early air-breathing fish became an amphibian, an animal able to spend part of its life on land, I suppose that I should not now be writing this book. The ancestor of the first amphibians was almost certainly one of the osteolepid tassel-finned fishes, many of which lived in pools that were liable to dry-up. And this is a danger that modern fishes also have to face.

Like the proto-amphibious tassel-fins, some of these modern air-breathing fishes use their fins and body to move over land, and at times may find a new habitat in place of a dried up pool. Certain of these walking-fishes, the climbing-perch, snake-heads, the cuchia and a few catfish such as *Saccobranchus fossilis* and *Clarias magur*, burrow their way into the mud or river banks at the approach of the dry season, and then aestivate until the rains come. Similar habits are also found in non-walking, air-breathing fishes, the best known being African lung-fishes, the South American lung-fish, the Asiatic gourami (*Osphronemus*) and the North American bowfin. The burrows in which the lung-fishes pass the dry season are neatly fashioned, one or more small openings being left at the entrance, through which they may breathe air. Whatever breathing is done by other aestivating species must also be aerial.

Before turning to fishes' means of air-breathing in water, some account of their habitat is necessary. Most of them live in shallow, stagnant waters in low-lying parts of tropical Asia, Africa and South America. Dr G. C. Carter, who with Professor L. C. Beadle made a long study of the swamps of the Paraguayan Chaco, has aptly described such kinds of living space. 'Some of the features of these swamps are common to most of them. Almost always the water lies under thick growths of aerial vegetation – trees in the mangrove swamps and forests, papyrus which may grow 12 to 15 feet high, and in the swamps of open country, grasses and other plants almost equally high. The water is highly coloured, it may have the colour of weak tea – and is almost or quite stagnant even in the mangrove and papyrus swamps on the borders of rivers and lakes.'

These waters contain very little oxygen, even quite close to the surface, but the content of carbon dioxide may be very high (as much as 70 cubic centimetres per litre). Dr Carter asks:

> How is this lack of oxygen in the swamps brought about? [and continues] I believe that it is the result of several conditions which are all present in these waters and not normally present in otherwise similar temperate waters. Oxygen can be introduced into a body of water by diffusion from the air, and produced in it by photosynthesis. It will also be removed by the respiration of plants and animals and by the chemical and biological oxidations of decay. In tropical swamps little oxygen is produced by photosynthesis owing to the weak lighting of the water, and decay, rapid at high temperatures, will actually remove any oxygen that gets into the water. Oxygen can only reach the water by diffusion from the air above it.

But not every kind of fish that lives in tropical stagnant waters has the means for breathing air. Indeed, only eight species out of twenty collected by Carter and Beadle [48] in the Paraguayan Chaco have air-breathing organs. Species not adapted for breathing air were seen to visit the surface film and hang below, drawing water over their gills. Here, very close to the air, must be adequate supplies of oxygen. Small cyprinodont fishes, such as the guppy (*Lebistes*) and the mosquito-fish (*Gambusia*), with jaws fitted for feeding at the surface film, are particularly adept at acquiring oxygen in this way. *Gambusia affinis* and *Mollienisia latipinna*, found in an anaerobic limestone spring in Florida, were seen to gulp air, and they died if denied access to the surface [49]. Larger fishes such as cyprinids and catfishes are also able to breathe at the surface film. But air-breathing fishes that can 'walk' and/or aestivate are ready for one of the great hazards of life in the tropics, the drying-up of small bodies of water.

Remembering that the swimbladder is essentially a specialized outgrowth of the foregut, the tissue systems that have been adapted for air-breathing in bony fishes are derived from the gut, gills and gill chambers. As air gulped into the mouth can only pass into the gut or gill chambers, the diversity of air-breathing adaptations is quite remarkable. Furthermore, freshwater eels and mud-skippers are able to breathe through the skin when out of water.

Certain of the fishes with a lung-like swimbladder – the bichirs, reed-fish and lung-fishes – have almost certainly inherited such an air-breathing organ from the long geological past of their kind. All these fishes are the survivors of once dominant groups, and their persistence today is partly due to their air-breathing habits. Except for the bowfin and the gar-pikes, survivors of the once diverse holostean fishes, the other fishes with a lung-like swimbladder are teleosts: tarpon (*Megalops atlanticus*), osteoglossids, featherbacks (*Notopterus*), a characin (*Erythrinus unitaeniatus*), the electric fish (*Gymnarchus niloticus*) and the mud-minnows (*Umbra*). In the teleosts, which evolved from the holosteans, the swimbladder is primarily a hydrostatic organ. The swimbladder of the air-breathing holosteans and teleosts and of many other species opens into the roof of the foregut through a pneumatic duct. In the lung-fishes, bichirs and reed-fish, the opening of the duct is in the floor of the foregut. In the South American and the African lung-fishes the swimbladder is divided equally into left and right lobes, but in the bichirs and reed-fish the right lobe is much the larger. The swimbladder of the Australian lung-fish, the bowfin, gar-pikes and teleosts is a single structure.

The possible evolutionary implications of these anatomical differences need not concern us here. The essential air-breathing adaptation is an alveolar conformation of the inner tissues of the swim-bladder wall. By the elaboration of hundreds of small, cell-like compartments, each with a rich blood supply, each having access to the air in the main cavity, the oxygen-catching surface of the swim-bladder is enormously increased (figure 29). And the air, just like water flowing over the gills, is brought very close to the blood [50].

When an air-breathing fish needs oxygen it swims up, breaks surface and gulps in bubbles of air, which are passed to the air-breathing organ. One kind is simply an aptly modified part of the gut, which becomes thin-walled, richly vascular and loses its digestive function. In certain of the loaches (Cobitidae), cyprinoid fishes that live in parts of Asia and Europe, most of the intestine has become a 'lung'.* After gulping in air, the Indian loach turns a complete somersault, '. . . at the same time driving out . . . the used-up gas from its intestine through the anus (eight to twelve small bubbles at a time) with such energetic force as to produce a distinct clinking sound.' [51] During the summer, when the water contains very little oxygen, the Chinese loach is an air-breather, but during the winter months, when the water is better aerated, the fish can use its gills. On resuming aquatic respiration, the oxygen-catching parts of the intestine regress and are replaced by regenerated digestive tissues. The intestine is thus a seasonal 'lung'.

An intestinal 'lung' is also found in certain of the South American, swamp-dwelling cat-fishes (*Callichthys* spp., *Hoplosternum littorale* and *Doras*) and in a very different kind of fish, the symbranchoid eel (*Monopterus javanicus*), which lives in marshy regions of Indonesia, China and Japan. But in the air-breathing mailed cat-fishes of South America (*Ancistrus* and *Plecostomus*; Loricariidae) the stomach has become a 'lung' [50].

Here then, in the cat-fish sub-order (Siluroidea) are two separate alimentary adaptations for air-breathing, one being 'copied' by certain loaches and a symbranchoid eel. Quite another kind of air-breathing organ is found in some species of the cat-fish genus *Clarias*, which live in both Africa and Asia. Each gill cavity is extended upwards as an air-chamber, having two branching structures, which develop from the upper parts of the second and fourth gill

* E.g. the spiny-loach (*Cobitis taenia*), stone loach (*Nemachilus barbatulus*) and *Misgurnus fossilis* of European freshwaters, an Indian species (*Lepidocephalus guntea*) and a Chinese species (*Misgurnus anguillicaudatus*).

arches (figure 29). Both the air-chambers, the lining of which may be folded, and the tree-like outgrowths, have a rich blood-supply, the whole complex forming an efficient 'lung'.

Even more curious is the Indian cat-fish (*Saccobranchus fossilis*), which belongs to the same family as the air-breathing species of

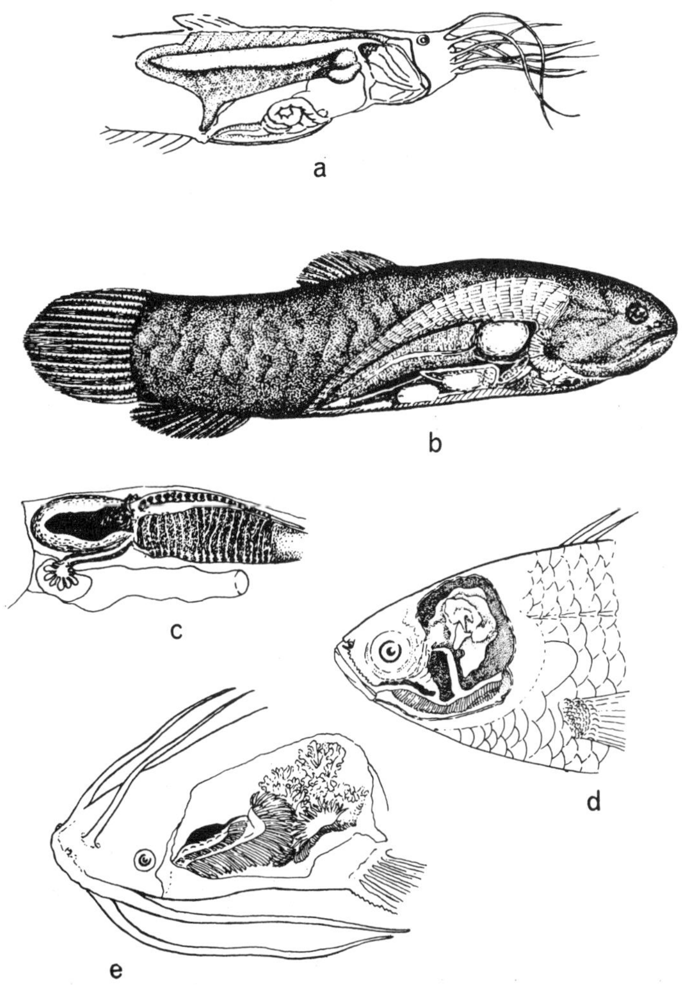

Figure 29. Air-breathing organs in fishes. (a) catfish, *Saccobranchus*, with a lung-like extension of the gill chambers down the back; (b) characin, *Erythrinus*, with (c) a lung-like swimbladder; (d) climbing perch (*Anabas*), with a labyrinthine organ above the gills; (e) catfish, *Clarias*, with respiratory trees. (b and c after Carter and Beadle [48].)

Clarias and like them lives in stagnant waters, but has adopted very different means of aerial respiration. During the early life of this species, the upper part of each gill cavity grows backwards through the muscle segments of the trunk to form an air-chamber, which in the adult extends to the middle of the tail (figure 29). The lining of the air chambers is folded into ridges, which unite in places to form air cells like those of a lung.

Like *Saccobranchus*, the Asiatic cuchia has a pair of air chambers, but these are developed from the roof of the pharynx just before the gill arches. It is thus very different from its relative the symbranchoid eel, *Monopterus*. Pharyngeal air chambers have also evolved in the snake-heads of Asia and Africa (figure 14), each chamber being constructed into an hour-glass shape. The air-breathing organs of the South American electric-eel, a cyprinoid fish, consist of vascular, papillate outgrowths of the lining of the mouth, the upper parts of the gill chambers and the inner surfaces of the gill arches.

To end this brief survey, mention must be made of the Anabantidae, spiny-finned freshwater fishes of Africa and South Asia.* As in air-breathing species of the cat-fish *Clarias*, each air-chamber is an upturned extension of the gill cavity, but the respiratory organ is labyrinthine, not tree-like in structure (figure 29). When a climbing perch is about half an inch long, the air-breathing complex begins to grow upward from the upper filamentous part of the first gill arch. Eventually, this outgrowth not only gives rise to the several saucer-like folds of the labyrinthine organ and the lining of the air-chamber, but also partly divides the chamber into upper and lower compartments. Each fold develops scalloped edges and becomes richly vascular, the whole elaborate structure, together with the air-chamber membranes, forming a serviceable 'lung' (figure 29).

Before a climbing perch gulps in a fresh supply of air, stale gases are fanned out of the gill cavities, which connect with the lower parts of the air-chambers through a special opening between the hyoid and first gill arches. Air taken into the pharynx enters each chamber through the first gill slit, which can be closed by a valve derived from the gill rakers of the first arch.

To live in stagnant waters, bony fishes have thus had to seek oxygen very close to the surface film or evolve means of breathing air. These means are quite diverse. A way for fishes with an open swim-

* Well known Asiatic forms are the Indian climbing perch (*Anabas scandens*) the Siamese fighting fish (*Betta splendens*), the gourami (*Osphronemus olfax*) and the paradise fish (*Macropodus opercularis*).

bladder, one opening into the foregut, was the development of inner air-cells like those of a 'lung' (p. 94). This way is barred to species with a closed swimbladder, such as snake-heads, symbranchoid eels and anabantids, and to those with an open swimbladder that is reduced in the adult stage, for instance loaches and various cat-fishes. These fishes have turned to an alimentary 'lung', or air-breathing extensions of the gill cavities, or to pharyngeal air-chambers. Related species, such as the anabantids, display their kinship to one another in the design of their air-breathing organs; but such common design is certainly not true of the symbranchoid eels *Monopterus* and *Amphipnous* and of the cat-fishes *Clarias* and *Saccobranchus*.

On the other hand, fishes that are not closely related, or even very remotely related, have acquired the same kind of air-breathing organ, the most striking instance being the development of an intestinal lung in various loaches, callichthyid catfishes and *Monopterus*. But congruence of functional design is a far-reaching outcome of teleost evolution and there is a thoughtful way of regarding this focus of living systems. The number of effective responses to an ambient challenge, such as lack of oxygen, must be limited by the nature and number of the parts that can be adapted to meet the challenge. Knowing the plasticity of living organisms, their capacity to develop structures, functions and habits that enable them to solve problems imposed by local conditions, but realizing that this plasticity is circumscribed, is it so curious that remotely related fishes may sometimes chance to master air-breathing and other problems by quite similar means?

A glance at a rarer kind of convergence leads to one remaining consideration. The bowfin not only has a lung-like swimbladder, but well developed, sieve-like gills, produced by the growth of cross connexions between the filaments. These elaborations provide the gills with extra surface area for the absorption of oxygen, which should facilitate breathing in oxygen-poor waters. Sieve-like gills are also found in adult sword-fish (*Xiphius*) and sailfishes (Istiophoridae) and in the wahoo, swift scomboid fishes that live in the turbulent, well-aerated, surface waters of the ocean [52]. But they are large fishes, and when moving at high speeds (20 to 30 knots) their oxygen requirements must be great: hence, no doubt, the significance of gills with increased surface area. Remotely related fishes living in very different waters have thus acquired similar kinds of gills, which meet similar needs but at very different levels.

The tarpon, which lives in the warmer coastal waters of the

Atlantic Ocean, also has a lung-like swimbladder and well developed gills. But even in a well-aerated aquarium, young tarpon will die if prevented from gulping air at the surface. In nature, however, young tarpon grow up in the poorly oxygenated waters of lagoons, estuaries, mud-flats and mangrove swamps, which may well explain their ingrained habits of air-breathing. On the other hand, the characin *Erythrinus*, an inhabitant of the swamps of British Guiana and of other parts of South America is perfectly able to use its gills in well aerated water. If the oxygen content is low or the carbon dioxide content high, it must then take to breathing air.

But many air-breathing fishes, notably the snake-heads, anabantids and the cuchia, have reduced gills, and it is understandable that they will die if denied access to the air. Here, as in fishes that always respire through the gills, there is a certain functional diversity. And, if we had a truly aquatic insight, we should discover that there are as many variations as kinds of fishes.

Chapter 7

Feeding and Growing

IN his Gifford Lectures, entitled 'Man on his nature', Sir Charles Sherrington introduced a striking image of life.

> The living energy-system, in commerce with its surround tends to increase itself. If we think of it as an eddy in the stream of energy it is an eddy which tends to grow; as part of the growth we have to reckon with its starting other eddies from its own resembling its own. This propensity it is which furnishes opportunity under the factors of evolution for a continual production of modified patterns of energy [53].

The 'eddies' of fish life are energized by many kinds of food. Taking first the scavengers, a good many fishes feed on detritus, particles of broken-down organic matter that fall to the bottom. Grey mullets, for instance, obtain much of their food by hovering head-downwards and browsing along the bottom, swallowing the deposits. Together with the detritus, bacteria, algae, protozoans and other small invertebrate animals may also be ingested. Just how much nourishment comes from the detritus and how much from the associated organisms is uncertain. Yet grey mullets have a muscular gizzard, the action of which causes food material to be ground against swallowed grains of sand and other hard substances, so rendering the food to a fine paste. Certain intertidal gobies and blennies also feed on detritus, but this sort of regime is less common in the sea than in lakes and rivers. In African lakes, various cichlids, such as *Tilapia nilotica* and *T. leucosticta*, feed on the soft bottom deposits. In the Grand Lac of Cambodia, a quarter of the fifty-six species that form most of the fishery are detritus-feeders; they include certain species of *Labeo* and *Cirrhina*, members of the carp-family (Cyprinidae). Related fishes that are aptly called suckers (Catostomidae), and are almost entirely confined to North America, grub along the bottom, drawing in the deposits and contained organisms through their highly protrusible jaws, which bear thick, fleshy lips.

A good many freshwater fishes feed on plants. Compared to the carnivorous species, they usually have a long intestine, suited to the slower digestion of plant material. 'The grass carp, *Ctenopharyngodon idellus* (plate 5),' to cite Dr C. F. Hickling [54], 'will clip the grass growing on the margins of their ponds as neatly as a pair of shears, and this fish is regularly fed on cut grass thrown into the ponds.' He also states that the giant gourami (*Osphronemus olfax*) eats great quantities of soft leaved plants. In the Malagarasi Swamps in Tanganyika, biologists from the Jinja Laboratory found that two characins, *Alestes macrophthalmus* and *Distichodus*, were '. . . feeding on, and partially digesting, the leaves, buds and seeds of the water lilies. It seems that the partially digested remains of the water lilies, after passage through the gut of these two fishes, form a digestible food for other fishes, including *Tilapia*.' [54] The productivity of the swamps, which contain very little plankton, seems to depend a great deal on these two characins.

Besides consuming flowering plants, many fishes feed on algae that grow attached to such plants or to other suitable sites. Filamentous algae and diatoms form the main diet of many food fishes caught in Indonesian West Borneo and this is also true in the Grand Lac region of Cambodia. *Pangasiodon gigas*, a large, toothless Asiatic catfish that grows to a length of over 2½ metres, feeds exclusively on filamentous algae. Cichlids, such as *Tilapia mossambica*, and a gourami, *Trichogaster pectoralis*, browse on attached algae and diatoms. The South American armoured catfishes (Loricariidae), which have a broad-lipped mouth under the snout, scrape algae from rocks and stones. And these are but a few instances of the great importance of plants in the diet of tropical freshwater fishes.

'In the sea and in the fresh waters of temperate climates, plant-feeding fishes are in the minority.' Dr Hickling [54] also draws attention to the part played by flooding in tropical regions. Herbivorous fishes move over inundated forests and plains, where they find grasses, decaying vegetation, leaf-debris, and the fronds and seeds of plants. The marine 'forests', where live the greatest diversity of herbivores, are made of calcareous algae and corals. More will be said of the food of coral fishes in a later chapter (p. 323), but we may note here that certain surgeon-fishes (Acanthuridae), rabbit-fishes (Siganidae), damsel-fishes (Pomacentridae), butterfly-fishes (Chaetodontidae) and trigger-fishes (Balistidae) are among the most important herbivores of coral reefs.

Plankton-feeding fishes are more prominent in the ocean than in rivers and lakes. The whale-shark (*Rhincodon*), basking-shark (*Cetorhinus*), manta-rays (Mobulidae), many herring-like fishes, flying-fishes, half-beaks and silversides (Atherinidae) depend largely on planktonic food. In all these fishes there is some kind of screen across the gill arches, which may be fine enough, particularly in conjunction with a curtain of mucous, to trap the microscopic plants. Menhaden (figure 30), pilchards (*Sardina* and *Sardinops*) and *Sardinella* species are among the herring-like fishes that gain much nourishment from planktonic plants and very small members of the zooplankton. Moreover, the larval stages of both marine and freshwater fishes feed on the phytoplankton and small crustacean larvae, such as the nauplii of copepods.

Certain freshwater fishes also depend on planktonic plants. A few species of the cichlid genus *Tilapia* are well fitted to deal with such food. In *T. esculenta*, Dr P. H. Greenwood found that the plankton, which enters in the respiratory stream, is trapped by a curtain of mucous in the mouth. The gill rakers bar the escape of the slimy boluses of food, which collect at the back of the throat, where they are raked into the gullet by the pharyngeal teeth [54]. Other cichlids depend on the animal plankton. In Lake Nyasa, for instance, there is a mid-water community of *Haplochromis* species which feed largely on copepods. The *dagaa* fishery of Lake Tanganyika, which consists mainly of a small anchovy-like fish (*Stolothrissa tanganicae*) mixed with an inshore clupeid (*Limnothrissa miodon*), is supported by the lake stocks of zooplankton. *Stolothrissa*, a mid-water fish, feeds on the animal plankton, following its up and down migrations during the day.

Keeping to fresh waters and turning first to the small carnivorous fishes, many characins, cyprinids, cyprinodonts and cichlids depend for much of their food on the aquatic larvae of insects, particularly on those of chironomid midges and mosquitoes. Indeed, in the economy of freshwater fish life, insect larvae almost take the place that the zooplankton occupies in the ocean. The surface film is also an important feeding place, especially for many cyprinodonts, which have small upturned mouths. Here they find small insects and other prey.

Some kinds of catfishes are among the larger predators. The sizeable species take frogs, fish, shrimps and so on. The best known predators among the characins are the South American piranhas (*Serrasalmus*) and the African tiger-fishes (*Hydrocyon*), which feed on other fishes. African lung-fishes also take fishes as well as molluscs, crustaceans and water beetles. Certain cyprinid fishes, which have pointed or hooked pharyngeal teeth, are also active predators. In the temperate fresh waters of North America, Europe and Siberia, pike (*Esox*), perch (*Perca*) and pike-perch (*Lucioperca*), are prominent carnivores. In North America, black-bass (*Micropterus*) bowfin and garpikes can be added to the list. Of the smaller species, sticklebacks are noteworthy. Thinking particularly of the three-spined species *Gasterosteus aculeatus*, A. F. Magri MacMahon [55] wrote thus:

> If I were asked which British freshwater fish I consider the most ferocious, I would not answer trout, eel or even pike: I would without hesitation award this undesirable pre-eminence to the stickleback, that fierce attacker of other creatures even bigger than itself, that voracious devourer of small fry. If the innumerable sticklebacks of our shores and rivers were as large as pike, they would make our waters more dangerous, even to man, than if they were inhabited by alligators.

In the surface waters of the open ocean, the main predators are sharks, dolphin-fishes, tunnies, marlins and sword-fishes. The sharks, such as the blue-shark (*Prionace glauca*), the man-eater (*Carcharodon*), mackerel-sharks (*Lamna*), makos (*Isurus*), hammerheads (*Sphyrna*) and threshers (*Alopias*) depend mainly on fishes and squid (plate 10). These are also the prey of tunnies, which also take a variety of pelagic crustaceans. Dolphins are avid pursuers of flying-fishes. Living below these large predators are hundreds of kinds of small mid-water fishes, mainly less than a foot long; most of them have needle-like or fang-like teeth.* They also seize fishes, squids and crustaceans and many, as we shall see in Chapter 17, are able to swallow relatively large prey, even larger than themselves. The smaller stomiatoids and lantern-fishes, which usually have many small teeth, depend largely on plankton.

In tropical coastal waters, garfishes, trumpet-fishes, groupers (*Epinephelus*), barracuda (*Sphyraena*), snappers (Lutianidae), bluefish (*Pomatomus*), amber-jacks (*Caranx* etc.) and certain drum-fishes (Sciaenidae), prey on small fishes and crustaceans. Moray-eels (Muraenidae), pike-eels (*Muraenesox*), lizard-fishes (Synodontidae), wrasses, hawk-fishes, scorpion-fishes and others find much of their prey on or near the sea floor. Rays with a crushing dentition, such as sting-rays and eagle-rays; also various wrasses and breams with molar-like teeth, feed largely on benthic molluscs and crustaceans.

* Stomiatoids, alepisauroids, giganturoids, Lyomeri (gulper-eels), angler-fishes (ceratioids), etc.

The skates, which are predominantly fishes of temperate seas, take fishes as well as crustaceans and other invertebrates.

The cod-like fishes (Gadidae) include many voracious predators. Atlantic cod (*Gadus morhua*) and Pacific cod (*G. morhua macrocephala*) eat fishes, crustaceans and molluscs, as do ling (*Molva*) and hake (*Merluccius*). Flat-fishes may be divided into small-mouthed, large-mouthed and halibut-like species. The first, such as dabs, plaice and flounders feed on small benthic organisms; worms, molluscs, crustaceans, echinoderms and so on. The large-mouthed kinds such as turbots and brill take fish as well as invertebrates. Halibut seize fishes, large crustaceans and cephalopods.

This is no more than an outline of the food of fishes (but see also pp. 293–343 on living spaces) and little need be said of the digestion of food. As in other vertebrates, preliminary digestion of proteins takes place in an acid medium produced by the stomach. But a number of fishes have no stomach; examples include chimaeras, lung-fishes, certain cyprinids and cyprinodonts, sea-horses and skippers (*Scomberesox*). The food passes straight from the gullet into the intestine, where all the digestion occurs.

The intestine is a simple tube, which may be long and coiled in plant and detritus-feeding species. In most teleosts its surface area is increased by a number of blind, tubular outgrowths known as pyloric caeca, which emerge close to where the intestine joins the stomach (figure 24). Protein digestion is completed in the intestine, together with that of fats and carbohydrates. The products of digestion are absorbed by certain cells in the walls of the intestine, and are thence circulated in the blood to all the cells of the body.

Food chains

The earth absorbs a minute proportion of the energy that streams from the sun. During the day time, chlorophyll-bearing plants intercept a small fraction of this energy, which is turned into the synthesis of food stuffs needed for plant maintenance and growth. As more and more humans are becoming aware, green plants are the ultimate basis of animal life on the earth. Now the sea covers rather more than two-thirds of the earth, but the microscopic marine plants, which can only synthesize their foods in the sunnier surface waters, are not necessarily twice as productive *in toto* as the plants on the land. The ratio between total plant productivity in the sea and on the land seems to be somewhere between two and one.

But one state of affairs applies equally in both environments: herbivorous animals have more food at their disposal than have carnivores. If, during any time, the herbivore digests more food than is needed for its activities and the repair of its tissues, the excess food can be turned into new protoplasm. Let us suppose that a fish can

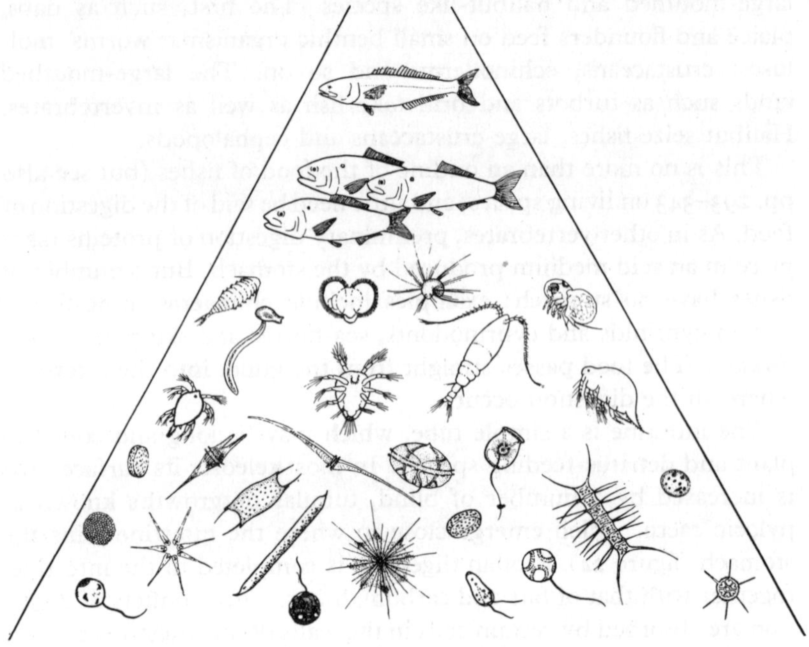

Figure 30. Different levels in the food chain may be represented by a pyramid. The base of the pyramid is supported by planktonic plants, which directly or indirectly support animal life. The second stage of the pyramid is represented by hosts of larvae and small planktonic animals, such as copepods, which feed directly on the plants and in turn are consumed by such fishes as menhaden (*Brevoortia*) shown as the third step in the pyramid. Above the menhaden is a blue-fish (*Pomatomus saltatrix*), a species which preys heavily on menhaden. The tapering of the pyramid expresses the diminished amount of food available between successive links or levels of the food chain.

use, on an average, one seventh of the weight of its food for growth and this is close to experimental indications. Over a given period, then, 1,000 pounds of plant food will turn into about 150 pounds of new flesh in a stock of herbivorous fish. Now one seventh of the weight of these plant-eating fishes will be available for the increase of a carnivorous species, which in turn will yield a seventh of its substance to a larger carnivore. There is thus less and less living material available down each step of the food chain. Thinking in terms of energizing and widening the eddies of fish life, about 15 per cent of the potential calories in growing tissues is available between each successive link of the food chain. The nearer the link is to the sun and plants, the more the available calories (figure 30).

But, except for the milk-fish (*Chanos chanos*), no marine fish seems to live wholly on plants. Larval fishes take a lot of plant food and so do certain clupeid fishes. To give two instances, juvenile Indian oil sardines (*Sardinella longiceps*) depend greatly on phytoplankton, which is also the main food of South African pilchards (*Sardinops sagax*). In tropical fresh waters, however, there are many herbivorous species. As Dr Hickling remarks: 'One of the reasons for the high rate of fish production in tropical waters, even when conditions seem unsuitable, is that there are a large number of plant-eating fish in the tropics.' [54] In terms of nutriment, one could also add that a fish will gain more from feeding on phytoplankton than on flowering plants. Compared to most land plants, the diatoms and various flagellates of the plankton contain high quantities of proteins and fats.

In ocean, lake and river, predatory fishes are larger and fewer than the species that eat plankton, which is to be expected, having in mind the decreased availability of food to carnivores. (Two exceptions in the matter of size are the large planktonic-feeding sharks and rays and many small deep-sea fishes.) In Lake Tanganyika, for instance, one night's work with a purse seine can yield 5 to 6 metric tons of fish. About two-thirds of this weight consists of *dagaa*, the small 2 to 3 inch clupeid fishes that feed on zooplankton (p. 101). Twenty per cent by weight are their predators, the Nile perch (*Lates*) and the mackerel-like *Luciolates*. Dr Max Poll (through Dr Hickling) has given a just account of the food chains.

> Towards the end of the afternoon, when his ship was stopped in mid-lake for a hydrographic station, he saw, at about five o'clock, the water change colour. Thousands of little fish were passing at the surface and he alerted the crew: There are the *Dagaa*. But their swarm was soon followed by bigger silvery shapes, those of the *Luciolates stappersii*, a fish of mackerel-like habit which feeds on the *Dagaa*. Finally, at the edge of the bank of fish could be seen the big shape of *Lates*, the wolves of Tanganyika, taking their toll of the *Luciolates* [54].

Turning to the sea, a vivid impression may be obtained of the great abundance of fishes that feed on plankton by considering the menhaden (*Brevoortia tyrannus*), which lives along the eastern coasts of North America. Menhaden are sturdy, tight-muscled clupeid fishes of great stamina and turn of speed. Besides man, they have a host of predators.

> Menhaden [as William A. Ellison observes] . . . is the prey to virtually all of those carnivorous fishes which inhabit the same waters. . . . In New England it is eaten by the whiting, cod-fish, pollack, dog-fish, shark, tuna, whale, and even the flounder. In more southern waters it is eaten by the pompano, cavally, bonito and bayonet-fish. Along the entire coast where-ever they are found in common with menhaden, the striped-bass, weak-fish, or sea trout and dolphin destroy and consume them. They are attacked and eaten in rivers, particularly southern rivers, by garfish and cat-fish.
>
> Of their enemies, the whales, dolphin, shark, tuna, blue-fish and weak-fish are the most destructive. As many as one hundred individuals have been taken from the stomach of one shark, and Dr Goode says that the whales and dolphins consume them by the hogshead. From the air, predatory sea birds attack the schools, and it is not unusual to see gulls riding the schools and enjoying a feast.
>
> The quantities in which they are eaten or destroyed by their natural enemies is prodigious. Dr Goode estimates that the number of menhaden annually destroyed by natural enemies amounts to a million, million of millions, or put differently, a thousand times the number taken by man in 1948, let us say, when the greatest capture on record was made [56].

Even more striking, nearly half the yearly world catch of fishes consists of herring and herring-like species. In 1951, the weight of such fishes amounted to 6,400 thousand tons, which may be compared with 3,400 thousand tons of cod-like fishes and 1,500 thousand tons of mackerel and tuna-like fishes. Since, *inter alia*, herring-like fishes are difficult to catch, these figures do not give a fair idea of relative abundance. All the same, one gets a vivid impression of the greater productivity of fishes that live near the sunny, plankton-producing end of the food chain.

The growth of fishes

Apart from the live-bearing kinds, such as sharks, rays, top-minnows, surf-fishes and so on, most fishes are born in a larval state. After using their yolk reserves, the larvae begin to feed and continue their growth and development. Now the eggs of any one species of fish are of much the same size, which is also true, as we might expect, of the newly born larvae. But soon after they begin to feed, as may be seen in rearing tests, some larvae are appreciably larger than others hatched at the same time. This may simply mean that the larger larvae chanced to find a better supply of food, or that they began to feed a little earlier than the smaller ones, which must often happen in nature. Yet some tests have indicated that there can be genetic differences: some larvae, so long as they find adequate food, possess gene combinations that give them a quicker growth rate than their contemporaries. Whether due to genetic or nutritional advantages, or both, a good start in a fishes' growing life is vital.

When a larval fish is ready to feed it must find food of a suitable size, such as phytoplankton organisms and very small members of the zooplankton. If it grows faster than its fellows, the sooner it will acquire larger jaws to take larger kinds of prey and it will go ahead even more.

Young fishes that outstrip their contemporaries may also have another advantage, at least in certain species. Trout fry and some salmon fry soon acquire a territorial sense, when the larger individuals become the dominant fish, feeding and growing better than the smaller ones.

> When food [as Dr Margaret Brown observed] is added to a tank with trout of various sizes, they all swim towards it and there is no obvious reluctance of smaller fish to feed in the presence of larger ones. When some food remains on the bottom of the tank the small fish have ample opportunity to feed after larger ones have satisfied their appetites, yet even when the foods consists of live and moving organisms, the small fish do not grow as large as the larger ones. It is possible that they are inhibited from feeding satisfactorily by the presence of the larger individuals. Another possibility is that the smaller fish suffer from stress in the presence of larger ones, and that their production of adrenocorticotropin (ACTH) is increased [57].

This substance, which is released by the pituitary gland, acts on the interrenal organ and an increased output, following stress, can lead to a decided decrease in growth. Whatever the causes of a slower growth rate, the disadvantages of the smaller fry are well known in trout hatcheries. A few weeks after hatching, they are separated into three size groups: large, medium and small. They are graded again after another few weeks, when some of the small group will have grown to the size of the large ones.

When young fishes are reared in tanks, the growth of the smaller ones may be affected in another way. The small cyprinid fishes,

Tanichthys albonubes and *Barbus tetrazona*, excrete a metabolic product that inhibits the growth of the smaller individuals. The females of both species laid about two hundred eggs, but in a 15-litre aquarium no more than fifteen to twenty of the young reached a length of 1 centimetre. Soon after the larvae began to feed, differences of size became evident. The larger ones continued to eat and grow, but the smaller individuals stopped feeding and died, even though there was an abundance of food. Changing the water in the aquarium several times a day made a great difference to the survival of young *Barbus*: instead of fifteen survivors or less, there were about one hundred and seventy. By drastically reducing the concentration of the metabolite, it would appear that the growth-retarding properties were largely eliminated [58].

Recent rearing tests with the medaka (*Oryzias latipes*), a small top-minnow from the rice fields of Japan, have been particularly illuminating. If young from the same parents are grown in isolation under the same conditions, some grow larger than others, which may well be due to genetic differences. These divergences in growth rate are no greater than when the fish are kept together at densities up to sixteen fish per litre, provided that they have plenty of food and the growth-retarding metabolites are removed.

Aggression among the medaka was greatest at moderate densities, not at the highest. If the food given was limited, a social hierarchy emerged and the larger fish became dominant and grew faster than their subordinates; but if food was abundant, the larger fish had no advantage. In a dense population given restricted and patchy supplies of food, the dominant fish was unable to defend his holding against all comers. When the food was evenly distributed, aggressive activities dispersed the medaka over the aquarium, but this did not happen if the food was patchy. These and other tests convinced Dr J. J. Magnuson [59] that aggressive behaviour is not so much a means of competing for space, but that it does reserve a greater share of food for larger and dominant fish under certain environmental conditions. This reminds us of Professor V. C. Wynne-Edward's contention [60] that it is through conventional forms of behaviour that animals conserve their food resources and control their numbers.

Earlier in this chapter (p. 104), we considered the proportion of its food that a fish can use for growth. The fraction one-seventh refers to favourable conditions. Looking more closely at a variety of conditions, Dr Otto Kinne [61] has investigated the growth of the desert pup-fish (*Cyprinodon macularius*), a small (20 to 40 millimetre) top-minnow that lives in the basin of the Lower Colorado River. The fishes tested came mostly from a population found near the mouth of a small creek that runs into the north-eastern part of the Salton Sea, California. In this region the desert pup-fish can endure considerable variations of temperature and salinity: indeed, it can run its entire life-span in salinities ranging from that of fresh water to sea water.

Rearing tests revealed that the rates of growth, food consumption and food conversion vary greatly within the ranges of temperature and salinity that occur in the fishes' natural surroundings. At the salinity of sea water (35 per thousand or 35‰) and on a restricted supply of food, young fishes grew quickest at 30°C, the rate decreasing with temperatures of 30° to 15°C. Regarding the effects of salinity, at a temperature of 30°C growth was maximal at the salinity of sea water and decreased in saline conditions in this order: 35‰, 15‰, 55‰ and fresh water. Food intake was highest at this combination of temperature and salinity: 30°C and 35‰.

The conversion of food to new living substance can be estimated by measuring the increase in weight of a fish over a given period and knowing the weight of food that was ingested. Dr Kinne expressed this in terms of dry weight (the increase in dry weight of the fish divided by the dry weight of food eaten $\times 100$). This conversion efficiency was highest at 20°C and it decreased in the following order of ambient temperatures: 20°, 15°, 25°, 30° and 35°C. Conversion efficiency is also influenced by salinity, being greatest at 15‰. At a salinity of 35‰ and temperatures ranging from 15° to 35°C, the conversion efficiency ranged from 3 to 32 per cent. On an average it was 15 per cent, which is close to our factor of one-seventh.

Few fishes live in so varied an environment as does the desert pup-fish. For most species the most uncertain factor is likely to be their supply of food, but fishes can adjust their rate of living to starvation conditions. Indeed, those of temperate regions, marine and fresh-water, do little or no feeding during the winter months, when their growth is almost at a standstill. This is reflected in the growth rings of the scales, ear-stones and bones: a wide zone formed during the favourable growing season is coupled with a narrow zone formed during the winter, the two making an annual ring. No one knows how the largest temperate fishes, basking-sharks (*Cetorhinus*), fare during the winter months but the few strays that have been caught during this season lack gill rakers. Now there is certainly not enough plankton in temperate waters to sustain these great beasts during the winter. Dr L. Harrison Matthews has calculated [62] that

> . . . a shark seven metres long would need to take each hour food with a calorific value of 663 calories to give merely the energy required in swimming to collect the food. Even if we multiply the recorded density of plankton in the North Sea in November by three, the shark's intake could only be 410 calories per hour under the most favourable conditions, and it would be losing on the deal. So it solves its problem by throwing away its rakers worn by a season's use, refraining from feeding, sinking to the bottom and hibernating. At this moment (he wrote in February, 1962) there are probably great schools of these enormous fish quietly resting on the bottom of the sea, perhaps in the heads of canyons at the edge of the continental shelf, with their metabolism running at its lowest level while they grow their new rakers ready for browsing on next summer's crop of plankton.

But fishes of Antarctic coastal waters, which I saw for myself, feed avidly during the winter months, when they live below the sea ice and at freezing sea temperatures of $-1{\cdot}9°C$. My measurements indicated that they grow during the winter, yet one can see annual rings on their scales. Their slower growing period is presumably at some other season. In the tropics, however, where there are no marked seasonal differences between the production of plankton, growth is more or less continuous. The scales of most tropical fishes have no annual rings.

Unlike birds and most mammals, fishes with a long enough life span continue to grow after reaching sexual maturity. Annual fishes, which are small top-minnows living in parts of tropical Africa and South Africa, last no more than one year. They grow to maturity, spawn and die soon afterwards. But some tropical freshwater fishes that live for several years, as do certain species of *Tilapia*, grow little after sexual maturity. Moreover, the males of some top-minnows, such as *Lebistes* and *Xiphophorus*, stay at a definite size on becoming mature, whereas the larger females continue to grow. Even so, most fishes that live on an average from five to ten years, such as goldfish, salmon, trout, freshwater bream, pike, haddock, cod, herring, pilchards and mackerel, keep growing after maturity. Now the larger the fish, the faster its swimming speed and the greater the size range of food that can be eaten. In this respect fishes are better off than mammals or birds, which after passing their period of active maturity, must give way to more vigorous individuals of the same size; whales are exceptional in continuing to grow after attaining sexual capacity.

Growth may be measured by the increase in length or weight over a given time. For instance, the length (L) and weight (W) of a fish can be related by the formula $W=aL^n$, a being a constant and n an exponent having a value between 2·5 and 4·0 [63]. In a local population size is a good criterion of potential growth. The larger fish, as we saw, tend to be dominant as well as having better powers of swimming and the ability to take more food. Growth is rapid during the early life of a fish, but slows down or stops after the attainment of physical maturity. In fishes that are born as larvae, there are also great changes in shape during the early life history. Larva, the Latin word for a mask, is just as apt a term for a newly born fish as for the early stages of invertebrate animals. The masks give little hint of the adult form, except that there is usually a complete series of muscle segments around the notochord, the early frame of the backbone. The eyes are also well formed (figure 64). Soon after, or even before the yolk is absorbed, the mouth and jaws are ready. Muscles, jaws and eyes must soon be put to use in finding and taking food. Fins, apart from the flimsy, fan-like pectoral fins and a medium fold along the trunk and tail, have yet to be formed. But the larvae grow quickly if they find enough food, and very soon the fins take more definite shape and begin to acquire their rays. At the same time, the sense organs, nervous system, gills, blood system and so on are becoming more elaborate. The young animals are looking more like fishes, but they have still far to go. Now follows a rapid metamorphosis; a marked refashioning of form, fin patterns, colour pattern and so on, at the end of which the larvae have shed their 'masks' and are adolescents. Despite continual but gradually slower growth, the fishes now keep much the same shape for the rest of their lives.

Growth depends on the pituitary gland. If this gland is taken from a common killifish (*Fundulus heteroclitus*), a top-minnow found along the eastern shores of North America, growth comes to an end. Though it still continues to feed, the fish stays the same size and no new circuli of bone are formed at the edges of scales. After injecting the fish with the pituitary growth hormone, somatotropin, growth is resumed. The pituitary gland, which probably acts together with the thyroid, stimulates and regulates growth; but whether the thyroid gland is involved in metamorphosis, as in amphibians, is still uncertain. Close study of hormones in fishes is only a few years old [64], and we can only await further developments, which are bound to enlarge our understanding of the complex processes of growth in fishes.

Chapter 8

The Internal Environment

THE LIFE of vertebrates is sustained by processes that preserve the stability of their internal environment, which is a name for the watery continuum that bathes the cells of the body. Water forms about four-fifths of the weight of most fishes: extreme values range from about 50 to 90 per cent. Most of the water is within or around the cells, but a small proportion, usually from 2 to 10 per cent of the body weight, is the fluid basis of blood and lymph. When these body fluids are analysed, they are found to contain salts in certain definite amounts and this balance of salts, no less than the balance of water, must be maintained if life is to continue. For the blood and lymph are in equilibrium with the internal environment, which forms, as it were, a microcosmic ocean, a salty medium that has to be balanced if cellular life is to flourish. Yet life will not proceed unless the waste products of excretion are eliminated from the body. Even so, one qualification is needed. Sharks, rays and chimaeras, as we shall see, are unique in being able to maintain high quantities of urea in their blood; quantities that would poison any other vertebrate, but which are vital to their wresting of water from the sea. Excretion, which may involve gills as well as kidneys, and the keeping of a proper water and salt balance, must be integrated activities. To these we may now turn.

The principal ions in the body fluids of fishes are also those of sea water: sodium, potassium, calcium, magnesium, chloride and sulphate. But, except in hagfishes, the total quantity of salts is less than that in the same volume of sea water. This means that freshwater fishes have more salts in their body fluids than occur in the same volume of the outer environment. Now, if a membrane that is much more permeable to water than salts separates two solutions containing different quantities of salts, water will pass from the weaker to the stronger solution; this process is called osmosis. Living membranes have this property and in fishes the most pertinent ones are the delicate epithelia covering the gill filaments and the mucous membranes of the mouth and pharyngeal cavities. Marine fishes with body fluids that hold less dissolved material than is contained in an equivalent volume of the sea will thus tend to lose water through these membranes; but in a freshwater fish the reverse will happen. Both tendencies must be countered if fishes are to preserve the inner stability on which their lives depend [65]. These, and related, activities will now be considered.

Marine Fishes

Sea water containing thirty-five parts of salt in a thousand parts of fluid (35‰), which may be taken as the mean salinity of the ocean, freezes at a temperature of $-1{\cdot}9°C$. The body fluids of fishes freeze at temperatures ranging from $-0{\cdot}38°$ to $-1{\cdot}9°C$. The more the quantity of salts and other substances in solution, the lower the freezing point. More specifically, the depression of a solvent's freezing point is proportional to the molecular concentration of the solute. The blood of hagfishes holds salts that are equal in concentration to those in sea water, which means that the freezing point of hagfish blood and the sea are much the same, and measurements have confirmed this. A similar correspondence, but it is not quite so close, is found in the cartilaginous fishes. Yet the blood of a shark, a ray or a chimaera, like that of a bony fish, is less salty than sea water, which has already been implied. We have also alluded to the presence of urea in the body fluids of these animals. It is this excretory product, together with the salts, that brings about the near match of freezing points.

By this unique biochemical adaptability, sharks and their relatives have evaded the physiological stress that faces a marine teleost. Holding less salt in its blood than exists in the same volume of sea water, a teleost loses water through its gills and the membranes of the mouth and pharynx. To offset this loss, it drinks sea water (figure 31), which can be shown by adding a known quantity of a dye, such as phenol red, to the water of an aquarium. The dye is taken up by the tissues of the gut and by measuring the amount absorbed, an estimate can be made of the volume of sea water that the fish has swallowed. In a period of twenty hours, for instance, a freshwater eel (*Anguilla rostrata*) absorbed 10 millilitres of sea water. A fish will also obtain some water, which will be salty, from its food.

This is one step towards restoring the balance of water in a teleost fish's body. To properly appreciate this balance, the water lost in the urine must also be taken into account. We must thus consider the activities of the two kidneys, which are suspended from the roof of the body cavity (figure 24). Each kidney consists of an elaborate system of tubules that discharge their excretory products into a ureter (figure 31). In turn the ureter empties these wastes into the urinary bladder or corresponding structure. When the kidney is fully developed, each tubule is headed by a capsule known as Bowman's

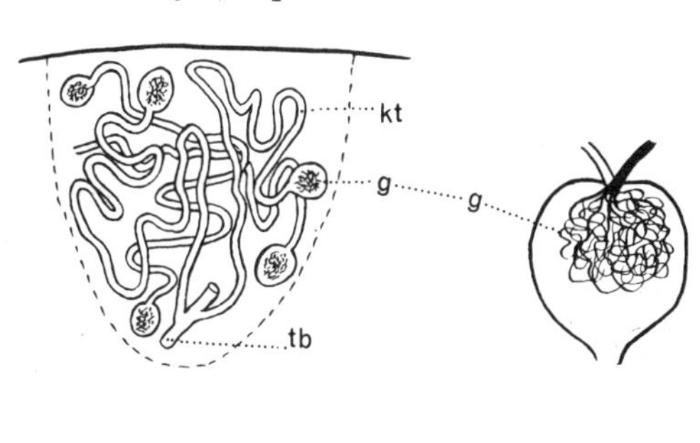

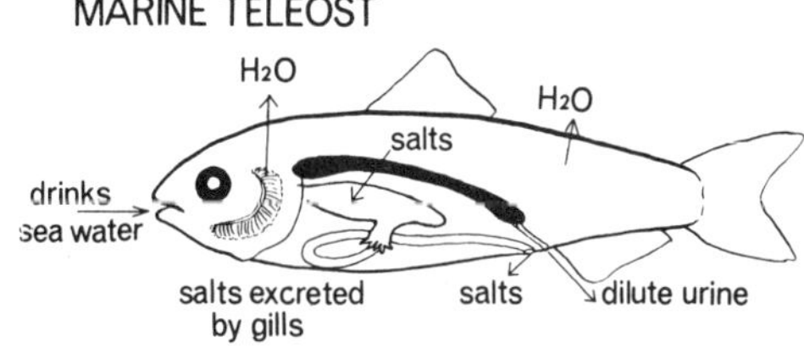

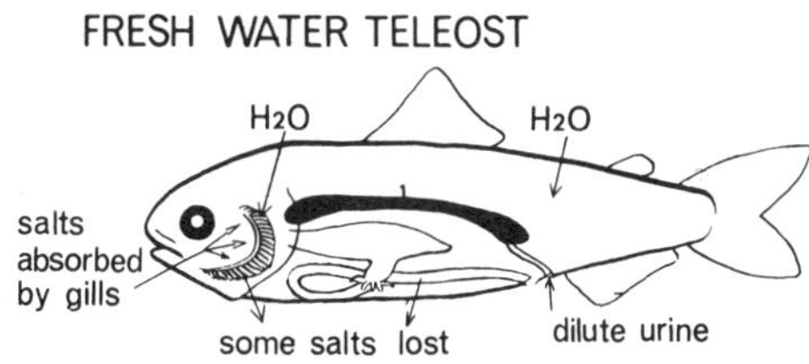

Figure 31. Above left: A few of the tubules in a kidney. The glomeruli (g) are about a quarter of a millimeter in size. The kidney tubules (kt) discharge into ducts (tb), which run to the ureter. Below: water and salt balance in marine and freshwater teleosts. The kidneys are shown in black.

capsule containing a small bunch of arterial blood capillaries, the glomerulus, which is connected to a small vein (figure 31). In some teleosts, as we shall see, these capsules are either incompletely formed or are absent. Now the function of each glomerulus, and there may be tens of thousands, is to serve as a filter bed for water, salts, nitrogenous wastes and sugars, which pass into the tubule. But much of the water is reabsorbed by the tubule, and so finds its way back into the body fluids. Experiments on a sculpin (*Myoxocephalus octodecimspinosus*) showed that it filtered 14 millilitres of fluid through the kidney glomeruli in one day, but excreted only 3 millilitres as urine. In fact the small amounts of urine discharged by marine teleosts hold a concentration of solutes about equal to that in the blood.

Though a teleost fish drinks sea water to make good the loss of water through its membranes, the kidneys must thus do work to preserve a healthy fluid balance. Above all, the capillaries of the glomeruli must be well supplied with fluid – needed to build adequate blood pressures – if they are to function efficiently as filters. Freshwater teleosts, which have plenty of internal freshwater at their disposal, develop greater filtering surfaces than their marine relatives. But, some teleosts seem to have found this provision of water a strain, for the glomeruli and their capsules have either become small and non-functional or have disappeared without trace. The angler-fish, for instance, has small inactive glomeruli, while the toad-fish and various pipe-fishes possess kidneys without glomeruli. In such fishes the kidney tubules must take over the functions of the glomeruli. But so far as present knowledge allows us to judge, the loss or regression of the glomeruli is a less common event than is their retention. Herrings, cod-fishes, flatfishes and mackerel are instances of successful families of teleosts that have well-formed glomeruli in their kidneys.

When a teleost fish drinks sea water, it takes into its system a fluid containing salts in greater concentration than those that circulate in its body fluids. To maintain its salt balance, the excess salts must be eliminated. Most of the calcium, magnesium and sulphate ions are concentrated in the intestine and pass out in the faeces. Sodium, potassium and chloride ions are absorbed from the gut by the blood and pass to the gills, where they are excreted into the water. Large secretory cells in the gill filaments are said to eliminate these ions and similar cells may also be found in the mucous membrane of the mouth and pharynx. In both marine and freshwater teleosts the gills also get rid of the simpler products of nitrogenous excretion, urea and ammonium compounds. The kidneys deal with the more complex wastes: creatine, creatinine and uric acid.

Compared to those of teleosts, the gills of sharks and rays are much less permeable to urea. By this holding provision and, it seems, through some reabsorption of urea by the kidney tubules, much of this substance is retained in their body fluids, which also contain small amounts of another waste product, trimethylamine oxide. It is these two substances, together with the salts, that give the body fluids of cartilaginous fishes a total solute concentration a little higher than that of sea water. This means that a shark or ray has no need to drink sea water, for water will enter its body through the gills and mucous membranes of the mouth and pharynx. The intake is evidently enough for the purposes of filtration in the kidney glomeruli – and no shark or ray has kidneys without glomeruli – for their filtering surface is comparable to that found in freshwater teleosts. By this curious biochemical aptitude, by playing safe, as it were, cartilaginous fishes have evaded the stresses of water and salt balance that occupy a marine teleost. It would be interesting to know how much the osmotic 'spirit of adventure' of the teleosts has contributed to their success. Has the possession of an osmotically strong kind of blood prevented the cartilaginous fishes from invading suitably sized freshwater habitats ? There is no fossil evidence to show that these fishes were ever important members of the freshwater faunas. Today, certain of the blue-sharks (*Carcharhinus* spp.) will sometimes venture upstream into brackish or even freshwater surroundings, while one species is found in Lake Nicaragua. Some of the sawfishes (Pristidae) also enter rivers and there are certain freshwater sting-rays (Pomatotrygonidae).

On entering a freshwater or brackish habitat, a shark or ray would seem to be in danger of becoming water-logged. The kidneys would have to work hard to cope with the greatly increased intake of water. Experiments on the sawfish *Pristis microdon* in the Perak River, Malaya showed that it produced more than ten times as much urine per unit of weight per day than marine dog-fishes. It had, however, lower salt and urea concentrations in its blood than did the marine species. Some adjustment to reduce the water intake is thus possible, but it is clearly a rare attainment.

Freshwater Fishes

Freshwater teleosts contain a less salty blood than do their marine relatives, the concentration of salts being nearer a quarter rather than a third of those in the sea. Even so, the body fluids of freshwater teleosts are rich in salts compared with the water in which they live. As might be expected, water freely enters their blood through the gill surfaces and mucous membranes. They thus acquire ample water for developing high blood pressures in the capillaries of their kidney glomeruli, which are always well developed. More precisely, the total glomerular surface for the filtration of wastes is greater than that found in marine teleosts. The kidneys are thus well fitted to eliminate water that is in excess of bodily requirements. In fact they produce a copious flow of urine. In one day, a rainbow trout, a gold-fish, a carp or a cat-fish (*Ictalurus nebulosus*) will produce a volume of urine that is ten or more times as much per kilogram as that produced by a marine teleost, such as a conger-eel, a moray-eel (*Muraena helena*), a sculpin (*Myoxocephalus*) or a scorpion-fish (*Scorpaena*).

There is, however, the problem of salt balance. Living membranes are not entirely impermeable to salts, which tend to diffuse through the gills to the salt-poor water that is continually passing over the filaments. Salts are also lost in the urine and faeces (figure 31). A freshwater teleost will derive some salts from its food, though this may not be enough to restore the balance. There is, in fact, evidence that the gills can actively absorb salts from the surrounding water.

Concerning this, and many other physiological aspects, we owe much to the late Professor August Krogh [66]. Since many freshwater fishes can go without food for very long periods, he thought it '. . . probable, *a priori*, that they possess, like so many of the freshwater invertebrates, a special mechanism for the absorption of ions from the water'. After washing out rainbow-trout, roach, gold-fish, cat-fish, stickleback, ruffe (*Acerina*) and perch in a slow current of distilled water, he found that these fishes could absorb salts (he studied chlorides) to make good the loss, but in varying degrees. Rainbow trout and gold-fish, for instance, were more active absorbers than roach. Krogh and other investigators have also shown that the salts are actually absorbed through the gills (figure 31). Besides their primary rôle in respiration, the gills of fishes may thus act as salt absorbers as well as salt eliminators (p. 116). In being actively and largely involved in keeping a healthy salt balance in both freshwater and marine teleosts, the gills have evidently facilitated the spread of these fishes into all manner of aquatic habitats. We have also seen that the gills are particularly active in getting rid of the simpler products of nitrogenous excretion. The versatility of teleost gills is almost a match for that of their swimbladder. Both kinds of organs act to keep a delicate balance, whether of salts or buoyancy. But we might expect an extremely adaptable group of fishes to have unusually adaptable organs.

The meeting of land and sea

Dr Lionel A. Walford writes as follows: 'There are long stretches of coast around the world where land and sea are not so sharply distinct as they are at rocky shores; where instead, the two merge gradually in an irregular and sometimes intricate edging of estuaries, sloughs, lagoons and mudflats, brackish swamps and fringing islands. This transitional area where the land reaches out to the sea and the sea to the land, is one of the most interesting of all marine ecosystems, and perhaps from a fishery point of view the most valuable. In some places it is richer than the richest farm country, for it is lavishly fertilized with inorganic nutrients which the land is continually pouring into it.' [67] The mixing of fresh and salt waters is particularly extensive and complex along the coasts of lands where rainfall is heavy and great rivers meet the ocean, as along the coast of north-east South America, West Africa, South-East Asia and Indonesia. Now in regions where the tides impinge on land waters, many invertebrate animals and fishes live in the brackish waters that are formed. Indeed, a good many species have spread from one medium to the other, but the movement is mostly from salt to fresh water. In North and Middle America, a hundred and six species of marine fishes enter freshwater habitats, almost nine times as many as the twelve freshwater species that can endure salt water.

Concerning the invasion of freshwater by marine fishes, let us first look at Thailand. Besides the many species, particularly those belonging to the order Ostariophysi (cyprinids, homalopterids, loaches, catfishes, etc.) that are confined to fresh waters, numerous marine fishes also live in the rivers. There are two saw-fishes, a shark (*Scoliodon walbeehmii*), two sting-rays, tarpon (*Megalops cyprinoides*), various herring-like fishes, certain anchovies, a pike-eel, a snake-eel, a garfish (*Stongylura*), a half-beak (*Hyporhamphus*), pipe-fishes, soles (e.g. *Synaptura* and *Cynoglossus*), numerous gobies, certain puffer-fishes (Tetraodontidae) and others.

On entering rivers, marine fishes have not only to change their ways of water and salt balance, but must face competition from the local inhabitants. (They have also to return to salt waters to spawn.) As estuaries can be rich feeding grounds, perhaps such competition is not too severe. Moreover, there are many volcanic islands, as in the tropical west Pacific Ocean, where the entire fish fauna of mountain streams that run down to the sea has come from the ocean. Living in the head-waters are freshwater eels, while gobies can endure the boisterous conditions in the cascade zone. Lower down, and still in freshwater surroundings, there are luminous perches (Leiognathidae), glass-fishes (Centropomidae), silver-basses (Kuhliidae) and pipe-fishes (Syngnathidae). Mangroves border the reaches of rivers influenced by the tides, and here are found certain puffer-fishes, archer-fishes (Toxotidae), half-beaks, barracudas, cardinal-fishes (Apogonidae), damsel-fishes (Pomacentridae), gobies, sting-rays and so on. In a stream on one of the Palau Islands, where the above survey was made, over fifty kinds of fishes are found in the estuarine mangrove region, while thirty kinds live in the high freshwater reaches.

This statistic suggests, and it seems understandable, that fresh waters are a greater barrier to invading marine fishes than are brackish waters. In the Baltic Sea, which is the largest brackish water basin in the world, there are numerous marine fishes, such as herring, sprat, cod, lump-sucker (*Cyclopterus*), plaice, flounder, sea-snails (*Liparis*) and sand-eels (*Ammodytes*). In the Kattegat, where the salt content is about half that of sea water, there are seventy-five species of marine fishes. On moving eastward into the Baltic, the salt content decreases. In the innermost reaches, the Gulfs of Bothnia and Finland, the salinity ranges from about one-fifth of normal sea water to near freshwater conditions, the surface waters being less saline than those overlying the sea floor. Off the southern coast of Finland, the fauna of marine fishes is reduced to twenty-two species. Indeed, the freshwater teleosts that live in this region, such as pike, bream, burbot and perch, are almost as diverse as their marine counterparts. In fact, quite a number of freshwater fishes can live in brackish waters. The fourth largest land-locked basin in the world is the Aral Sea, which has a salinity of rather less than a third of average sea water. Three quarters of its fish fauna consists of carp-like fishes and perches and there are important fisheries for carp, bream and roach.

Even so, the main invasion has been landward rather than seaward, which is also true of the aquatic invertebrates. Diverse marine fishes are thus able to live in waters that range in salt content from oceanic to freshwater conditions. Certain species are still more adaptable. Along the western side of the sea of Azov, grey mullet (*Mugil cephalus*) live in surroundings that contain three times as much salt as oceanic waters. Flounders (*Platichthys flesus*) can tolerate salinities that are about twice that of average sea water. But we should remember that most fishes, whether marine or freshwater, are strictly confined to one medium or the other.

Of the fishes able to live in waters of widely ranging salt content, (euryhaline species), most physiological attention has been paid to killifishes and sticklebacks. When these fishes are transferred from fresh to sea water, they lose water and gain salts, which might be expected. But in forty-eight hours their water and salt contents return to normal. On being moved from sea water to fresh water, the reverse occurs: they gain water and lose salts, but again they soon master their new medium. Their adaptability is largely dependent, it seems, on the low permeability of their body surfaces to water and salts and on extraordinary activities of their gills and kidneys. When a marine fish moves into fresh water its kidneys must be able to rid the body of excess water and to conserve salts. On entering the sea, the body fluids will become saltier following the loss of water and this change, it seems, automatically ensures that urine formation will be reduced.

Concerning the gills, activity resides in secretory cells that are formed on the filaments. In the killifish (*Fundulus heteroclitus*), these cells appear to be responsible for both the extrusion and intake of salts, the first process being vital in sea water, the second in fresh water.

Just how all these activities are co-ordinated is not yet clear, though there is evidence that the endocrine system is involved in some way or other. In sticklebacks, for instance, the thyroid gland is vigorously active in migrating and spawning fishes, and during this time they are much less able to tolerate sea water, which is also found when thyroid extract is given to fishes outside the breeding system. Pituitary activities may also be important, for if this gland is removed from killifishes they are unable to survive in fresh water or diluted sea water. No doubt the nexus of endocrine and nervous activities is rather complex and will require much more study before their means of co-ordinating water and salt balance in a fish are understood. If we could appreciate why it is easier for a fish to move from the sea to fresh water rather than the reverse, another aspect would become plainer. Estuaries, as we have seen, may bear more life than the richest farming land. Are all but a few freshwater fishes prohibited from taking advantage of these fine feeding grounds because of their highly developed glomerular kidneys, adapted to deal with a copious osmotic intake of water by the body? This seems unlikely, for there are reaches in estuaries where the salt content is less than in the body fluids of freshwater fishes. Perhaps the gills of most freshwater fishes cannot change from absorbing salts to excreting them. If so, why should the reverse change be more readily managed by so many marine fishes? Here is an intriguing but puzzling problem.

Perhaps fishes are led to estuarine waters because they are rich in food. The certain attraction of the ocean for freshwater eels or of rivers for salmon is concerned with their breeding. Besides the Atlantic salmon (*Salmo salar*) and Pacific salmon (*Oncorhynchus*), certain lampreys and sturgeons, shads, sea trout, steelhead trout, certain char and smelts also seek fresh waters when the time is approaching to reproduce their kind. When they leave fresh waters for the ocean, the gonads of freshwater eels are growing but are still far from maturity. Both the European eel (*Anguilla anguilla*) and the American eel (*A. rostrata*) face a long oceanic journey before they reach their spawning grounds in the Sargasso Sea.

Regarding the changes that occur as these fishes leave, or are preparing to leave, one aquatic medium for the other, most study has been given to lampreys, salmon and European eels. In the sea the lampern (*Lampetra fluviatilis*) evidently regulates its water and salt content like a teleost fish. Fresh-run fishes are capable of living in half diluted sea water. They keep up their water content by swallowing sea water and they probably get rid of excess salts (chlorides) by way of secretory cells on the gill filaments. After lamperns enter fresh waters, there is evidence that these cells disappear and are replaced by smaller kinds of cells. As these cells arise when marine modes of water and salt balance are being replaced by freshwater ones, they are believed to be responsible for salt uptake. Concerning their water balance in rivers, lampreys have well-formed glomerular kidneys, which produce a copious flow of urine.

Young Atlantic salmon living in a particular reach of a river may not leave for the sea at the same age: some are a year old when they start to migrate; others are two years old, while a few wait until they are three years old. Before a salmon can migrate, certain physiological changes are necessary and some of these concern the endocrine glands. Study of Atlantic and Pacific salmon has shown that before they migrate the thyroid gland, which is controlled by the pituitary, becomes very active and it continues to function during their migration. The fish become restless and their senses seem to be sharper. There is also a change in their requirement for salts. During the spring, two kinds of Pacific salmon, chum (*Oncorhynchus keta*) and pink (*O. gorbuscha*) migrate seawards as fry and they will not survive if they are kept in fresh water beyond their normal migration period. Just after hatching, the larvae prefer fresh water, but near the time for their spring migration the fry begin to prefer water with a higher salt content. The fry of coho salmon (*O. kisutch*) migrate seaward but do not enter the sea, and it is only for a short period in the spring that they have a partiality for salt water. Like sockeye salmon (*O. nerka*), coho usually take to the ocean as one- or two-year old smolts, but they can survive in fresh water for many years if they do not reach the sea or have become landlocked. Sockeye salmon can even mature and reproduce when land-locked. In keeping with this, both the coho and sockeye salmon, unlike chum and pink, can change preference from salt to fresh water if kept in fresh water at the end of their seaward migration period. There are thus trenchant differences in the lives of these five species of salmon. But they all migrate seaward in spring, when the increasing length of day controls the time – probably through the pituitary and thyroid glands – when they begin to prefer salt water for river water [68].

Before they take to the ocean, freshwater eels change from yellow eels to silver eels. Apart from acquiring silvery guanine pigments, the skin becomes thicker and fatter; the eyes are enormously enlarged and develop golden retinal pigments like those of deep-sea fishes; the thyroid and pituitary glands become more active, and the gonads begin to develop. Silver eels also stop feeding. They are changing to deep-sea fishes, and one mark of their readiness to live in the ocean is an increase in the number of chloride excretory cells in the gills. Silver eels are more active than yellow eels; and while they lose chlorides during their transformation, they are better able to regulate their water and salt content when put in salt water than are yellow eels. So far as we can see, they are well fitted to enter the ocean.

Leaf-like larvae emerge from the eggs that are laid in the Sargasso Sea. Drifting and feeding in the surface waters, they move across the Atlantic for periods up to three years to European continental waters, where they change into elvers. Besides becoming smaller and acquiring a cylindrical form, the elvers lose water and gain salts. They swarm round the mouths of rivers and in due course some develop a very active thyroid gland, which is probably controlled by the pituitary. These forms grow well and have a marked tendency to swim into currents, which take them into the rivers. Elvers with a less active thyroid grow slower and are not so sensitive to currents; they enter the rivers at a later date. The earlier entrants are believed to grow into females, the later ones into males. At all events, elvers with an active thyroid gland consume more oxygen in fresh water than in the sea. They are prepared for their life in the rivers and, provided the mucous coating of the skin is intact, they can adapt to sharp changes in salinity.

The internal environment and the geological history of fishes

Life is most likely to have begun in the sea, which marvellous event seems to be imprinted in the ionic make-up of animal body-fluids. Indeed, A. B. Macallum concluded [69] that blood plasma is, so to say '. . . the heirloom of life in the primeval ocean'. Protoplasm, having begun to exist in salty surroundings, thus needs the intimate and constant support of an inner microcosmic ocean if it is to continue to be the seat of life. Moreover, the ionic composition of the body fluids of some invertebrates is close to that of sea-water, but the resemblance is less close in crustaceans and molluscs, for instance, and the same is true of fishes. Even though having blood that is just as salty as the ocean, a hagfish has less calcium and magnesium in its plasma than is contained in the outer medium. These two ions are similarly less prevalent in cartilaginous and bony fishes.

While the ionic balance in the plasma of a shark is rather like that of a teleost, the shark has the saltier blood. Macallum attributed this to the longer geological history of the cartilaginous fishes: they had been exposed to the sea since Devonian times, whereas teleosts arose much later in the Jurassic period. His contention was that once the blood stream had become isolated from the ocean, the ionic make-up of the plasma had somehow remained unchanged. Divergences in this make-up in different groups of animals could be explained by two considerations: first, that the primeval ocean had a different ionic balance (e.g. more potassium and less magnesium) than later oceans, and secondly, that the blood system of different groups had become separated from the sea at different times in geological history.

But blood cannot remain 'fossilized' in this way, for there is a continual exchange of fluid between an animal and its environment, while its living membranes are not equally permeable to all the ions in sea water. Moreover, all available geochemical evidence indicates that the seas of Ordovician times were quite similar in salty composition to those of modern times. Macallum's elastic, but essentially static, concepts must thus be discarded, though this does not mean, of course, that life did not begin in the ocean.

There is also the idea, based partly on the structure of the fish kidney in relation to its preserving a correct water balance, that the ancestral fish was a freshwater animal. The exponents of this hypothesis argue that kidneys were evolved as organs that were primarily concerned in ridding the body of excess water, which would flow from a salt-poor, freshwater medium to the more salty fluids in the fish. In particular, it is contended, the glomeruli of the kidney arose to cope with this hazardous dilution of the body fluids. Now it is true that the glomeruli in the kidneys of freshwater fishes are well developed and one could go on to argue that when fishes took up life in the ocean the glomeruli would be something of an embarrassment. For the osmotic stress facing a fish is now reversed: somehow or other it must acquire water from a medium that is saltier than its blood. The sharks and rays seem to have avoided this stress by acquiring urea-laden blood that is a little stronger in solutes than sea water, but marine bony fishes have either had to reduce their glomerular surface or to dispose of it entirely.

This seems an attractive idea, but like Macallum's, it is not in accord with the facts. The earliest known fishes, jawless ostracoderms, are found in Ordovician deposits and they are joined in Silurian times by the jaw-bearing placoderms. They are found together with marine invertebrates, nor is there any other evidence that they lived other than in the sea. Moreover, the three groups of invertebrates that are nearest to the vertebrates – Hemichordata (acorn-worms, etc.), Urochordata (sea-squirts, etc.), and Cephalochordata (lancelets) – are marine animals. Like these organisms, hagfishes have inner fluids that are as salty as the sea. They also have well formed glomerular kidneys. Again, the kidneys of sharks and rays have an overall glomerular surface that is equal to that of freshwater fishes, but their body fluids have but slight osmotic powers of attracting water from the sea. There is, in fact, no direct linkage between the extent of such powers and glomerular development in fishes. All available geological and physiological evidence thus points to a marine origin for fishes [70]. Kidneys bearing glomeruli were evolved in the sea. Indeed, it looks as though the primary rôle of glomeruli is not to control the water balance of the body but rather to preserve a correct balance of ions in the blood. It was when fishes took to fresh waters that their kidney glomeruli evolved to serve the further function of guarding against excessive dilution of the body fluids.

Invasion of fresh waters certainly took place after Silurian times, for in the Devonian period ostracoderms and bony fishes are found in both marine and freshwater deposits. Recent detailed study of the Lower Devonian rocks of the Welsh Borderland, which were thought to have been deposited in an arm of the sea, are particularly revealing. It now appears that there were gradual changes in this region.

Marine conditions prevailed at first, but afterwards the sea retreated. The now shoaling coastal waters began to be diluted by river water, and in these shallow, slightly brackish waters marine ostracoderms still lived. Later they were joined by brackish water species. Later still, when the sea was in full retreat, the rivers formed an extensive flood plain, where lived freshwater ostracoderm fishes (cephalaspids and pteraspids). Many of the freshwater species are also known from Eastern Canada, Spitzbergen and the Ukraine. How did these species, which are not found in marine or brackish deposits manage to spread so widely? As L. B. Tarlo says, the only reasonable explanation seems to be that '. . . their larval life was spent in the sea, and that like the modern eel, they were able to travel long distances during the period of their life, so that on the onset of metamorphosis they would be ready to invade fresh waters and colonize their own particular environment, going back to the sea only to spawn'. [71] This reconstruction accords well with the idea that the life of fishes began in the sea. Like modern freshwater eels, jawless fishes living in early Devonian times had to return to their ancestral medium in order to spawn. Seeking the exactly appropriate medium for spawning is the most critical migration that fishes must make and on this depends the proper unfolding of a healthy internal environment in the growing fish. Such conditions and migrations were evidently just as vital about 300 million years ago. There is, then, 'no new thing under the sun'. The Preacher also said that 'time and chance happeneth to them all'. But may we not conclude that the inroads of chance can better be faced if there are means to steady the internal environment? There is, after all, Claude Bernard's celebrated dictum: 'La fixité du milieu intérieur est la condition de la vie libre.'

Chapter 9

The Sensory World

HOW FAR can we enter the sensory world of fishes? Living in a transparent atmosphere, we can see for miles; but for a fish the water is very clear if he can distinguish objects fifty feet away at the same level. In turbid waters, this critical distance may be less than an inch. Yet many kinds of deep-sea fishes must be much more sensitive than men to very dim light, while certain species may well have just as marked a binocular vision. Moreover, most kinds of teleost fishes, together with lizards, turtles and birds, appear to be the only vertebrates, besides ourselves and other primates, to have a sense of colour.

While water is much less transparent than air, it is a better medium for transmitting sounds. Moreover, we are beginning to appreciate that fishes exist in a world of sounds, part of which may arise as they swim. Certain crustaceans, fishes and cetaceans also have special means of producing sounds; 'voices' that mingle with the noises made by water in motion. Like man, but unlike bats and dolphins, fishes are virtually deaf to supersonic sounds, that is those above a frequency of about 20,000 cycles per second. At frequencies up to 1,000 cycles per second, about two octaves above middle C, 'specialist' fishes may have sharper hearing than humans. But all fishes are less sensitive to high-frequency sounds than we are, though their hearing is not inferior to that of land animals in any essential way.

The sense organs of smell are diversely developed in fishes. We human beings, for instance, are better placed to appreciate the sensory world of a flying-fish or a pike, a world of vision rather than one of olfaction. More imagination is needed to enter the sensory life or a shark or an eel; for in such fishes the olfactory organs are highly developed (figure 32). Even so, land animals with a keen sense of smell, and even human beings, may be alerted to the source of a smell more quickly than is a fish. Odours spread much more rapidly

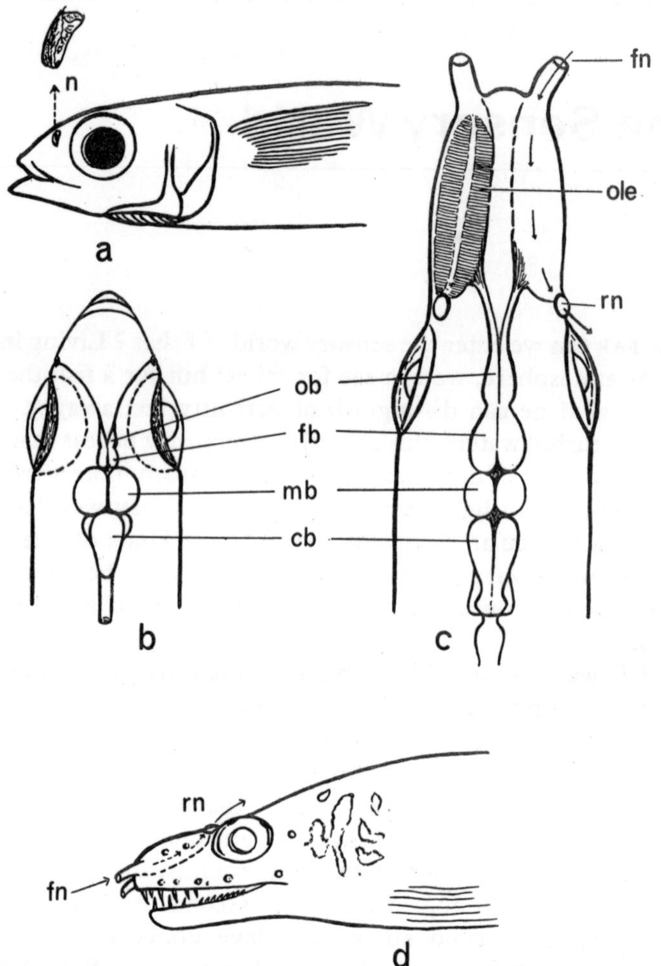

Figure 32. Nasal organs and brain of two telost fishes. (a and b) side and top views of the head of a flying-fish, *Oxyporhamphus micropterus*. The very reduced nasal organs (n) lie in a rectangular pit. Note the small forebrain (fb), which in fishes is largely concerned in olfaction, and the small olfactory bulbs (ob) – a striking contrast with the large optic centres (mb) of the brain. (c and d) top and side views of a moray-eel (*Gymnothorax*). Note the tubular front nostril (fn) which leads to a large nasal capsule containing a large nasal organ with numerous nasal lamellae (ole), the receptors of odours. The rear nostril (rn) is the exit for the water flowing over the nasal organs. In sharp contrast to the flying-fish, the olfactory bulbs (ob) and the forebrain (fb) are very large, surpassing the volume of the optic centres (mb). (Drawn from specimens in the Museum of Comparative Zoology, Harvard University.)

in air than in water. Again, when considering taste, we must remember that the taste-buds of fishes are not only inside the mouth but may be in the skin covering the lips, barbels and fins or even the body (figure 33).

In the lower vertebrates, the diversity of sensory nerve endings in the skin is not so great as in man, who has receptors sensitive to touch, pressure, heat, cold and pain. Yet the sense of touch appears to be well developed in fishes, which are also keenly sensitive to slight changes in temperature, salinity and pressure. Lastly, the skin

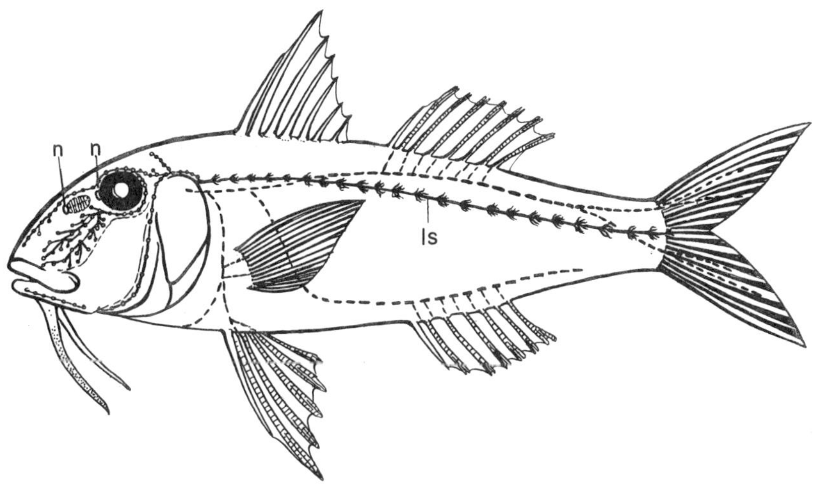

Figure 33. The sense organs of a red mullet (*Parupeneus*). Note the course of the lateral line canals and openings (shown as small circles) over the head and the continuation of this system (ls) along the body. Taste-buds are shown as dots on the barbels. The courses of nerves that supply taste-buds on the fins are shown as broken lines. The nasal organs and nostrils (n) are just before the eyes.

of fishes and certain amphibians houses lateral line sense organs, which respond to disturbances in the water and to changes in flow patterns near the swimming animal. This 'distance-touch sense' or *Ferntastsinn*, as it is called by Germans, is a special attribute of the lower aquatic vertebrates. When the vertebrates became masters of the land, these sense organs must have been lost. Nothing like them is found in the reptiles and mammals that have returned to the water, not even in whales and dolphins, which rely largely on their sense of hearing.

Before turning to a closer look at each sense, certain aspects should be stressed. One aspect is best left in the words of Sir Wilfrid le Gros Clark:

> Our knowledge of the world around is entirely dependent on multitudes of nervous impulses which stream into the brain from sensory nerve endings, or receptors, scattered over the surface of the body. Each of these receptors is susceptible to stimulation by certain forms of energy – mechanical, chemical, light, etc. – and it is this stimulation which initiates the nervous impulses that are ultimately translated into conscious perception [72].

After suitable changes in the first seven words and the omission of the word 'conscious', this statement is equally true of fishes. But the sense organs will only respond to certain ranges and intensities of light or sound energy etc. The sense organs abstract and the fish may act. This 'may' is essential, for a fish is not bound to respond to every perceivable change in its surroundings. It will act if the need arises. Moreover, a fish is not a passive receiver of sensory information concerning its outer world. From time to time it will explore its living space, which may well have been chosen, and it is certainly capable of learning.

Vision

The eyes of all vertebrate animals are essentially alike, and they are strikingly similar to the eyes of squids and octopuses. Each eye has an outer fibrous tunic called the sclera, which changes into a transparent cornea in front of the iris, pupil and crystalline lens. After passing through the lens, the light rays transverse a vitreous humour which fills and distends the eye-ball on their way to the retina, the light-sensitive part of the eye. The retina is separated from the sclera by a choroid layer, which contains an elaborate system of blood vessels to nourish the many millions of retinal cells (figure 34).

When fish eyes are compared with ours, certain structural differences are immediately apparent. The lens of a fish eye is perfectly spherical and it bulges well through the pupil (figure 34): the human lens is more or less oval in cross section and it lies just behind the plane of the pupil. The salient feature is that the cornea is optically absent in fishes and other aquatic vertebrates, for its tissues, being watery, have the same refractive index (1·33) as water. Light rays are not bent and focused until they reach the lens of a fish, which has the highest effective refractive index (1·65) of all vertebrates. In

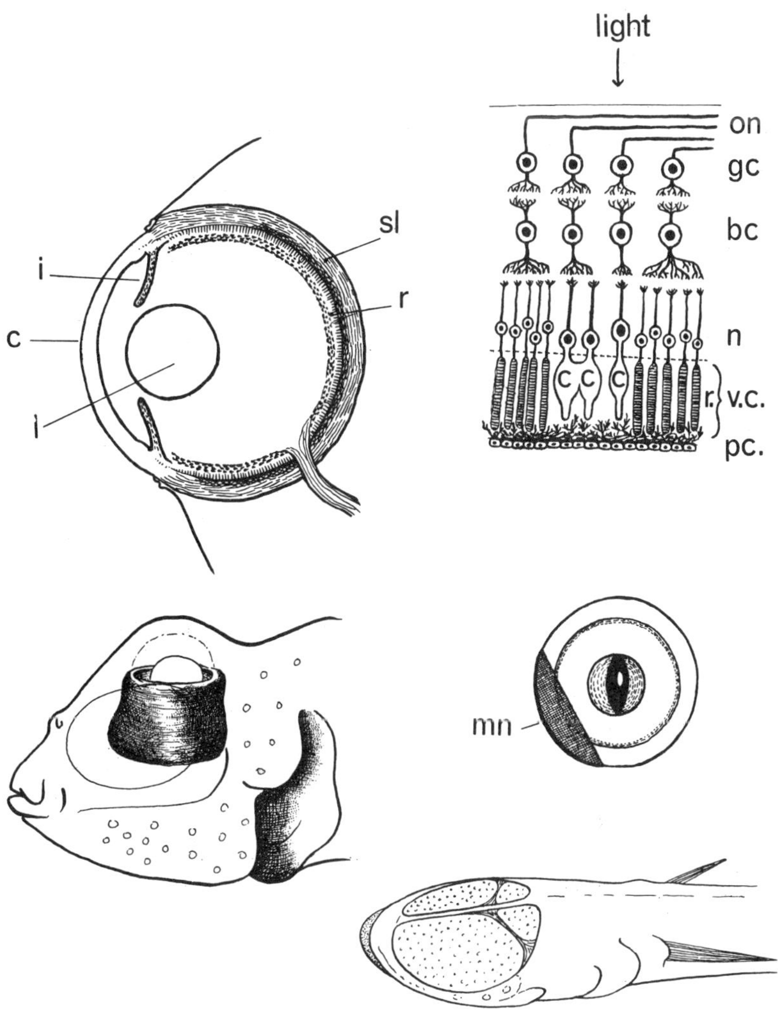

Figure 34. Eyes. Top left: a transverse section of a salmon's eye, showing the cornea (c), the iris (i), the lens (l), the retina (r), the choroidal blood vessels and the sclerotic layer (sl). Top right: Schematic detail of a retina, showing cones (c), rods (r) and their nerve cells (n). These cells are linked through bipolar cells (bc) and ganglion cells (gc) to the optic nerve fibres (on). At the base of the rods and cones are pigment cells (pc). Below this figure is the eye of a carcharinid shark, showing the nictitating membrane (mn). Bottom left: the large, upturned tubular eye of a deep-sea salmonoid fish (*Macropinna microstoma*). Bottom right: the head of a deep-sea, bottom-dwelling fish, *Ipnops murrayi*. Note the two plate-like bones on top of the head. Below these bones, which are transparent, are the flattened-out eyes containing visual cells (shown as dots).

land-dwelling vertebrates, however, both the cornea and lens play a part in refracting and focusing light on to the retina. If each eye of a fish is to have a wide visual field, the lens must not only protrude through the pupil, taking the place, so to speak, of the absent cornea, but also stand out from the surface of the head itself. In this way all-round vision can be obtained by fishes, most of which have eyes on the sides of the head (figure 35). The significance of all-round vision to a fish, as Dr G. L. Walls has written [73], '. . . is not only seen ecologically, in his increased awareness of near-by prey and increased difficulty of approach by enemies, but is also seen anatomically in his lack of a neck. Despite his buoyancy and rotability on his vertical axis, a fish would need a neck almost as badly as a land animal, were it not for his full visual field.' A fish lens is both spherically and optically perfect, which must mean that images are transmitted to the retina without distortion. Being a relatively large sphere with a high refractive index, the lens also has a short focal distance and good resolving powers. But if the image of an object is to fall, as it must, within the layer of light-sensitive cells, the focus of the lens may sometimes need to be changed for near or distant vision. Such optical accommodation in human eyes is attained by alteration in the shape of the lens, which is moulded by ciliary muscles. In fishes the lens is simply moved as a whole, and the means, which are muscular, are differently contrived in lampreys, elasmobranchs and teleosts.

Fishes are sometimes said to be short-sighted, but a recent study has confirmed that quite the reverse is true [74]. Retinoscopic tests on the eyes of living, immersed fishes – on dog-fishes, sharks, skate, lantern-fish, flounders, dolphin-fish, jacks, silversides and others – showed that they are markedly long-sighted or hypermetropic. Moreover, the tests indicated that fishes are more long-sighted in their lateral field of vision than in their forward one. When looking forward they use, of course, the backward parts of the retina, where the density of visual cells is greatest – much greater than that over the middle and forward parts of the retina. In brief, fishes should see the clearest images in their forward field of vision. The finest visual grain of the retina thus being linked to their forward, least long-sighted vision, fishes evidently have the most detailed sight where they need it – as when taking prey, avoiding obstacles and enemies and so forth. The coarser grained parts of the retina associated with their long-sighted, lateral vision, must be more useful in detecting movements than in mediating clear images (figure 35).

Fishes, like other vertebrates, have two main kinds of visual cells in the retina: rods and cones (figure 34). These are elongated cells, and being packed neatly and closely together, they look like the pile of a regularly woven carpet. The tips of the rods and cones point away from the light, which must traverse a clear layer of nerve cells and fibres in order to reach them (figure 34). This neural layer – really layers – connects with the rods and cones through elaborate circuits and with the brain through the fibres of the optic nerve.

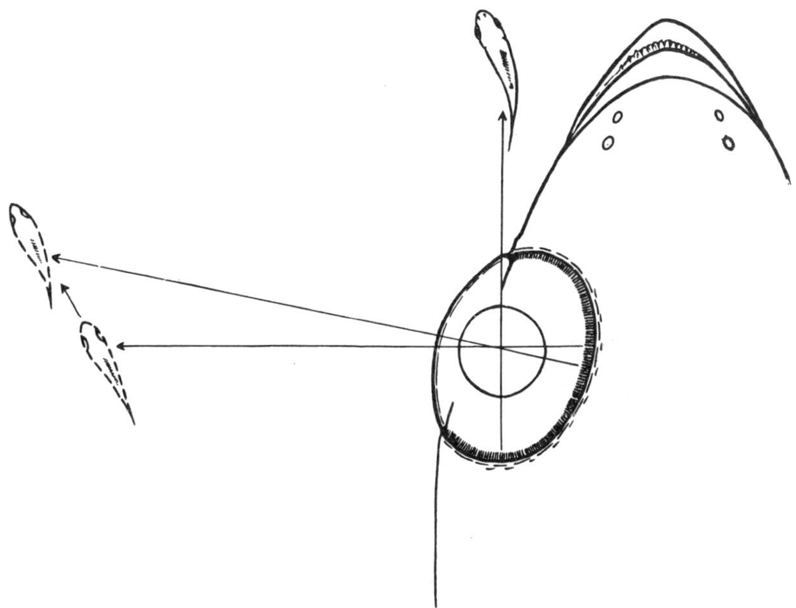

Figure 35. Showing how a fish sees more clearly in its forward visual field, where it is less long-sighted than in its lateral field of vision. The former field is covered by many more visual cells than is the latter.

The rods and cones hold pigments that absorb the light of the focused image, and are thereby chemically changed to substances that stimulate their possessors to fire trains of nerve impulses that are led to the brain. It is now, we presume, that the fish sees. The rods of marine fishes contain rose-coloured pigments or rhodopsins, while visual purples, known as porphyropsins, are the corresponding pigments of freshwater species. The pigments of the cones have proved more elusive, but we do know that these cells will only respond to bright illumination and that they are responsible for colour vision.

The rods, being much more sensitive, are able to mark the falling of very dim light on the retina. In fact the retinal image is registered by groups of rods, each group forming a visual unit; each is linked, by way of the nerve layers, to a single fibre of the optic nerve. Each fibre is thus fired by the combined responses of the rods in the group, and by this means the retina can continue to picture the most shadowy images. If, like many of the cones, each rod had its own optic nerve fibre, this would not be possible. Thinking now of the retina as a whole, it will be evident that the sharpness of the image will depend on the number of visual units that have been concerned in its reception. The finer the grain of the retinal film, the finer the grain of the image.

Fishes that are most active during the hours of daylight have retinae with relatively high numbers of cones compared to the complement of rods.* In nocturnal fishes the rods greatly outnumber the cones. As the cones, in sharp contrast to the highly sensitive rods, will only register the play of bright images and colours, this is to be expected. But no fish, like so many lizards, snakes, birds and mammals, has retinae composed entirely of cones. Again, this fits one's expectation, for underwater light is much dimmer than light in the atmosphere. The eyes of most fishes have both cones and rods, but some species, as we shall see, have dispensed with cones.

The pike, which hunts from dawn to dusk, has eyes to match its activities. For every three or four rods in the retina, there is one cone; but as the cones are very large, the total mass of these cells is about equal to that of the rods. When a pike is looking straight at its prey – and in clear lake waters visibility may range to about 50 feet – binocular vision is enhanced by two sighting grooves that converge towards the tip of the spatulate snout. The fish darts, sometimes for 20 to 30 feet, and the prey, which must be moving, is seized. Clearly, accurate day-time vision is the very life of the pike, a fish that 'thinks' through its sight. This may be appreciated by looking at its brain, which has very large optic lobes. By contrast, the nasal organs and olfactory centres of the brain are quite small. Smell, as experiments have shown, means much less to a pike than vision. But let us not forget that a blinded pike can still find its prey by means of its lateral line sense.

Like the pike, the burbot (*Lota lota*), lives in rivers and lakes of Europe, Siberia and North America. It is the only cod-like fish to

* Many teleosts have double cones, adapted to function, it seems, in rather dim surroundings.

inhabit fresh waters. Leading a secretive kind of life, it attracts much less attention than does a pike. In England, A. F. Magri MacMahon says:

> It is found only in the rivers of the east coast from Suffolk to Durham, and it is not particularly well known, even to the local inhabitants, because it prefers rather deep water, where during the day it lies hidden among the weeds, or under stones, or in cracks and holes, rather like the Eel; and again like the Eel, it goes about at night, voraciously eating worms, fish-spawn, small fish or frogs [55].

One square millimetre of its retina houses about 810,000 rods and 3,400 cones, a ratio of more than 200 to 1. The eyes, like those of other nocturnal vertebrates, whether aquatic or terrestrial, have many more rods than cones. But a burbot's eyes, and consequently the optic centres in its brain, are rather small. In fact the olfactory centres in the forebrain are about one and a half times the weight of the visual centres, a factor that is still greater (over two and a half) in the freshwater eel. In both burbot and eel these large brain centres receive olfactory 'intelligence' through nerves that come from large nasal organs, which are much more useful to these species than are their eyes. Yet an eel eye has about three million rods in a square millimetre of its retinal surface, which seems quite excessive for a fish that is so led by its nose. But the eyes of a freshwater eel are 'awaiting' its marine life, when it becomes a silvery deep-sea fish and heads for its breeding grounds in the Sargasso Sea. Before this migration the eyes become greatly enlarged, and the rods, which now bristle densely within an extensive retina, acquire the golden, light-absorbing pigment that is typical of deep-sea fishes.

Diurnal eyes, those with balanced complements of rods and cones, are much commoner among freshwater and coastal fishes than are nocturnal eyes. However, a good many fishes that live in turbid, tropical waters – fishes such as the mormyrids, knife-fishes and numerous cat-fishes – have small eyes. Even in clear surroundings, such eyes must have limited visual powers, serving more to detect movements and distinguish light from shade, rather than to register images. At all events, the mormyrids and the knife-fishes are able to navigate and find their prey by means of electric fields of their own creation. Cat-fishes have keen hearing and many have well-developed nasal organs that work together with long probing barbels that bear hundreds of taste-buds. All such fishes are well enough able to move around and find their food without normally developed means of vision.

Cave-dwelling fishes, those that exist in lightless underground streams, must certainly turn to senses other than sight, for their eyes have become quite degenerate. The remnants of their eyes, which may be almost unrecognizable, are usually buried below the skin of the head. Yet in some blind fishes, at least, the early development of the eyes is quite normal. It is only as the fish becomes an adult that the eyes regress, as do the visual centres of the brain.

Reduction of the eyes and the visual centres is drastic enough to make one suppose that cave fishes would have lost all sensitivity to light. But if the Mexican cave characin (*Anoptichthys jordani*) and the blind Congo barb (*Caecobarbus geertsi*, figure 36) are given a clear-cut choice of light and darkness, they tend to move away from the light. A blind cave cyprinid from Iraq (*Typhlogarra widdowsoni*) seems to have almost lost this reaction. Experiments on the Mexican fish and other blind characins have revealed that the relics of the eyes are still light-sensitive, so long as they are connected to the brain by the reduced optic nerves. The pineal organ of the brain may also be sensitive to light and it is even probable that some cave fishes may have light receptors in the skin.

Green plants need light to live and grow and cave-dwelling animals are ultimately dependent on the fruits of plant activity. Streams that disappear underground carry down plant and animal remains that become the food of cave-stream crustaceans, which in turn may fall to the fishes. But how do cave fishes find their food? Their success in making a living depends on a dogged scanning of their surroundings by olfactory, acoustic and lateral line senses. In the most specialized cave fishes, these senses, judging by their sense organs, are much better developed than in their relatives with normal vision. The three cave fishes mentioned above seem to be always on the move, *Caecobarbus* and *Typhlogarra* alternately swimming and gliding, *Anoptichthys* appearing to be in a frantic hurry. When they almost literally bump into something edible, they are quick to seize it. Having lost their visual means of scanning the space around them, they have come to rely on random and persistent exploration. It looks as though they cannot afford to stop and await distant 'intelligence' by way of nostrils, ears and lateral line canals.

Blind cave-fishes have been found in the Americas, Africa, Iran, Iraq, Indonesia, Australia, Japan and the Philippine Islands. More than forty species are known, about a third of which are cat-fishes (figure 36). There are six blind cyprinids (three being African) and six blind amblyopsids (figure 36). All but one of the latter live in streams running through limestone caves in the central United States.

Most cave-fishes range in length from 1 to 6 inches. Some species, having nearly or entirely lost their pigment cells and scales, have a

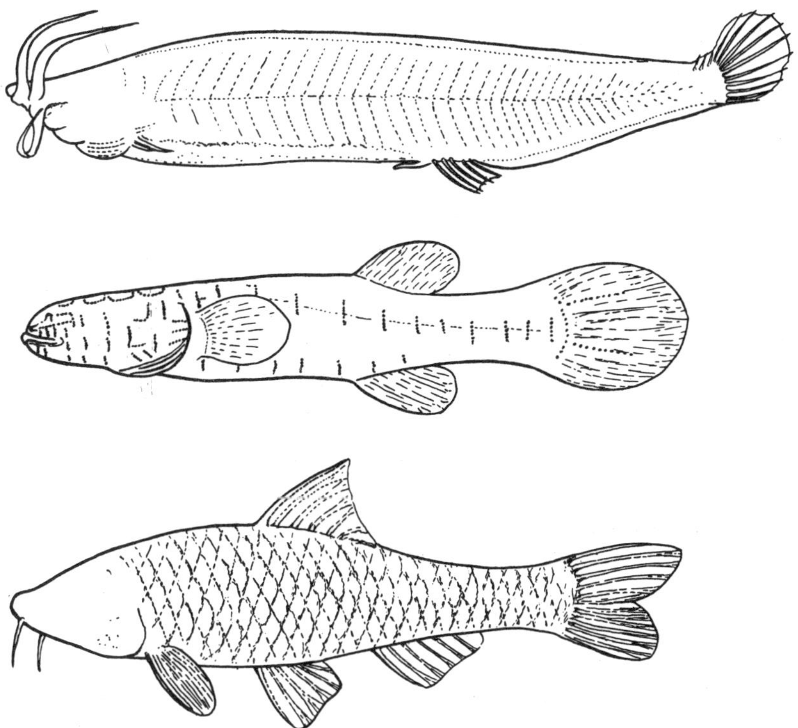

Figure 36. Three blind fishes. Above: pygidiid cat-fish (*Pygidianops eigenmanni*) from the Rio Negro, South America. Middle: amblyopsid (*Typhlichthys subterraneus*) from limestone caves in the south-eastern United States. Below: the barb, *Caecobarbus geertsi*, from caves in the Congo. (Top figure after G. S. Myers.)

curiously naked and pink appearance, for the blood vessels now show through the transparent skin.

'Down to a sunless sea' is an apt description of a descent into the deep ocean. Even in the clearest tropical waters, the most deeply penetrating blue rays of the spectrum are not detectable much beyond a depth of 1000 metres. Between this level and one at about 150 metres there is underwater twilight. Yet few deep-sea fishes have degenerate eyes like those of cave-fishes. On the contrary, most of these fishes, whether they live in the mid-waters or near the bottom, bear large and elaborate eyes (plates 18 and 19 and figure 34). We can only presume that such ocular evolution is a response to the presence of two kinds of underwater illumination, the twilight given by the sun and the luminescence produced by the animals themselves. Concerning the latter, flashes have been detected 3,000 metres below the lower limit of the twilight zone (see also p. 173).

Large eyes are found in most groups of deep-sea fishes.* There are no cones in the retina but vast numbers of rods (100,000 to 20 million in a square millimetre), which being long and slender are efficient catchers of light. Even more significant, the rods contain golden pigments that are most sensitive to the blue rays of the spectrum, those that travel farthest into the ocean. But these highly sensitive retinae, probably the most sensitive to be found in any vertebrate, must be matched by aptness of optical design. To let twilight flood into the eye, a relatively large pupil is needed, which means a corresponding increase in the size of the lens. To perceive spots and patterns of luminescence efficiently, the pupil, and hence the eye, must be as large as possible. In varying ways, and in ways that we are only beginning to appreciate, these conditions are met in the eyes of deep-sea fishes (figure 34) (plates 8, 9, 13 and 17).

The populations of most kinds of large-eyed fishes, both pelagic and benthic, are centred between depths of 200 and 1,000 metres. These levels not only encompass the twilight zone of the ocean but also the region of the greatest luminescence, where flashes of living light are particularly frequent. The deeper-living pelagic fishes (ceratioid angler-fishes, gulper eels, snipe-eel and so on) and the abyssal kinds of bottom-dwelling fishes (such as various kinds of tripod-fishes, rat-tails, brotulids and sea-snails) – those that live below the reach of the sun and in regions of sparser luminescence – tend to have smaller or much smaller eyes. Judging by their structure, such eyes can only be used to make their owners aware of living light, which may come, of course, from prey or predator. Some of the deeper living fishes, notably certain kinds of brotulids, have very reduced eyes. One genus of abyssal fishes called *Ipnops* has the most curious eyes of any vertebrate. The head of *Ipnops* is markedly flattened, and lying below thin, transparent roofing bones is a pair of large, plate-like organs (figure 34). That these are modified eyes can

* Deep-sea salmonoids; clupeoids (alepocephalids and searsids); stomiatoids (hatchet-fishes, etc.); lantern-fishes, giganturoids, macrouroids (rat-tails), and so on.

be told from their microscopic structure, for each organ bears about 250,000 rods, which are linked to the optic nerve. Each eye is little more than a flattened fibrous tunic and retina, all trace of the optic parts, other than relics of the iris, having vanished. Perhaps these abstracts of eyes enable an *Ipnops*, which lives well below the twilight zone, to keep in touch with luminescent events near the deep-sea floor.

Deep-sea sharks and the chimaeroid fishes, which live in dimly lit waters, have, as we might expect, pure-rod retinae. But so have nearly all the elasmobranch fishes that live in shallow waters or near the sea surface. Cones are known only in the eyes of certain sharks, smooth-dogfishes (*Mustelus*) and the eagle-ray. Now nocturnal eyes may well match the habits of some sharks, dog-fishes and rays that live on or near the sea floor, yet it is odd to find such eyes in sharks that swim in the well-lit surface waters: in blue sharks (Carcharhinidae), basking-sharks (*Cetorhinus*), and so on. But elasmobranchs, unlike most bony fishes, have means of contracting their pupils, so shielding the retina from bright illumination. Blue sharks and hammerhead sharks also have a nictitating eyelid that can be flicked across the eye (figure 34). And there is evidence, judging from their attacks on man, that sharks tend to be more active at dusk. Even so, the eyes of sharks, like their owner's habits, are something of a mystery.

Hearing and the lateral-line sense

Fishes may virtually lose their eyes, but never their ears and lateral-line organs. Ears not only hear but are concerned in maintaining and regulating the tone of the body muscles, in keeping them in a state for instant action. The ears also respond to angular accelerations that a fish may make away from a straight course and to movements involving gravity. The lateral-line organs make a fish aware of nearby water disturbances and of distortions in the field of flow around its moving body.

The inner ears are housed in capsules on either side of the skull behind the eyes. Each ear or labyrinth, which rests in a lymph bath and is itself full of lymph, has an upper part called the utriculus bearing one to three semicircular canals (one in hag-fishes, two in lampreys and three in the jaw-bearing fishes). In the jawed fishes the utriculus communicates through a narrow neck with a lower part or sacculus, which has an annex known as the lagena (figure 37). The

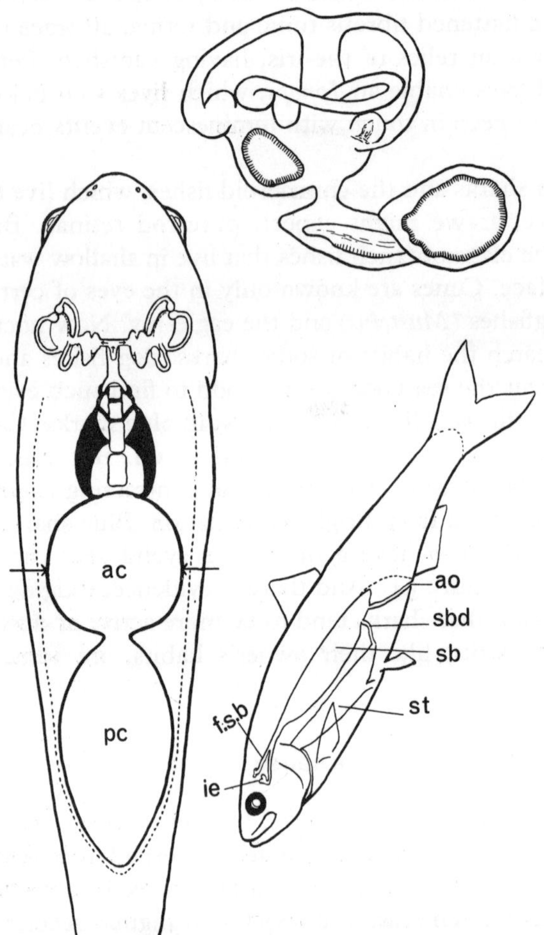

Figure 37. Linkage of ears and swimbladder in carp-like and herring-like fishes. In the former (*Ostariophysi*), shown on the left, the swimbladder is usually divided into anterior and posterior chambers (ac and pc). The anterior chamber is linked through a large pivoted ossicle (the tripus) and two smaller ossicles with a Y-shaped extension of the canal that joins the two ears. In the herring, a Y-shaped forward extension (fsb) of the swimbladder runs to the two ears (ie). The swimbladder (sb) has a duct (sbd) to the stomach (st) and also opens through a duct (ao) that has an exit near the anus. The top right figure represents the ear of a minnow (*Phoxinus*). Note the forward, rear and horizontal semicircular canals, each ending in a bulbous ampulla. Below these canals is the utriculus, which connects with the sacculus and lagena. Note the earstone in each of these three parts.

labyrinth is completely closed in the bony fishes, but is open to the sea through a narrow duct in the elasmobranchs.

Each of these ear chambers, utriculus, sacculus and lagena, holds a calcareous ear-stone or otolith that is coupled to a cluster of sensory hair-cells. A cap of hair-cells, crowned by a large gelatinous cupula in which the sensory hairs are embedded is also found in a widened part, the ampulla, of each semicircular canal. In the jawed-fishes, two of the semicircular canals are set fore and aft in the vertical plane, while the third traces its arc in the horizontal plane (figure 37). Lastly, the responses of each group of sensory hair cells, whether to movements of otolith or gelatinous cupula, are led to the brain by branches of the auditory nerve.

Except for the lack of a cochlea, or the beginnings of such, the inner ears of fishes are like those of other vertebrates. Though called labyrinths, the ears are not particularly labyrinthine, but in one way they would please a Daedalus, for they are cunningly contrived. Many well-planned experiments have revealed that the upper part of the ear, the utriculus and semicircular canals, is concerned with balance, while the lower part, the sacculus and lagena, is the seat of hearing.

Like the instruments of an aircraft, the semicircular canals of a fish provide the central nervous system or 'pilot' with information – in the form of codes of nerve impulses – concerning movements that are made in the yawing, pitching and rolling planes. As the body rotates in any one of these planes, the flow of lymph in appropriate canals rocks the cupulae, which stimulate the sensory hair-cells. The otolith in the utriculus is also shifted with respect to its cushion of hair cells.

On receiving this sensory information, the central nervous system brings into play nerves that control muscles moving fins and eyes. If a fish is rolling, as it may well be when swimming in boisterous and turbulent waters, the fins are flexed so as to bring the fish to an even keel. At the same time the eyes are moved so as to maintain a constant visual field. G. L. Walls writes: 'If the eyes always turned with the head instead of automatically "against" the head, the swimming of the visual field in a wholesale "apparent" movement would conceal from the animal small real movements within the field. So many more parallactic relative movements would take place, that actually-moving objects would be harder to spot. The gyroscopic stabilization of the eye is a means of combating the relativity of motion – by keeping the visual field still, the animal can better know what moves when and where.' [73] By keeping the visual field still, a fish is better fitted to keep its life.

Many fishes have special means of making sounds and while swimming in certain ways they may also generate sounds. The play of water itself is a further source of underwater noise. Water noise is probably audible to fishes, which are certainly quick to respond to their own 'voices' and *concerted* swimming sounds. There are experimental grounds for supposing that sounds are received by the lower part of the ear. Consisting mainly of water, the tissues of a fish will be much more 'transparent' to underwater sounds than are the very dense otoliths in its lower ear. The consequent vibratory play between sensory-cell cushion and otolith will lead to impulses that pass to the brain through the auditory nerve.

This is the way that sharks, rays and most kinds of bony fishes receive sounds. But in a good many teleost fishes – in at least five thousand species – the inner ears are acoustically coupled to the gas-filled swimbladder, the walls of which are set vibrating by waves of underwater sound. These vibrations will then be transmitted to the ears, the arrangement being analogous to that in a hydrophone.

In characins, carp-like fishes and cat-fishes (plates 4 and 5), the forward swimbladder chamber is linked to the ears by two chains of Weberian ossicles. The largest and hindmost pair of ossicles pivot on an adjacent vertebra, and convey vibrations by way of two other ossicle pairs to the forks of a Y-shaped lymph sinus, which is in contact with a lymph-filled canal joining the saccular chambers of the two ears (see figure 37). In the other fishes the swimbladder bears a forked, forward extension, each branch ending in a close contact with a membrane covering a window in the outer wall of the ear capsule. This direct kind of coupling is found in the tarpon, feather-backs, deep-sea cods (*Moridae*), various squirrel-fishes (Holocentridae) and sea breams. The arrangement in the herrings and mormyrids is more elaborate but essentially similar. The forward tip of each tubular fork in the herrings actually enters the corresponding ear capsule, where it expands into two vesicles (occasionally one) that are closely connected with the upper part of the inner ear. The entire arrangement is reminiscent of a doctor's stethoscope (figure 37). In the mormyrids a closed air sac is closely applied to each ear capsule and the two sacs are actually the elaborated terminal parts of an earlier forked outgrowth of the swimbladder. Lastly, the air-breathing chambers of anabantoid fishes also act as hydrophones.

These tympanic kinds of hydrophone endow their owners with

better hearing than is found in fishes without such extra auditory channels. They hear a wider range of frequencies (up to 10,000 cycles per second in some characins); they have sharper hearing; and they have a better discrimination of pitch. The hearing of minnows, gold-fish and cat-fish (*Ictalurus nebulosus*), as well as certain characins and mormyrids, has been compared with that of non-hydrophonic fish such as guppies, the freshwater eel and gobies. This is very intriguing, but what is the significance of good hearing in the day-to-day life of a herring, a mormyrid, a featherback, a carp or a deep-sea cod? We know very little but our knowledge is beginning to improve, as may be appreciated by turning to pages 163–73.

The lateral-line sense organs, which look like those in the semicircular canals, are usually housed in mucous-filled canals that extend under the skin of the body (figure 38). Both kinds of organs are also physiologically alike. Quite spontaneously, when in a resting condition, both send continuous trains of nerve impulses down the sensory fibres that connect them with the central nervous system. When the fluid filling the canals is troubled, either in the ears by certain rotary movements of the body or in the lateral-line canals by disturbances near the fish, this pattern of impulses is changed; the frequency is either increased or decreased, depending on the direction of flow of the lymph. However, there are no lateral-line canals in deep-sea fishes such as the gulper-eels, the bob-tailed snipe-eel (*Cyema*) and the ceratioid angler-fishes – fishes in which the sense organs are carried at the ends of skin papillae that are free to wave in the water (figure 38). Certain loaches and the minnow have both freely exposed sense organs and canal-covered ones (see figure 33 for the general arrangement of these organs).

Facing a wind, we feel its pressure where we slow down its motion. More precisely, according to Bernoulli's Principle, as the velocity of air or water decreases, the pressure increases, and vice-versa. The changes of water pressure along a model ship can be measured by connecting a series of inlets in the hull to tube manometers. The lateral-line canals are also open to the water through pores or pits, but do the sense organs act as pressure detectors? Water pressure will be high just in front of the fish's snout, where the flow of water is impeded, and low around the head and shoulders, where water is flowing more quickly. If this pressure field could be 'watched' by the lateral-line sense organs, the fish might detect any local disturbance of the field and thus, like an electric fish, have means of finding or avoiding the disturbing source, which might be prey, predator or

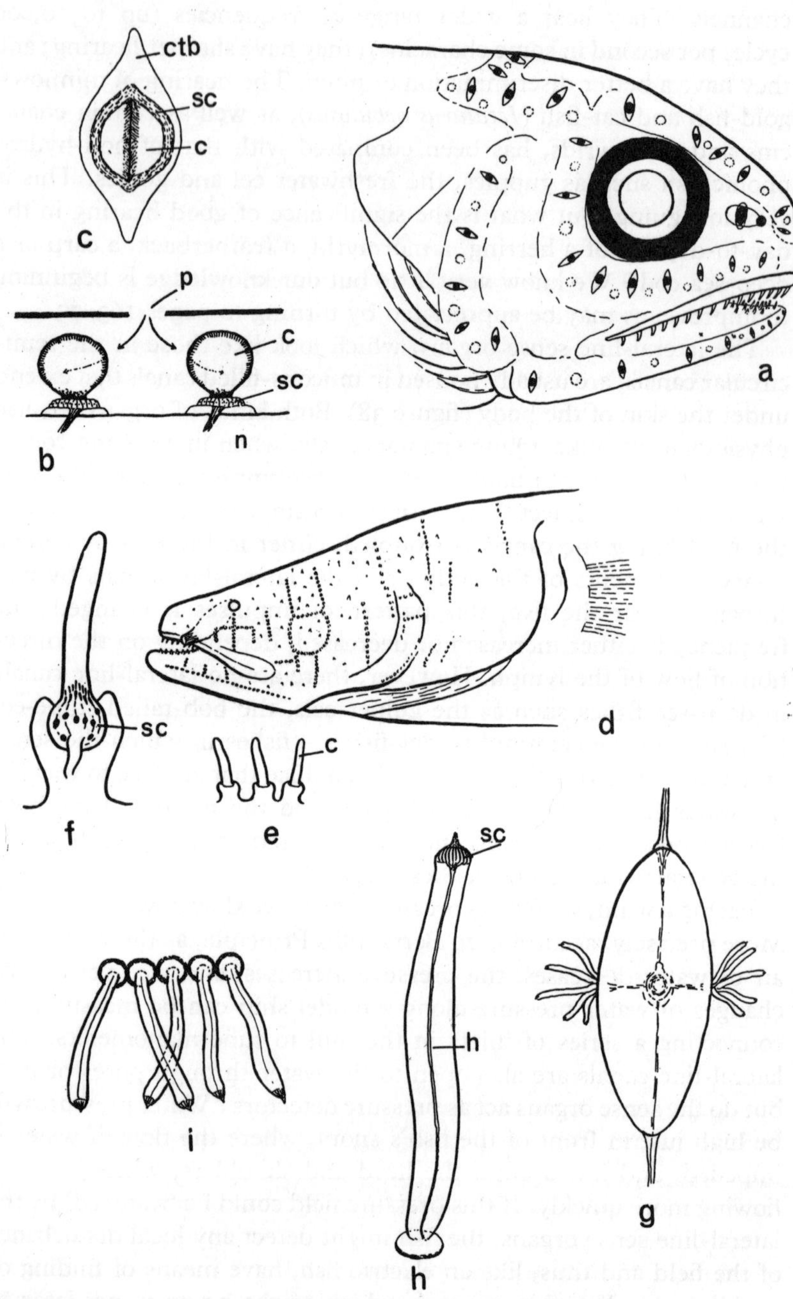
ctb
sc
c
c
p
c
sc
b
n
a
sc
d
c
f
e
sc
h
i
h
g

some obstacle. But Dr S. Dijkgraaf [75] has evidence that it is the diverging flow pattern of water – the hydrodynamic shadow moving before the head of a fish – that is used as a short range warning system. Distortion of this pattern by an obstacle is detected by the lateral-line sense organs of the head. In this way a fish avoids bumping into the glass walls of an aquarium.

A migrating salmon is quick to use any aid it can get from the water. It will rest in pools and behind rocks, where the water is calmer, and swim near the bottom, where the flow of water is slower. Is it using its lateral-line sense to detect local variations in the velocity of the water it meets? Fishes certainly use their sight to keep station while swimming in a current (p. 204), but can they, by visual means, appreciate small scale variations in water velocity? How do they fare on a dark night? Perhaps the answers to these questions reside in the lateral-line sense, though the ears may be expected to play a part if water movements are strong enough to upset the equilibrium of a fish. Through conditioning experiments, as Professor O. Lowenstein remarks, we at least know that: 'Blind individuals learn fairly quickly to associate small water disturbances in their vicinity with the presence of either food or punishment and – what is more – they locate the centre of the disturbance with astonishing accuracy.' [76] The lateral-line sense may thus enable a fish to track current-like water disturbances caused by its fellows, as in a school, or those stirred by prey or predators. But one has the feeling that even more subtle experiments will be needed before the living rôle of this sense is fully revealed. When this has been done we shall come close to gaining good insight into the life of fishes.

Figure 38. The lateral line system. The head of a deep-sea fish, *Melanonus*, showing the arrangement of the sense organs (neuromasts) in the lateral line canals. (a) Each sense organ, shown as a black dot, rests on a shield-shaped bolster of connective tissue. The canals open through pores, shown as open circles. A neuromast, much enlarged, is seen in (c). Note the gelatinous cupula (c), which carries a vane and rests on the sense organ (sc). Below this two sense organs are shown in part of a canal. (b) When current-like disturbances pass through the pore (p) so that they run along the axis of the vanes, the sense organs respond. Middle: (d) The head of a blind fish (*Amblyopsis spelaeus*) from the Kentucky caves. The free sense organs, shown as dots, are packed closely together in a definite pattern. In e and f are shown enlarged neuromasts (sc). Each sense organ, which bears a long cupula, is seated in a stalked cup. Below: deep-sea gulper-eels (*Eurypharynx*), have free neuromasts (sc) at the end of long stalks (h) which carry a nerve (n). These organs are arranged in groups (see g and i) along the fish. (Drawn from specimens in the Museum of Comparative Zoology, Harvard University.)

Smell and taste

As in air, odours are spread in water by diffusion and movements of the medium. On reaching the nostrils of a hovering or resting fish, the animal may try to find the source of the smell, or it may turn tail and flee. To taste something the fish must make contact with it. Just as the lateral-line sense is a long range kind of touch, so is the olfactory sense, compared to taste, a means of gathering chemical information from a distance. At the same time, a fish may stumble upon a source of odours as it moves around its habitat.

The nasal sacs of fishes are lined with a sensory epithelium that is linked to the brain through the olfactory nerves. Odours in the water are wafted to this lining by way of the nostrils. If an odour is detectable, the sensory buds fire quickened trains of impulses down the nerves to the forebrain. Now the fish becomes aware of the smell.

Lampreys and hag-fishes have a single median nostril and nasal sac (figure 79), but in the jaw-bearing fishes each of the two nasal sacs, which are housed in capsules on either side of the snout, communicates with one or two nostrils (figure 32). The nostril of a lamprey may be seen on top of the head just before the eyes and it leads to a nasal sac and a hypophysial pouch (figure 79). This pouch is squeezed and relaxed by the pulse of the breathing movements and thus acts like an aspirator. Water drawn into the nasal tube enters the sac through an aperture guarded by a valve.

Hag-fishes breathe through the nostril, which opens at the tip of the snout and leads to a passage that continues past the nasal sac and enters the foregut (figure 79). A hag-fish thus smells as it breathes and it has good need of this sense, for its eyes are degenerate. Lampreys have well-formed eyes, the parasitic species depending on vision for finding their prey by day. Yet their olfactory centres form the most conspicuous part of the brain. The perception of odours must be an important part of a lampreys' sensory life, and recent tests do indicate that *Petromyzon* can nose out food.

The cartilaginous fishes – sharks, rays and chimaeras – bear large nasal sacs and well-developed olfactory centres in the forebrain. The two nostrils are usually on the underside of the snout, each being more or less divided by a cartilage-strengthened flap of skin into a forward inlet and a backward outlet, the latter sometimes leading through a groove to the corner of the mouth. Sharks soon gather round sources of smells. When whalers cut up their catch in the sea, Dr P. Budker says, the blood '. . . attracts sharks, who hurl themselves on the whale carcase with incredible force, so much so that flensing has sometimes had to be interrupted. Guns fired from ships at the sharks hardly disturb them. They would continue to shear off mouthful after mouthful of meat and blubber.' [77]

Once I was able to watch the food-tracking tactics of a young blue shark in the Red Sea. On throwing a dead fish into the water, the splash startled the shark and it soon began to make quick weaving turns across the line between its original position and the dead fish, which was about 20 yards away. After several such turns, it streaked towards the fish and snapped it up. The shock of the splash, which may have been perceived through the lateral-line sense, evidently aroused the shark to make tracking movements like those of a bloodhound.* If the fish had been seen by the shark – and the water was very clear – one would have expected a more direct approach. It seems likely that the fish was found by scent.

When both nostrils of a dog-fish are plugged with cotton-wool, the fish loses its power of finding food; but if one nostril only is blocked, the animal keeps turning towards the side of the open nostril. This suggests that dog-fishes and their relatives find sources of interesting odours by wheeling towards the side that is getting the stronger scent. The comparison between the responses of the two nasal sacs is made in the brain. If the scent is being wafted by a current from a distance, a shark may need to make a number of tracking movements before it can draw an olfactory bead. The greater the distance between the nostrils, the more readily should a shark be able to detect a particular gradient of odours in the water. A large shark ought thus to have better tracking powers than a small one. Hammerhead sharks, in which the nasal organs are near the ends of hydroplane-like extensions of the head (plate 10), should be particularly good ocean bloodhounds; they are said to be among the first sharks to arrive when there is blood in the water. The great, wide-headed whale-shark and giant devil rays might also be suspected of having a keen, directional sense of smell, one that should lead them to good feeding in the ocean.

The nasal sacs of teleost fishes are housed in bony capsules on either side of the snout. As in other fishes, the sensory lining of a sac

* Sharks are attracted to struggling fish. Low frequency (20 to 60 cycles per second) pulsed sounds resembling those made by a speared grouper attracted bull sharks, hammerheads, lemon sharks and tiger sharks, which swam directly towards the transducer. The investigators [78] consider that particle displacements associated with these broadcast sounds might well be large enough to be detected by the lateral line organs of sharks at a distance of 250 metres from the sound source.

is pleated, thus increasing its surface area. There are usually two nostrils to each sac. Water enters the forward one, and after irrigating the sac, is expelled through the backward nostril (figure 32). This circulation may depend on the compression and relaxation of the nasal sacs, which respond to the breathing movements; or it may depend on currents created by the beating of cilia in the nostrils. In other species water is deflected through the nostrils as the fish swims. Where each sac has a single nostril, as in damsel-fishes, cichlids, sticklebacks and so on, the circulation must depend on the aspirator-like action of the sac [79].

All the sharks and rays have large olfactory organs, but these are very diversely developed in the teleost fishes. At one extreme there are eels with very large nasal sacs containing elaborate rosettes; at the other, there are flying-fishes, half-beaks, garfishes, and skippers, an entire order of fishes (Synentognathi) with curiously reduced olfactory organs. Freshwater eels are quick to find food enclosed in a bag, but a hungry pike, which has small olfactory rosettes, is not interested in unseen food. An eel's sense of smell is so keen that it is able to detect pure chemicals, such as β-phenyl-ethyl alcohol, at concentrations that would admit no more than a few molecules of the substance to each nasal sac at any one instant. Moray-eels, which are particularly common in coral reefs, hunt by night and depend largely on scent for tracking their prey (figure 32).

Most teleost fishes have moderately developed nasal organs but it would be wrong to suppose that all their olfactory powers were proportionately moderate. Some fishes may be remarkably sensitive to certain odours. Water that had been rinsed over human hands proved to be very repellant to coho salmon ascending a salmon-ladder, whereas urine and tomato juice had no such effect. Salmon, as many tagging experiments have proved, usually return to the rivers where they were born. It has also been shown that particular streams and creeks have characteristic odours, due probably to volatile organic substances, which are remembered and used by migrating salmon to find the home stretch.

Fishes also recognize the body scents of other species. By means of conditioning experiments, minnows were found to discriminate between the odours of fifteen species of fishes belonging to eight different families. Blinded blunt-nose minnows (*Hyborhynchus notatus*) were also trained to distinguish between the odours of various aquatic plants. Lastly, Karl von Frisch has shown that the skin of cyprinid fishes contains alarm-substances (*Schreckstoffen*). When an injured minnow was introduced into a school, nothing happened for about half a minute, but afterwards the fishes drew together and suddenly dispersed. By means of apt experiments, von Frisch was able to show that alarm-substances diffuse out of lacerated skin and lead to a panic reaction on being perceived through the nasal organs. More recent tests with twenty-one species of European and Asiatic cyprinid fishes have revealed that each species reacts to the alarm-substances of its own kind. Though the alarm-substance is present in the skin of young fishes, panic flight is not developed until later in life, after schooling has been established. Attack by a predator causing injury to a member of a school thus leads to appropriate responses [79].

Taste-buds look rather like the sensory buds of the olfactory epithelium, but they appear to be much less sensitive to chemical substances. Moreover, contact is needed before a fish can taste its food. The taste-buds are found in the mouth and pharynx and on the lips and gill-arches. Fishes with barbels, such as bichirs, sturgeons, various cyprinids, cat-fishes, red-mullets, certain cod-like fishes and drum-fishes, have these organs studded with taste buds. In fishes such as sturgeon, carp, American cat-fish (*Ictalurus*), cod, ling, red-mullet and grey-mullet, taste-buds are also found in the skin covering the body (figure 33).

When a piece of meat is brought into contact with a barbel or the flanks of an American cat-fish, the animal is quick to seize the food; but the fish no longer responds in this way if the nerves to the taste-bud are severed. Fishes with barbels may well find their food by a combination of taste and touch. A red-mullet will fan the mud with its pectoral fins and explore and test the bottom with its highly mobile barbels (figure 33), which in some attitudes remind one of the forked, flickering tongue of a snake. The slender pectoral and pelvic fins of an African lung-fish, which bear taste-buds, are also used as 'gustatory antennae'. When one of these fins touches food, the fish whips round and grabs it. Rocklings (*Gaidropsarus*) and hake (*Urophycis*) search for food with their filament-like pelvic fins. Gurnards feel their way along the sea floor on the tips of the free pectoral rays, which may bear taste-buds. On touching interesting food, one gurnard (*Trigla lineata*) suddenly wheels round and swallows it, or the prey may be subject to further tactile scrutiny. Other gurnards (*T. cuculus*, *T. gurnardus* and *T. hirundo*) also finger their way along the bottom, but these species depend more and more on the visual sense for the recognition of their prey, and dart on it from a distance [80].

The sense of touch

The skin of fishes, both the epidermis and dermis, contains myriads of free nerve-endings, many of which must be sensitive to touch. Fishes that swim freely at mid-water levels flee at the slightest contact with a strange body. To fishes that live on the bottom or burrow, a certain positive sense of touch is clearly very much a part of their lives. Many fishes live or shelter in branches and crevices of coral and the seeking of appropriate tactile satisfaction is likely to be important to their well-being. Fishes with adhesive, sucker-like pelvic fins, such as gobies, lumpsuckers, sea-snails and cling-fishes, are no doubt guided by touch when they begin to fasten on to a rock or weed. The same may well be true of the shark-suckers (*Echeneidae*), which use their suction-disc to cling to other animals and objects besides sharks. Appropriate contact between the sexes during the breeding season is yet another way in which the tactile sense is involved in the life of fishes.

To take a well-investigated example, recent tests with the rice-fish (*Oryzias latipes*), a small egg-laying toothed-carp, showed the relevance of appropriate tactile stimulation to successful egg-laying [81]. The male finds his mate by sight, and after embracing the rear part of her body with his dorsal and anal fins, proceeds to beat her with these fins until the eggs are extruded. Males deprived of such fins were not able to induce egg-laying, nor could females lay their eggs successfully after being immersed in weak anaesthetic solutions. It seems that tactile stimulation, which will be registered in the central nervous system, induces the pituitary gland to produce hormones that trigger-off the egg-laying process. Many other instances could be given of sexual activities that must involve the sense of touch, but here we are concerned with minute sensory units, free-nerve endings that are widely scattered in the skin. To prove definitely that these are involved when a fish seems to be relying on the tactile sense is a delicate undertaking.

Certain other senses of fishes

By means of conditioning experiments, both freshwater and marine fishes have shown that they are able to respond to a momentary rise or fall in temperature and the minimal differences they can detect are as little as 0·03° to 0·07°C [82]. The sensory elements involved seem to be free-nerve endings in the skin. Similarly conducted experiments have also revealed that marine fishes can discriminate between salinity changes of no more than 0·5 part per thousand and between small changes in hydrostatic pressure. The sensory receptors for changes of salinity have yet to be identified, but in the minnow and probably in other ostariophysan fishes, the Weberian ossicles linking the swimbladder to the ears can function in detecting changes of pressure. However, fishes without a swimbladder are sensitive to quite small alterations in pressure.

While the structural basis of these kinds of senses are more elusive than the ones already considered, it will be evident that we are dealing with faculties that are vital to the well-being of fishes. Marine fishes are likely to make subtle use of their means of temperature and salinity discrimination when seeking their spawning grounds, which are usually in a circumscribed region of the ocean. The sword-fish, for instance, migrates well to the north and south during the northern and southern summers respectively, to seas where temperatures are as low as 12° to 13°C. But during the breeding season it must seek tropical waters of temperatures not lower than 23·5°C, and it seems to favour regions where the temperature is between 25° and 29°C. Indeed, when any kind of marine fish is closely studied, its spawning grounds are not only found to lie in water masses having a restricted range in temperature and salinity, but in regions of a certain limited range in depth. Fishery biologists, in particular, spend much time in trying to discover the physical and biological requirements for spawning in fishes. And just as fishes use all the refinements of their senses to test and explore their surroundings, so are fishery biologists endeavouring to refine their methods of exploration, which are gradually leading to a greater overlap between the ways of fish and man.

Life has been likened to a flame. A. J. Lotka writes: 'If the lamp of life is a poetic symbol, it is an image essentially true to fact. Not only is life, in particular animal life, largely a combustion process: like the flame, life reaches out for fuel, and with the power gained, strains again for more. Like the flame it consumes, and it spreads. And as the fire sends out sparks, of which many die, but a few, falling on favourable ground, flare up as a second generation, in reproduction of the parent flame; so the living creature scatters its seed, some to die, but some also to live again the life of the parent.' [83] While this reminds us most of the need for food, we should not forget that the flame of an animal is kept alive by its senses. Before reaching out for fuel, the flame that is fish scans its surroundings with the senses at its command. To reproduce the flame, breeding grounds are sought, where the sparks have a good chance of catching fire. Much is still hidden, but it is as well to know that fishes have been exploring the hydrosphere for many millions of years. Man, whose means of underwater intelligence are still comparatively crude, has been an explorer of aquatic space for less than a hundred years.

Chapter 10

'Physical' Fishes

'Just as the beauties and structural variety of animals and plants are almost infinite, so it is becoming clearer every year that wherever one looks in nature there are functional inspirations of the utmost significance.'

D. R. GRIFFIN, Listening in the dark, 1958

The production of heat, light, sound and electricity

TWO PHYSICAL signs of life are the evolution of heat and the maintenance of minute charges of cellular electricity. Heat from the oxidation of sugars and fats sustains the high body temperature of birds and mammals, but most fishes are as cool as their liquid living space. Certain fishes are unique, however, in possessing definite organs for generating series of electrical pulses. In such organs, the small quantities of electricity that are normally associated with nervous and muscular activity have been enormously magnified. At any time, the human brain emits a minute fraction of a volt, which can be recorded continuously through electrodes attached to the scalp. A large electric eel (*Electrophorus electricus*) can deliver lightning shocks of 550 volts.

Sounds, as we shall see, may also be associated with the activities of fishes. Again, certain of these sounds can be transformed, this time by a resonantly vibrating swimbladder. This gas-filled sac may also have its own muscles to set it vibrating and roll out its 'voice'. But luminous light is not a by-product of vertebrate metabolism. The production of living light requires special enzymes and proteins, which in fishes are produced in glandular parts of their light organs; and these may be very elaborate structures. Even so, the light organs of certain fishes, which may be no less elaborate, contain cultures of luminous bacteria. Whether or not the light of these simple organisms plays a part in their own lives, it has been put to good use by many kinds of fishes.

The body temperature of fishes

Fishes are cold-blooded or poikilothermous animals. The Greek adjective *poikilos* means 'many-coloured', which in one sense of the word 'colour' is quite apt. The blood of most fishes has the same temperature as that of the environment: it is, so to say, 'coloured' by its surroundings. The blood of a tropical fish may thus be more than 25°C warmer than that of species from the polar regions. The salient feature is that fishes have no means of keeping a constant body temperature.

Most of the heat comes from muscular activity: for muscles not only burn plenty of carbohydrate fuel, they are easily the heaviest organs in the body of fishes. Any heat in excess of the ambient temperature must soon be lost through the surfaces of contact between the fish and water, through the skin and the membranes of the mouth, gill chambers and gills. The smaller the fish the more surface it has in relation to weight; for the surface area varies as the square and the weight as the cube of the body dimensions. A goby or a guppy will thus lose excess heat much more quickly than a whale-shark.

The body temperature of a whale-shark has yet to be measured, but it does seem that the larger scombroid fishes, such as tunnies, marlins and so on, may at times generate a temperature higher than that of the sea. Just after being caught, a long-finned tunny (*Thunnus alalunga*), which had no doubt been struggling hard, was found to be 9°C warmer than the ocean. Five striped marlin (*Makaira mitsukurii*), weighing from 184 to 283 pounds and ranging in length from 8 feet 3 inches to 8 feet 10½ inches, had their body temperatures taken by a thermopile-harpoon. They were from 1° to 6°C warmer than the New Zealand waters where they were caught, but they were played for half an hour or more before being landed. All this indicates that actively swimming tunny and marlin may become 'warm-blooded' fishes. Their massive, close-grained muscles must generate heat quicker than it is lost to their surroundings.

On the other hand, a 269-pound ocean sun-fish (*Mola mola*), which was also caught off New Zealand, was found to have a body temperature of 18°C, whereas the temperature of the surface water was 21°C. But between depths of 15 and 30 fathoms the sea temperature was equal to that of the fish. The two investigators, J. E. Morrow and A. Mauro [85], were led to wonder if the sun-fish had recently come up from these cooler subsurface levels, the time being too short for the temperature of the body to become equal to that of the sea. After all, if a large fish loses heat slowly it will also gain heat slowly.

More measurements of this kind are needed, but it does seem that large and active fishes are bound to become 'warm-blooded' when swimming at full stretch. After a burst of activity, the time that muscles take to recover depends on the rate at which lactic acid, a by-product of metabolism, can be eliminated. Fishes are slower to recover than mammals, but the steady body temperature of a bird or a mammal, being higher, will allow lactic acid to diffuse more rapidly into the blood and so be removed from the muscles. By becoming warm-blooded, a tunny or marlin might thus be better placed to regain its full swimming powers after strenuous activity. But how long does it remain warm-blooded? Perhaps more can be learned by studying tunnies in captivity.

Electric fishes

Electric organs are found in several hundred kinds of fishes. Those able to generate substantial shocks are in the minority and they live in the ocean, lake or river, as do those with weaker electrical powers. Nearly all the marine electric fishes are rays, the torpedo-rays (Torpedinidae) of the temperate and tropical ocean having greater powers than their more diverse relatives the skates, most of which are temperate species. Certain of the stargazers (*Astroscopus* spp.), which are the only marine teleosts known to have electric organs, have a capacity equal to some of the torpedo-rays. Like the skates and electric-rays, they are bottom dwellers.

Except for the electric-eel of tropical South America and the electric cat-fish (*Malapterurus electricus*) of tropical Africa, all the freshwater species are weakly electrical. The mormyroid fishes of African lakes and rivers, comprising about a hundred and twenty species, are easily the most diverse. Their counterparts in Central and South America are the knife-fishes, which are relatives of the electric-eel (figure 39).

During a field trip to Ghana, Dr Hans Lissmann [20] found that the muddy waters of the Black Volta and its tributaries could be alive with electricity. This activity was readily detected by an amplifier coupled to two copper wire electrodes, which were suspended in the river. The output of the amplifier could be displayed on a small cathode-ray tube or heard through headphones. Indeed, on some occasions the discharges of electricity could be heard without amplification.

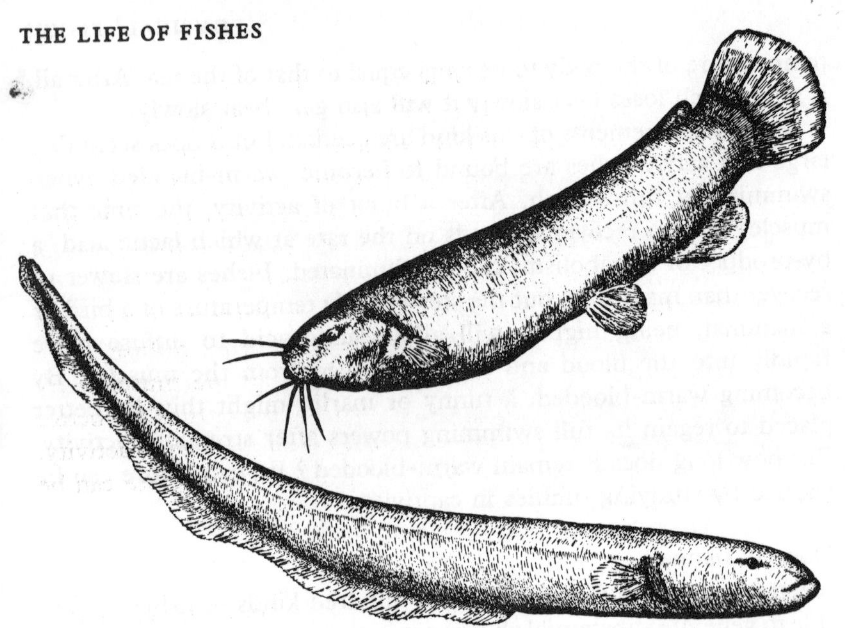

Five kinds of discharges were heard and viewed; a study hum of pulses, about 300 per second; rather regularly spaced pulses 1 to 6 per second; similar pulses of higher frequency, usually between 20 to 50 per second; bursts of pulses with a frequency of about 120 per second; and powerful and brief discharges consisting of a train of pulses. The last kind of pulse may well have come from an electric-cat-fish: the other four kinds must have been due to certain mormyroid fishes, for the first was very reminiscent of the electrical activity of *Gymnarchus niloticus*, while the second type resembled the pulses produced by *Mormyrops boulengeri*.

Study of the activities and behaviour of mormyroid fishes is revealing the unique part that electricity plays in their lives. The electric organs lie below the skin of the after-body, those of *Gymnarchus* consisting of four tubular structures on each side which extend forward from the tip of the tail. The rest of the mormyroid fishes belong to another family, the Mormyridae, and these have a pair of elongated electric organs on each side of the tail (figure 40). Each organ, as in all electric fishes, is composed of many electroplates, which are derived from muscle cells. A single electroplate is a disc-like structure, one surface of which is innervated. The opposing surface is deeply folded into finger-like processes (figure 40). As each electroplate is embedded in a gelatinous matrix, and as the cell contents are rather transparent, an electric organ has a clear, jelly-like

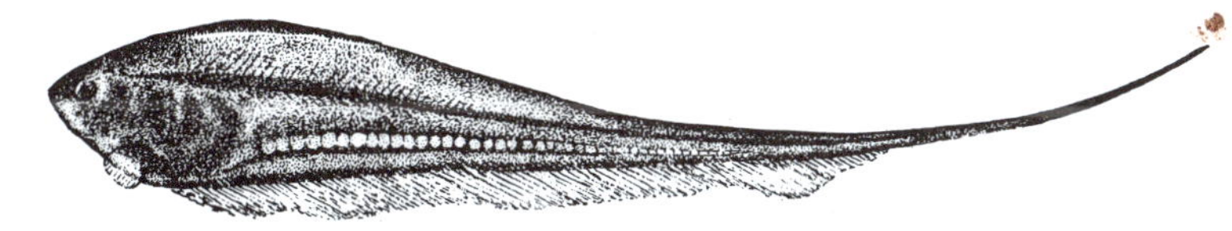

Figure 39. Electrical fishes. Left: electric cat-fish (*Malapterurus*) and electric-eel (*Electrophorus*). Right: knife-fish (*Eigenmannia*) (above), and a mormyrid (*Mormyrus*).

appearance. Like a pile of coins, all the electroplates usually fit closely together. They also face the same way. The electroplates of the mormyroids have fore and aft faces, and there are from 150 to 200 in each electric organ, the total complement thus being from 600 to 800.

When the electric organs are active, they emit series or bursts of pulses, so surrounding the fish with an oscillating electric field. The nature of each pulse and the frequency of discharge may vary from one species to the next. During part of its life, each mormyroid fish thus swims within its own electrical world, but to what ends is the world created? The beginnings of an idea came to Dr Lissmann [20] through study of the habits of *Gymnarchus*. It feeds on other fishes and lives, as do the mormyrid fishes, in waters that are rarely other than turbid. Dr Lissmann reasoned thus: '. . . a specimen of *Gymnarchus niloticus* aroused interest through its navigational abilities. This fish, which frequently swims backwards, appears to avoid obstacles, to find its way through crevices in rocks, and to locate its prey from a surprising distance.' The weakly developed eyes could hardly peer through turbid waters to this extent, nor did it seem likely that the lateral-line sense was involved. The implication was that the fish might well be using its electric field (figure 41) both as a means of navigation and of detecting its prey.

Like its mormyrid relatives, *Gymnarchus* is extremely sensitive to electrical charges. It will react visibly to the motion of small electrostatic charges, such as can be induced by running a vulcanite comb through one's hair. When a concealed rod magnet is moved, the fish is quick to respond. More spectacularly, it will follow a magnet that is moved up and down the sides of the slate tank, the fish seeming to be drawn along by a sorcerer's wand. Tests on responses to electrical currents showed that *Gymnarchus* was about five hundred thousand times more sensitive than a minnow.

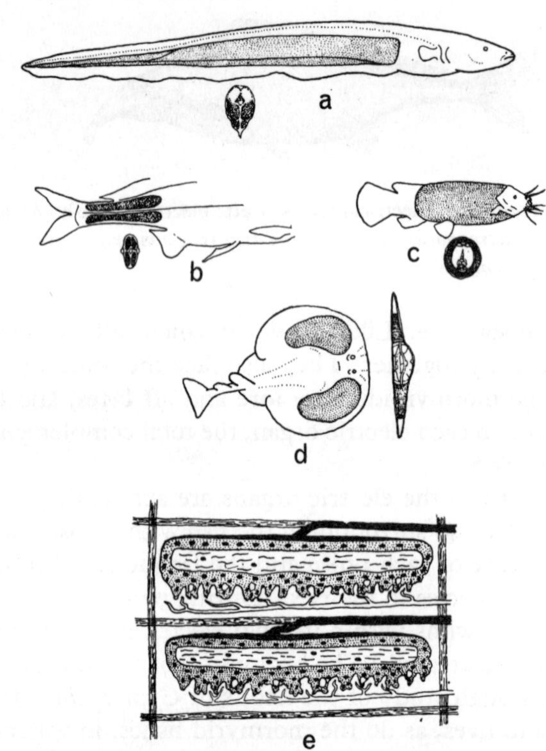

Figure 40. Electric organs, shown dotted, of (a) electric-eel, (b) mormyrid, (c) electric cat-fish and (d) torpedo-ray. Two electroplates are shown (e). Note the much folded surface (a to d after H. Grundfest [87].)

After experimenting with an electrical model of *Gymnarchus* and the fish itself, Dr Lissmann and Dr Machin [86] proved that electric fishes set up flow patterns of current in the surrounding water (figure 41). Any body with a conductivity unlike that of the water will distort this electrical pattern, so changing the field of potential around the fish. In view of the very refined electrical sensitivity of the fish, perception of this charge can be used to detect the position of the body. A *Gymnarchus* can thus avoid an obstacle, steer through a crevice, or make for its prey, all of which are decidedly useful powers in muddy tropical waters. This kind of electrical sense is a parallel to the lateral line sense, by means of which a swimming fish seems to be able to detect a distortion due to some nearby object in the flow patterns of water around its snout.

Pulses from electric organs may also serve as signals. One *Gymnarchus* will sense the presence of another, even at some distance, and it may then attack its neighbour. The attacker could have homed to the electrical signals of the victim, as the following experiment suggests. When the discharges of a *Gymnarchus* were picked up and fed to any one of six pairs of electrodes immersed at intervals around a tank, the fish always attacked the pair in circuit. Perhaps an exchange of pulses has a very different significance for a pair of courting fishes, but this is something for future study.

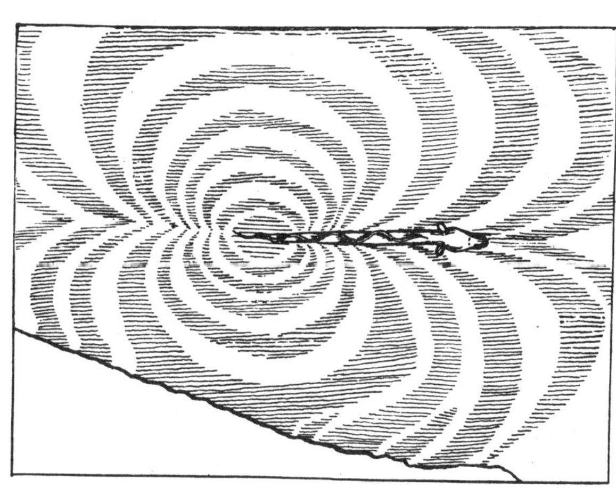

Figure 41. An impression of the electric field surrounding *Gymnarchus*, a mormyroid fish. (After J. Gray.)

Electric fishes are highly sensitive to electrical changes, but where are the sense organs ? In the mormyroid fishes the skin is pitted with pores, each leading through a jelly-filled canal to a sense organ known as a mormyromast. A much enlarged part of the cerebellum forms a brain centre for these sense organs. If, as seems likely, the mormyromasts are electric sense organs, this centre could compare the responses from all parts of the body, so making the fish aware of any distortions in the electric field, distortions that may mean food or an obstacle to be avoided.

The mormyroid fishes form a subdivision of the Isospondyli. The knife-fishes, which live in rivers and lakes of Central and South America, belong to another order of teleosts, the Ostariophysi. But though these two groups are only distantly related, they are very alike in electrical and other respects. The electric organs of the knife-fishes are derived from outer parts of the trunk and tail muscles and they generate pulses at frequencies between two and one thousand per second. Like *Gymnarchus*, one knife-fish (*Gymnotus carapo*) can detect the whereabouts of a stationary magnet and discriminate between conducting and non-conducting bodies in its surroundings. Again, the knife-fishes have a very large cerebellum in the brain and the electrical sense organs are like those of the mormyrids. One of the most striking convergences between the knife-fishes and *Gymnarchus* is in the way of swimming. The body is not undulated but held stiff and straight: the driving force comes from waves passing down the long anal fin in the knife-fishes or down an equally long dorsal fin in *Gymnarchus*. By this means the long axis of symmetry, both of the fish and its electric field, are kept coincident, which, judging from trials with a model fish, is essential if the field is to be effective in electro-location of nearby bodies.

Powerfully electrical fishes, as we might expect, have larger electric organs than have the mormyroids or knife-fishes. In the very long tail of an electric-eel, the lower parts of the muscle segments have been turned into three electric organs, a main one followed by a smaller terminal one known as the organ of Sachs. Beneath these two is a second weak element, the organ of Hunter (figure 40). The surfaces of the wafer-like electroplates face fore and aft, those of the main organ being packed closer together than are the others. There are about seventy columns of electroplates along each side of the body and each column contains from six thousand to ten thousand plates, whose innervated surfaces face the tail. When the electric organs are firing, these nerve-receiving surfaces are negative with respect to the opposite, deeply-folded surfaces, which is the rule in nearly all electric organs. As most of the electroplates are connected in series, their charges, like those of a series of batteries, add up to produce a large voltage. Each electroplate contributes 150 millivolts. Current thus flows from the tail to the head of the electric-eel, and in the reverse direction in the surrounding water, so completing the circuit. The series arrangement of the electroplates enables the eel to overcome the high resistance of its freshwater habitat. But certain of these tiny batteries are connected in parallel, a linkage enabling brief pulses of high amperage (about 1 ampere) to be generated.

Like all powerful electric organs, those of the electric-eel produce brief bursts of pulses, each pulse lasting three milliseconds, the bursts being repeated several times in one second. The nerves supplying the electric organs are fired by a command centre in the brain, so releasing the charges of the electroplates. The shock produced may be more than 500 volts.

The living batteries of an electric-eel occupy nearly half the bulk of its body, which may reach a length of eight feet. A shock from one of these fishes will severely shock a man, which shows that they have powerful means of defence. In an aquarium an electric-eel will use its electric organs to stun other fishes, which can then be devoured at leisure.

Like its relatives the knife-fishes, an electric-eel swims in a stiff-bodied way, driven by undulations of its long anal fin. This suggests that it uses its electric organs for guidance and prey detection, for it has very small eyes and lives in muddy waters. In fact, the electric-eel uses its weaker electric organs to such ends. The organ of Sachs has been found to emit low frequency pulses, which seem to be used just like those of a knife-fish, a *Gymnarchus* or a mormyroid. The electric-eel enjoys the best of both electrical worlds. It is a truly remarkable fish.

The electric-cat-fish, which grows to a length of three feet or more, will also stun other fishes in an aquarium, for it can deliver shocks of up to 350 volts. The translucent electric organ is found just under the skin of the trunk and the fore part of the tail (figure 40), and it is divided into a number of dove-tailing compartments that contain the electroplates. The nerve-bearing sides of the electroplates face towards the tail and in each half of the organ they are innervated by the branches of a single large nerve cell that is housed in the grey matter of the spinal cord.

Again, the large torpedo-rays and the electrical stargazer can at least be suspected of switching on their batteries to shock their prey in the surrounding sea. Individuals of *Torpedo nobiliana*, which reaches a length of 6 feet, have been found to contain a 4- to 5-pound salmon, a 2-pound eel, summer flounders, red mullet, plaice and spotted dog-fish. As torpedo-rays appear to be the laziest of fishes, we may presume that their prey was devoured after being stunned.

Torpedo-rays have two large, kidney-shaped electric organs, one on each side of the head (figure 40). They are derived from the hypobranchial muscles. Each organ is composed of a number of vertical columns of electroplates which are circular to hexagonal in cross section. The number of electroplates in a single column varies from one hundred and forty to over a thousand and in a large ray the total number may be about half a million. The largest recorded dis-change, from *Torpedo nobiliana*, is 220 volts.

As the electrical resistance of the sea is much less than that of fresh water, substantial currents must be produced if the electric organs are to deliver powerful shocks.* At least two-thirds of the electroplates of the torpedo-rays are, in fact, connected in parallel, as are some of those of the stargazers, whose electric organs are derived from the muscles that move the eyes.

Do the torpedo-rays have a direction-and-prey-finding system? One species, *Narcine brasiliensis*, has a small electric organ behind each of the main batteries. Here, at least, is an exciting indication.

The weak electric organs of the skates (*Raja* spp.) are formed from parts of the tail muscles. But these fishes seem very reluctant to display their electrical powers, although the thorn-back ray (*Raia clavata*) can produce a discharge of about 4 volts. Dr Harry Grundfest thinks as follows:

> . . . of all the electric fishes thus far studied, only the electric skates remain an evolutionary puzzle. Their organs, located in their whip-like tails, and producing low voltages at low current, discharge only with great provocation, and they would seem to have little, if any, adaptive value. Considering the vast number of electrical experiments tried by natural selection, however, it is not surprising that one of them should be inconclusive [87].

It will be interesting to see if he is right.

All kinds of electric fishes were a puzzle to Charles Darwin, who was well aware that the different groups are far from being closely related. There are also different kinds of electric organs, formed in different parts of the body. Clearly, then, electric organs could not have been inherited from a common ancestor. In any event, he wrote, '. . . it is impossible to conceive by what steps these wondrous organs have been produced' [88].

Concerning these steps, Dr Lissmann [20] feels there might have been three. When a fish – in his experiment a freshwater eel – makes a sudden movement, an electrical disturbance, perhaps a muscular action potential, can be recorded in the water nearby. If the perception of such a disturbance came to have a survival value, as it might in a gregarious species, the way would be open to the development of weak electric organs. From these, more powerful organs could be evolved. But whatever the steps, it is now clear that electric organs must have developed quite independently in six separate lines of

* Power (watts) equals i^2 (current in amperes) times r (resistance in ohms).

evolution – in the skates; torpedo-rays; mormyroids; knife-fishes and the electric-eel; the electric cat-fish; and the stargazers. Evidently there is only one way to make an effective electric organ – by the transmutation of nerve and muscle; but there have been at least six variations on this theme [89].

The independent acquisition of like structures for similar functional ends is known as convergence, which is just as basic a part of evolving life as divergent descent from a common ancestor. Considering only the substance of this chapter, convergences in the design of sound-producing swimbladders and light-producing organs are no less striking than those displayed by electric organs. Our understanding of evolution will remain unbalanced until all kinds of convergences have been properly investigated.

Sound production

Far away air-borne sounds may be heard quite clearly on a calm, cool day, but gusty days bring gusty hearing, even of nearby sounds. Given perfect conditions, however, air is not such a good conductor of sound as water. Loss of sound energy through frictional effects is much less in water. Except for boisterous aquatic habitats, water is also more stable than air and much less broken up by turbulence, which means that sound paths in water will be subject to less deviation.

Since fishes live in this acoustically superior fluid, is it so surprising that many species have acquired the means of producing definite sounds? But before these are considered, I must mention that swimming fishes can make sounds which may be picked up by a hydrophone. James Moulton [90] found that an idling school of an anchovy (*Anchoviella choerostoma*) was noiseless, but that sounds were produced when the school was streaming or veering. During veering movements there were sharp increases in sound lasting from one to three-fifths of a second. Amberjack (*Caranx latus*) made a thumping noise as they veered. But even during steady streaming movements the sounds produced are not continuous, which suggests that the noises are struck from the sea by the quick flexures of thousands of swimming bodies. All the sounds recorded were below 2,000 cycles per second, the pitch range of the more intense sounds tending to be lower in larger fishes, which swim by fewer tail beats in a given time than do smaller individuals.

Swimming sounds are simply a by-product of certain kinds of

movements, but they are not without significance in the lives of fishes. A blinded anchovy paid no attention to twelve normal individuals when they were resting or quietly milling in an aquarium; but as soon as they were startled and began to stream and veer, the blind one joined them and behaved as a member of the school. Sight certainly plays a part in the schooling of fishes, yet here is evidence that sudden concerted swimming sounds may bring the school together and that visual means of station-keeping are not the only ones.

Bursts of swimming sound may also arouse predators. When the sounds made by a large school of anchovy were played back into an aquarium containing young amberjack, they began to quicken their swimming movements. Man, who is a great fish predator, is also aware of swimming sounds. Ghanian 'herring' fisherman* use three-pronged paddles about 5-feet long to listen for schools of fishes. The paddle is lowered into the water over the stern of the canoe and rotated slowly, the fisherman holding his ear to the end. Those skilled in this art are not only able to judge the direction of the school, but also to get an impression of how far away it may be [90].

Incidental sounds are also made by fishes with a swimbladder opening into the foregut through a pneumatic duct. When such a fish expels a gas-bubble through this duct, the escape of the bubble produces a mouse-like squeak. Carp, minnow, chub, barbel, loaches and eels, and no doubt many other such fishes with an open swimbladder, will release gas bubbles if they are suddenly disturbed, but no one knows what part the incidental mouse-calls may play in their lives. Professor Sven Dijkgraaf [91] tried to play Pan to minnows by blowing small underwater bubbles of the size they produce through a fine glass tube. Such a charming experiment deserved success, but the minnows were quite unresponsive. Herring, pilchard and other clupeid fishes must release bubbles as they move towards the surface. Mediterranean fishermen spot a rising school of sardines by the altered appearance of the sea surface, due to the bursting of myriads of gas-bubbles. As such a school is ascending it must produce a Bedlam of mouse-like squeaks. If nothing else, one can imagine that these noises will attract the attention of predatory fishes and cetaceans.

A good many fishes are noisy eaters. Marie Poland Fish [92], who

* They catch *Sardinella aurita*, *S. cameronensis*, shad (*Ethmalosa dorsalis*), long-finned herring (*Ilisha melanota*) and a drum-fish (*Cynoscion senegalla*).

has studied sound-making in many fishes, has this to say of feeding noises:

> Such bottom feeders as the rays have pavement-like teeth of a grinding type, and their activity on feeding grounds is marked by the crackling sound of crushing shells. The trigger-fish has a powerful set of chisel-like teeth that drill noisily through molluscs. Flounders and wrasses use blunt, molar-like pharyngeal teeth for mastication of similar hard food. Even vegetarian fishes may produce unusual noise with pharyngeal grinding when a piece of sea weed, scraped from a rock or piling, happens to have hard-shelled animals attached. Listening in shallow water that is well populated by cunners (*Tautogolabrus adspersus*, a kind of wrasse) may reveal constant clicking and chirping, which are chewing sounds.

Coral crunchers, such as parrot-fishes and puffer-fishes, are also noisy feeders.

Teleost fishes with pharyngeal teeth are particularly interesting, for these teeth, and especially the muscles that move them, are close to the front of the swimbladder. As the lower pharyngeal teeth bite against those on the roof of the throat, rasping or grinding sounds are produced which set up answering vibrations in the swimbladder, at frequencies depending on the source. Much of the sound produced by feeding amberjacks, yellow-tails, grunts (Haemulidae), trigger-fishes and file-fishes seems to stem from the swimbladder. More significantly, it is becoming clear that these incidental sounds have their biological uses. On hearing their own pharyngeal tooth-rasps, which were played back by a transducer, amberjacks (*Caranx latus*) began to behave as though they were feeding: 'The fish became exceedingly active, swimming about furiously, and facing the transducer to nibble at its rubber surface.' [90] One can imagine a group of foraging fishes gathering round an individual that had chanced to find food, the discovery of which would be plainly audible. In this way feeding would become a group activity and food resources would be more thoroughly explored. But there is another side to audible feeding activities. When playing back the sounds from *Caranx latus* into the sea '. . . an adult barracuda (*Sphyraena barracuda*, *Sphyraenidae*) came abruptly to a spot about eight feet from the suspended transducer and lay quietly facing it for about three minutes.' [90] Even at best, underwater vision is very restricted, but if a predator can hear his prey and find the sound-makers, which are preoccupied with filling their own stomachs, what a fine opportunity he has.

Turning now to structures that are set aside, as it were, for the production of sounds, there are two main kinds. Like crickets, fishes make sounds by stridulation, that is by the friction of one part of the skeleton against another. These sounds have a rasping, scratching or whining quality; they are usually higher pitched than those made by the swimbladder, the second special kind of sound-producing organ. Swimbladder sounds, to quote Marie Poland Fish, '. . . are variously described as thumps, grunts, groans, growls, knocks, thuds, clucks, boops and barks. Typically they have a hollow quality, like the sound of a distant tomtom or of hammering on a wooden wall or like the sound produced when a wet finger is rubbed along the surface of an inflated balloon' [92]. Such sounds are generally from 50 to 1,500 cycles per second (middle C on a piano is 256 cycles per second), but most of the energy is concentrated below 300 cycles per second. Stridulatory sounds range from 50 to 10,000 cycles per second.

All known fishes with special sonic organs are teleosts. Sharks and rays may well make sounds as they swim and we have seen that some rays are noisy eaters; but these fishes have no swimbladder and their gristly skeleton, even if well calcified, is unlikely to be hard enough for stridulatory sound.

Sound-producing fishes in freshwaters

In most freshwater fishes,* the swimbladder is linked in some way to the ears, endowing its possessor with enhanced powers of hearing. The air-breathing, suprabranchial chambers of the Anabantidae (see figure 29), which lie close to the ears, can also act as hydrophones. But what do these fishes hear in nature and what part does sound play in their lives? What of the other species with lesser capacities of hearing? Very little is known.

Recent studies of North American cyprinid fishes are particularly welcome. H. E. Winn and J. F. Stout [93] found that the satinfin shiner (*Notropis analostanus*) makes two kinds of sounds: a high-pitched sound when gas-bubbles, presumably from the swimbladder, are released from the mouth, and single knocking sounds. The latter sounds are made when male fishes are fighting. When males are courting females, they make a purring sound. The pitch of the knocks is from 85 to 11,000 cycles per second and the duration of each is 12 to 60 milliseconds. In male fishes the sounds are most noticeable at the beginning of the breeding season, when the water temperature is from 25 to 30°C. Related fishes, such as *Notropis spilopterus* and *Semotilus margarita*, also make knocking sounds, and again, these are

* Mormyroids, notopteroids and Ostariophysi (carp-like fishes, characins and cat-fishes).

emitted as the males are fighting. But the sound-making organ has yet to be identified.

Using the blacktail shiner (*Notropis venustus*) and the red shiner (*N. lutrensis*), E. A. Delco [94] recorded the activity and responses of male fish to calls coming from ripe females. By playing back these records, he found that males of both species were perfectly able to distinguish the calls of their proper partners, even when offered female calls of the other species at the same time. But the trilled calls of the red shiner are much longer (0·84 second) than those of the blacktail shiner (0·047 to 0·07 seconds), which are not trilled. Here, then, is definite evidence that sounds may serve as distinct recognition signs between the sexes of two related kinds of freshwater fishes, so being a means of isolating one species from another and concentrating the breeding stock.

Numerous cat-fishes* have an extraordinary sound-making swimbladder, which is vibrated by an 'elastic-spring' apparatus (figure 42). In the cat-fish *Auchenipterus nodosus*, for example, the two transverse processes of the fourth vertebra are bent downwards and backwards, each ending in an oval bony plate that is embedded in the interior wall of the swimbladder and each being springy and flexible. To the forward part of each spring is attached a powerful muscle that runs forwards to an insertion on the back part of the skull (figure 42). The activity of the two muscles, coupled with the spring of the processes, makes sounds through the swimbladder. A South American cat-fish, *Doras maculatus* has '. . . a very deep, grunting tone, that is so intense that it is still heard very distinctly at a distance of 100 feet when the fish is out of water.' [95] Certain other cat-fishes, such as members of the tropical American family Pimelodidae, are without an elastic-spring device; instead the swimbladder has drumming muscles that are attached to the skull. Some kinds also produce sounds by stridulation, by rubbing the basal parts of spiny pectoral rays or of strong dorsal rays against neighbouring bony structures.

Sound-making is thus well developed in cat-fishes, and we can be sure that our knowledge is far from complete. Most cat-fishes, as Dr C. T. Regan says, '. . . are carnivorous fishes that live at the bottom, many of them in stagnant water or muddy rivers; the eyes are often small, but these fishes find food by means of their barbels, which are generally six or eight in number, and may be very long.' [96] Nearly

* For instance, those belonging to the South American genera, *Auchenipterus*, *Doras*, *Oxydoras*, *Rhinodoras* and *Euanemus*, the African genera *Synodontis* and *Malapterurus* (the electric cat-fish) and the Indian genus, *Pangasius*.

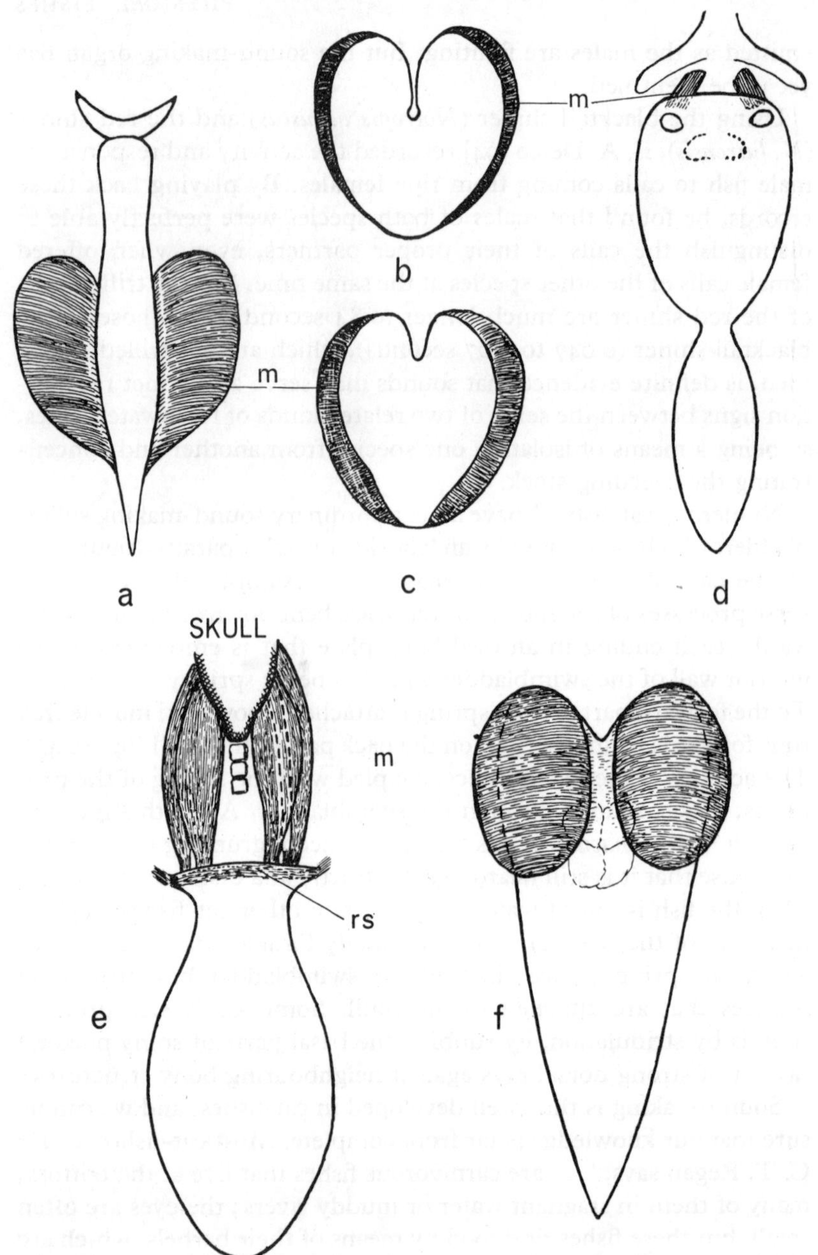

Figure 42. Swimbladder with drumming muscles (m). (a) drum-fish (*Bairdiella chrysura*), (b) and (c) toad-fishes *Opsanus tau* and *O. beta*, (d) cat-fish (*Pangasius djambal*), (e) deep-sea brotulid (*Dicrolene nigra*) and (f) rat-tail (*Macrourus berglax*).

all the sound-makers certainly live in turbid, tropical waters and it may well be that their 'song' is particularly important during the breeding season, enabling dispersed members of a species to assemble and reproduce – congress that might otherwise be difficult.

Sound-producing fishes of the ocean

Most kinds of marine fishes live in the seas that overlie the shelves of the continent. Many are sound makers and more are likely to be discovered, for exploration with hydrophones and recording equipment is still in an early stage. Deep-sea sound has yet to be properly explored, but there is evidence, as we shall see, that sounds must be made by numerous kinds of rat-tailed and brotulid fishes, most of which live near the deep-sea floor.

Many coastal teleost fishes have a sound-producing swimbladder with special drumming muscles (figure 42). When these muscles contract, the swimbladder is twitched into vibrations, the fundamental frequency corresponding to the frequency of muscular contraction.

In one kind of sound-producing swimbladder,* the red drumming muscles are closely attached to the wall of the swimbladder (figure 42). In a second type, one attachment of the muscles is to thc swimbladder wall, the other to a neighbouring structure. This may be the lateral wall of the body cavity in the case of the ling, burbot and certain drum-fishes (figure 42); or to the ribs in the case of cod and *Raniceps*; or to the back of the skull in sea-perch (*Therapon*), three-spined fish (*Triacanthus*) and *Ophidium* [32].

During the second world war, underwater listening for enemy craft, which became possible with the development of hydrophones and associated electronic equipment, showed that the seas were far from silent. Along both coasts of the USA, some of the more chronic interferers with hydrophone networks proved to be drum-fishes (Sciaenidae). In eastern coastal waters there was the croaker, with its rapidly repeated bursts of drumming; the black drum and red drum, with their isolated groans; the spot, uttering a series of raucous honks, and so on. Close by Scripps Institution of Oceanography at La Jolla, California, listening from 1942 to 1947 revealed that there was an underwater chorus from April or May to September. Dr

* Found in such fishes as the pollack, haddock, coal-fish, John Dory (*Zeus*), the spade-fish (*Chaetodipterus faber*), the porcupine-fish (*Diodon hystrix*), gurnards belonging to the genera *Trigla* and *Prionotus*, the singing midshipman (*Porichthys notatus*) and toad-fishes (*Opsanus*).

M. W. Johnson writes: 'The chorus begins about sunset and increases steadily to an uproar of harsh, froggy croaks, with a background of soft drumming. This continues unabated for two to three hours and finally tapers off to individual outbursts at rare intervals.' [97] The chorus was largely provided by drum-fishes called the black croaker, spot-fin croaker and yellow-fin croaker.

Daily and seasonal rhythms of fish sounds were also discovered along the east coast of the USA. During the day, croaker (*Micropogon*) noise was mostly confined to their feeding time, which begins towards sunset. In Chesapeake Bay the principal frequency of drumming was 600 cycles per second in early June, but had fallen to 250 cycles per second in early July, an indication that older, deeper-voiced fish had migrated into the area. Dawn choruses have also been heard. Like birds, certain fishes (notably drum-fishes and squirrel-fishes) and marine invertebrates (snapping shrimps) make more noise at dawn and at dusk [92].

Besides being associated with feeding, sound-making in the drum-fishes is evidently important during the breeding season. In some drum-fishes, the males alone are provided with drumming muscles on the swimbladder, although both sexes of certain species make noises by grinding their pharyngeal teeth. On the other hand, both male and female croakers have sound-making muscles. But closer appreciation of the biological rôle of these specific and sexual sonic capacities has yet to be gained. However, certain drum-fishes use their voices during aggressive and defensive activities, which is also true of the singing midshipman and the related toad-fishes.

Toad-fishes, which have a heart-shaped swimbladder with a band of red muscle along each side (figure 42), utter very loud calls. According to Dr W. N. Tavolga, 'The sounds produced by the submerged fish are loud enough to be heard out of water, but underwater, if one is close enough, the blast can be almost deafening. Measurements have shown that at a distance of 2 feet the sound output of a single toad-fish may reach an intensity of 100 decibels, a value comparable to the noise of a riveting machine or a subway train.' [98] Of the several species of toad-fish, one (*Opsanus tau*) is common along the east coast of the USA: the other (*O. beta*) lives in the Gulf of Mexico. Both species utter two main sorts of sound, powerful foghorn-like 'boops' and short grunts, which may be associated with deep growling tones. If the nerves supplying the swimbladder muscles are stimulated electrically, grunts are produced. 'Boops' have yet to be elicited by this means.

The fundamental frequency of the single foghorn calls of *Opsanus tau* is 140 cycles per second (almost an octave below middle C), and there are harmonics at 140 cycles per second intervals up to nearly 2,000 cycles per second. Grunts are lower pitched. The Gulf toad-fish (*O. beta*) is a little smaller than the Atlantic species, and has quite a different voice. The fundamental pitch of the foghorn call is more than an octave higher (350 cycles per second) than that of *O. tau*, and instead of uttering single calls it emits a double 'boop', the second being shorter than the first. 'Each of the boops is introduced by a short grunt and the entire sequence is announced by a longer grunt.' [98]

Toad-fish live on the sea floor and are decidedly territorial in habit. A resident fish will see an intruder off, including one of his own kind, by threatening postures and grunts. A male guarding a nest of eggs is angrily vocal on being disturbed. But the significance of the foghorn calls is not yet clear. While they may well be important during the spawning season, they are certainly not confined to this period. Perhaps, like bird song, they are linked to territorial activities.

Sound-making has a sexual significance in sea horses (*Hippocampus*), which produce snapping or clicking noises by tossing their heads. At the back of the head is a star-shaped bony crest known as the coronet and this has a loose articulation with the rear edge of the skull. When a sea horse suddenly lifts its head, this edge slips under the coronet, and then, as the head is lowered, snaps out, an action which must presumably produce the sound. Perhaps the swimbladder resounds to these snaps, but at all events there is no doubt of their association with mating activities. Marie Poland Fish [92] found that the mating of a pair of captive sea horses '. . . consisted of slow swimming, either together or apart, accompanied by occasional noisy snapping of the head. Clicks were often produced alternately by the fishes, and during their actual embrace, these sounds were loud and almost continuous.'

In quite a different way, the males of an eastern Atlantic goby (*Bathygobius soporator*) produce low-pitched grunts, which are synchronous with sudden downward thrust of the head. At the same time, the mouth is snapped shut and the gill covers opened; through the upper parts of these emerge strong jets of water. Just how the sounds are made is not yet clear, but again they play a definite part in courtship. The male calls only when he is courting a female and he becomes particularly vocal when she is following him to his refuge.

When the recorded sounds are played back to males in an aquarium, they emerge and move towards the speaker. Females only became interested if they could see another goby as well as hear the sounds. Yet sound is but one means of ensuring that pairing and spawning are successful in this goby: sight and smell are also involved [99].

Deep-sea sound

No part of the ocean is perpetually silent. The deep sea, no less than the seas that fringe the land, may be full of sounds. During cruises of research ships various kinds of deep water sounds have been heard through hydrophones suspended in the sea. Some of these sounds have been traced to cetaceans, but others might easily have been made by fishes. Two other kinds of sounds arise from the play of waves and the grinding of ice.

Many of the Lilliputian luminous fishes that swim in the mid-waters of the ocean have a gas-filled swimbladder, but not one of the species that I dissected – and I looked at more than sixty – has special drumming muscles attached to the sac [38]. Recalling earlier remarks (p. 170), this need not mean that these are silent fishes. But I did find that numerous kinds of rat-tailed fishes (Macrouridae) and brotulids have a capacious swimbladder with large drumming muscles (figure 42). While these fishes may well spend most of their life swimming close to the deep-sea floor, they can be taken at mid-water levels. They are most diverse and common over the continental slopes of the temperate and tropical ocean. In their living space sun-light is reduced to twilight or is entirely extinguished. But, as we shall see (p. 173), there is a background of luminescent flashes, produced by the light organs of deep-sea animals. Moreover, many kinds of rat-tails have their own lights, yet water is a poor transmitter of light compared to the atmosphere. Remembering the superior acoustic properties of water, we can imagine that sound signals may well play a useful rôle in their lives [100 and 101].

At least thirty kinds of rat-tailed fishes must be able to make sounds. The large, red drumming muscles are either completely attached to the swimbladder wall or one attachment is joined to the adjacent wall of the body cavity (figure 42). So far, I have found these muscles only in male fishes, which would suggest that sound production is somehow linked to sexual activities. Perhaps the males begin to call at the beginning of the breeding season and by this means assemble the local breeding stock, which is likely to be dispersed over a large area of the deep-sea floor. How far the calls of the males might travel is unknown, for we have yet to measure the intensity of these sounds. At all events, rat-tailed fishes have extraordinary ears. The part concerned with hearing, the sacculus, contains a very large ear-stone. But what this signifies is something more for future research.

Fishes with lights

Even in the clearest parts of the ocean there can be little or no sunlight beyond a depth of about 750 metres. But above and below this threshold of light, which separates underwater twilight from darkness, the sea is full of luminescent animals. If a sensitive photomultiplier tube is lowered into the ocean, flashes from these organisms can be detected to a depth of at least 3,750 metres [102]. The number of flashes recorded can vary from one to one hundred and sixty in a minute. At the most luminous depths, which are around 800 metres in parts of the Western North Atlantic, there is at times a nearly continuous background of living light.

The living space of luminous deep-sea organisms, many of which are fishes, thus comprises most of the entire volume of the ocean. There are, however, some luminous fishes in coastal seas, which will be considered after the deep-sea species. Furthermore, this review of luminous deep-sea fishes will be concerned with the structure, pattern and performance of their light organs. Discussion of the biological rôle of these organs is reserved for a later chapter (pp. 302–306).

Fishes with light organs or photophores live near the deep-sea floor as well as in the middle bathypelagic regions of the ocean. The most diverse bathypelagic groups of luminescent fishes are the Searsidae, the Stomiatoidea (hatchet-fishes, bristle-mouths, etc), Myctophidae (lantern-fishes) and the Ceratioidea (angler-fishes) (plates 7, 8, 9, 10, 16, 17, 18 and 19). In fact, nearly two-thirds of the species of bathypelagic fishes, and there are close to a thousand, have some kind or kinds of light organs. In numbers of individuals, this proportion is likely to be nearer four-fifths. Fewer of the bottom-dwelling, deep-sea fishes bear light organs, the only known luminous species being numerous members of the rat-tail family (Macrouridae) and a few deep-sea cods (Moridae). The light organs of these fishes, and perhaps those of the angler-fishes, contain cultures of luminous bacteria. The other deep-sea fishes make their own light in special glandular cells.

The Searsidae, which are classified in the herring suborder Clupeoidea, have a unique kind of luminescent organ. On each shoulder is a darkly pigmented sac that opens to the sea through a narrow duct (figure 43) (plate 9). Within the walls of the sac are strands of connective tissues bearing small cells, which probably give rise to larger kinds of cell that fill the cavity of the sac. On being ejected into the sea, the largest of these cells are believed to be broken up and be transformed to sparks of luminous light. At all events, when Dr J. A. C. Nicol handled fishes of the genus *Searsia*, the shoulder glands gave a splendid display. 'When gently stimulated by being lightly pressed with the fingers, it (a specimen of *Searsia schnakenbecki*) suddenly shot forth myriads of blue-green sparks into the water. These brightly illuminated the whole dish with a glow lasting about two seconds.' [103]

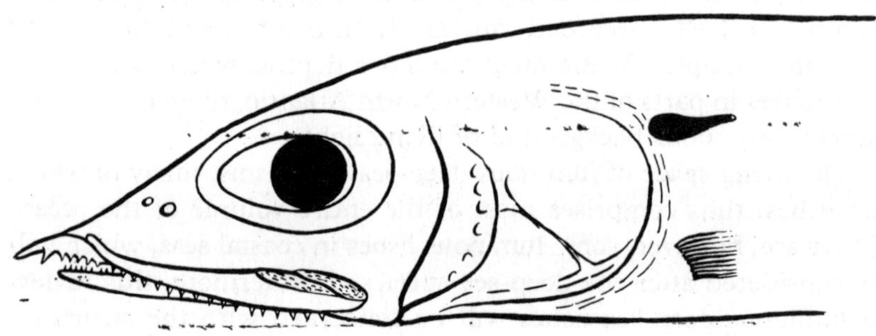

Figure 43. The luminous shoulder organ (shown black) of a searsid fish (*Holtbyrnia*). (Drawn from a specimen in the Museum of Comparative Zoology, Harvard University.)

Besides this pair of shoulder organs, some of the searsid fishes also have photophores on parts of the head and on the underside of the trunk and tail. Each such species has an individual pattern of lights, which may be used as recognition marks. The discharge of the shoulder organs may mislead an attacker, the fish escaping to leave its aggressor worrying a cloud of sparks.

Nearly all of the three hundred-odd kinds of stomiatoid fishes have four rows of large light organs, two along each side of the underbody (plate 8). Each of these lights has a black pigment cowl faced by a silvery reflector and a lens (figure 44). The largest light organs are found on the predatory stomiatoids.* Behind or below each eye is a

* Astronesthidae (star-eaters), Melanostomiatidae (black dragon-fishes), Stomiatidae (scaly dragon-fishes), *Chauliodus* (viper-fishes) and Malacosteidae.

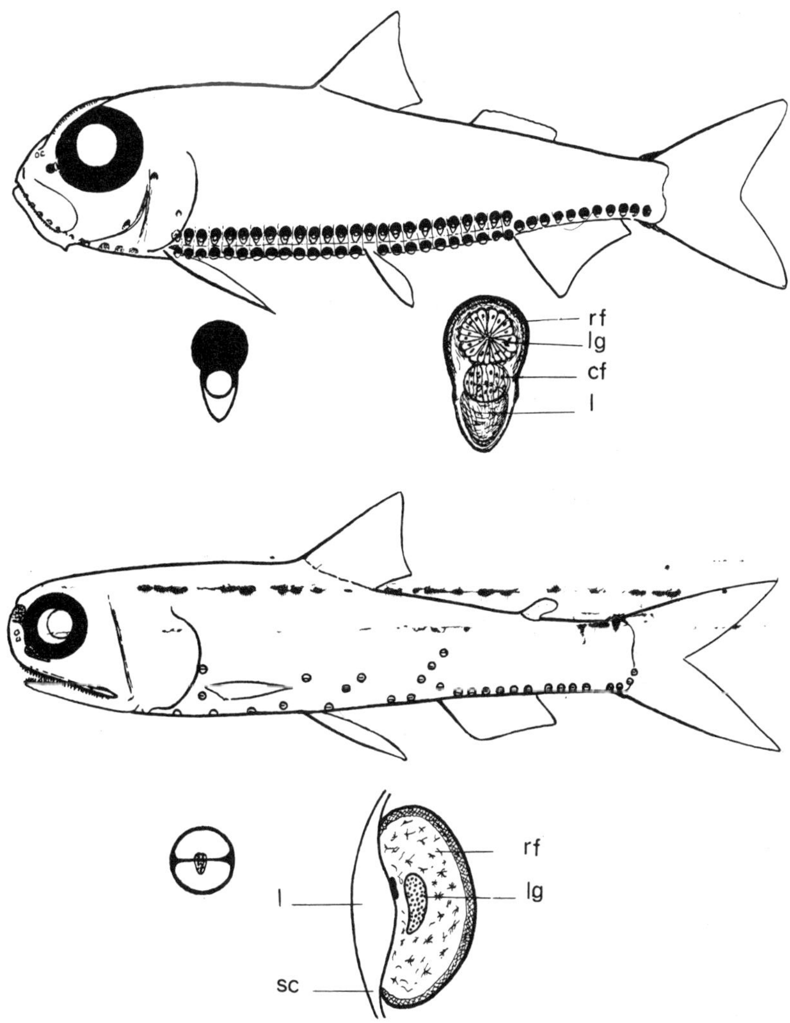

Figure 44. Light organs of (top) *Ichthyococcus ovatus*, a stomiatoid fish and *Diaphus rafinesquei*, a lantern fish. In the former note the light organ before the eye, which shines on the retina, and the double row of lights along the underparts. One of these lights is shown below the fish. Note the reflecting layer (rf), the large light gland (lg), the colour filter (cf) and the lens (l). The lantern fish has light organs (shown dotted) before and under the eyes. These lights shine forward. Note the pattern of lights over the sides of the trunk and tail, which have a much smaller light gland (lg) than do the corresponding organs of a stomiatoid fish. The lens (l) is a thickening of the scale (sc) overlying the light organ. (Drawn from specimens in the Museum of Comparative Zoology, Harvard University.)

large cheek light (figure 45) that has its own muscle, the action of which is to pull the luminous face of the light organ downward and out of sight. Fishes belonging to the first three families listed in the footnote (page 174) also bear a chin barbel, which may be fantastically elaborate in certain black dragon-fishes. At the tip or along other parts of this barbel are patches of luminescent tissue. This luminous barbel, as we shall see (p. 303), may well be used as a lure. In this respect it is significant that viper-fishes, which have no barbel, bear a

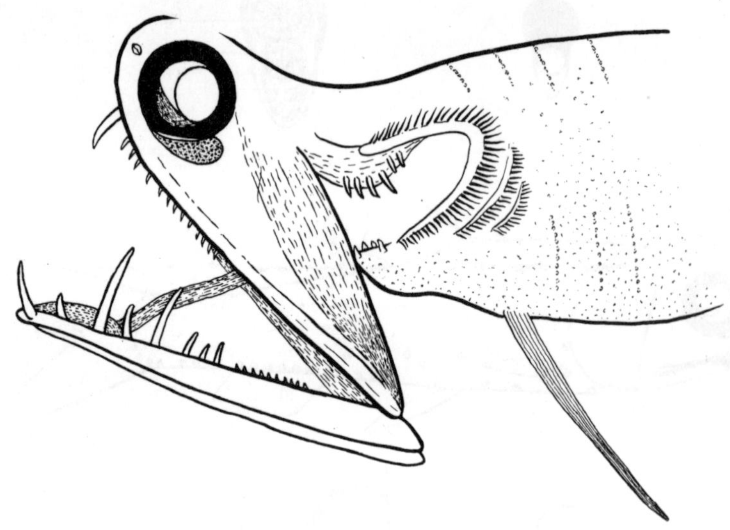

Figure 45. The large subocular, cheek light (shown dotted) of a predatory stomiatoid fish, (*Malacosteus niger*). Note also the small light organs over the body. (Drawn from a specimen in the Museum of Comparative Zoology, Harvard University.)

specially long second dorsal fin ray and this is tipped with luminous tissue. This ray can be swung forwards so that the lighted tip hangs as a lure in front of the mouth (plate 9). Lastly, but having done scant justice to the intricacies of structure and pattern of stomiatoid photophores, most of these fishes have small orbital lights, one or more to each eye. These organs are so placed that they shine directly into the eyes. Clearly they must have some special function, and one will be suggested in Chapter 17 (p. 303).

Though there are more than two hundred kinds of lantern-fishes (Myctophidae), each species bears an individual constellation of light organs over the flanks (plate 16). Among closely related species, the difference in the disposition of these lights can be subtle. Each of these photophores looks like a dwarf pearl-button, the nacreous lustre being largely due to a silvery reflecting layer at the back of the organ. The lens of each light is formed by an appropriate thickening of an overlying scale (figure 44), not by a mass of hyaline cells as in the main body lights of stomiatoid fishes. Lantern-fish lights also differ from those of stomiatoids in containing a relatively small mass of luminous cells.

Besides these pearl-button lights, some lantern-fishes, particularly those belonging to the genus *Diaphus*, have patches of luminescent tissue over the snout (plate 17). Certain kinds also bear plate-like photophores along the dorsal and ventral surfaces of the tail, which may be disposed differently in the sexes. Usually the male has these plates on the tip of the tail whereas the female carries them along the underside. How these lights may be used is discussed in Chapter 17 (p. 302).

These groups of luminous fishes have their headquarters in the twilight zone of the ocean. In the sunless depths below live about a hundred species of angler-fishes (plate 7). All but two (*Neoceratias spinifer* and *Caulophryne jordani*) bear a light organ, but this is confined to females. The light hangs at the tip of a modified dorsal fin ray, the illicium, which is hinged to a basal bone set on the front of the head. Inserted on the basal bone and connected to the illicium are two pairs of muscles. One pair lowers the illicium so that the luminous lure dangles in front of the mouth; the other pair swing the illicium backwards. In this way the bait can be fished and retracted [104].

The fine structure of the light lure has been studied in two species, *Gigantactis vanhoeffeni* and *Dolopichthys niger*. A globular light gland is backed by a stratum of spindle-shaped cells that form a reflector. Outside the reflector is a thick mantle of black pigment cells. The gland cells discharge a granular secretion into a central cavity, which opens through a canal into a small chamber at the front of the bulb. This chamber is covered by transparent skin and is open to the sea through a central pore. Just how this light organ functions is not yet clear, but one suggestion is considered on p. 305.

All but a few kinds of rat-tailed fishes swim near the deep-sea floor and many have a light organ along the mid-line of the abdominal wall. This organ has an inner cavity that opens near the anus. In all species of *Hymenocephalus* and some of *Coelorhynchus* the light organ is in the form of a long tube (figure 46), but in *Malacocephalus*, *Nezumia* and *Ventrifossa* it is more compact and pear-shaped.

The light has an inner lining of glandular cells. Within the folds of this gland are other cells that contain capsules and in these are masses of luminous bacteria which can be removed and cultured. In *Coelorlynchus*, the light of the bacteria is reflected downwards by a layer of cells along the roof of the organ and, when it is unveiled, a luminous line can be seen along the abdomen. There are two lenses at the front of the light organ of *Hymenocephalus*, an inner one and an outer one, the last being set in the skin before the pelvic fins. In the other three genera the light has one or two lenses. The outer lens – or lenses– is faced by skin containing many black pigment cells, and when these contract light is emitted, provided, of course, that the bacteria are active [105].

Elsewhere (p. 308), the suggestion is made that the light organs of rat-tailed fishes are most likely to play a part in courtship and mating. The production of light by luminous bacteria is thought to be a vestigial system of organic evolution, developed when there was no free oxygen on the earth. W. D. McElroy and H. H. Seliger write:

> The first organisms, therefore, were anaerobes. When in the course of the millenniums free oxygen slowly appeared – as a result of solar decomposition of water vapour, augmented, perhaps, by primitive photosynthesis – it would have been highly toxic to anaerobic organisms that could not quickly get rid of it. Chemically the most efficient way to remove oxygen is to reduce it to form water. In the forms of life then present, the most likely reducing agents would have been these organic compounds that were already part of the hydrogen-transport system of the primitive anaerobes. When oxygen is converted to water by such compounds, enough energy is liberated in single packets, or quanta, to excite organic molecules to emit light [106].

Luminous bacteria live near the surface of the deep sea oozes. If their light is simply a relic of the anaerobic phase of organic evolution, it has been put to good use by numerous kinds of rat-tailed fishes.

Lastly, certain kinds of deep-sea sharks are luminous. One, *Euprotomicrus bispinatus*, which grows to about a foot long, is the smallest known shark. This species, like another small luminous shark *Isistius brasiliensis*, has thousands of tiny light organs studding its underparts. Both species emit a vivid, bluish-green light. A third species, *Etmopterus spinax*, also carries many small light organs, each consisting of light-producing cells backed by black pigment cells. When the species, *E. lucifer* was stimulated, the light gradually appeared and grew brighter. Since the light organs of *Etmopterus* do not seem to be innervated, the emission of the light is thought to be controlled by the expansion and contraction of black pigment cells around each organ, the cells acting like the iris of an eye or camera. This is also true of the genus *Centroscyllium*. At all events the light organs of sharks are not so highly organized as are those of the teleosts.

Of the luminous fishes that live in coastal seas, two have been known for some time. These are *Anomalops kaptotron* and *Photoblepharon palpebratus*, which belong to a family (Anomalopidae) of berycomorph fishes (p. 373) and are found in Indonesian Seas. A third luminous fish (*Krytophanaron alfredi*) of this family occurs in the Caribbean. Under each eye is a large oval light organ containing a gland made of many parallel tubes. The tubes are well supplied with blood and contain luminous bacteria. According to E. Newton Harvey,

> As *Anomalops* swim through the water the light is turned on for about ten seconds and off for about five seconds. *Photoblepharon* shows a light for much longer time periods (thirty minutes), but as a result of excitement or of partial asphyxiation the light becomes intermittent. In *Anomalops*, by means of the hinge (*and muscle*) at the antero-dorsal edge, the light organ can be turned downward until the light surface comes in contact with black pigment tissue forming a sort of pocket. The light is thus cut off. In *Photoblepharon*, a fold of black tissue has been developed on the ventral edge of the organ socket, and can be drawn up over the light surface like an eyelid, thus extinguishing the light. Why these two fish, so similar in most respects, and especially in the general structure of the luminous organ, should have developed such totally different means of extinguishing the light is one of the mysteries of evolution [107].

In other berycomorphs called the pinecone fishes (*Monocentris*), there is a pair of light organs under the lower jaw. As in the Anomalopidae, glandular tubes within the organ hold luminous bacteria. The emission of the light, which may be seen by daylight as well as in darkness, seems to be controlled by the movements of mantles of black pigment cells.

Lastly, certain cardinal-fishes (Apogonidae) and silver-perches (Leiognathidae) are luminous. The light gland, which again holds luminous bacteria, is housed within the body and is provided with a reflector. The muscles near the light are usually translucent and act as a lens. Curtains of black pigment cells appear to regulate the brightness of the display (see figure 46).

The light organs of deep-sea and coastal teleost fishes are thus elaborate structures. Except for the rat-tails, the light glands of the former make their own light. But the coastal fishes depend on the

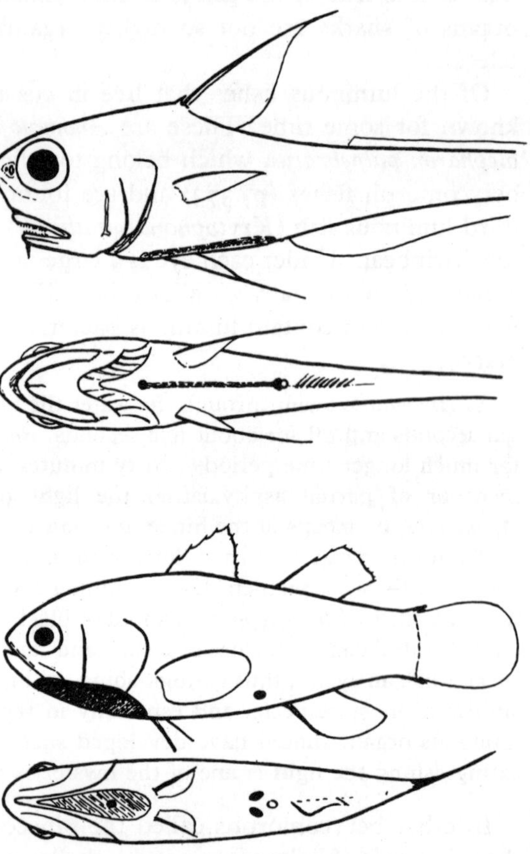

Figure 46. The tubular light organ of a rat-tail, *Hymenocephalus cavernosus* (above) and the lights (black) of a cardinal fish (*Apogon ellioti*). The forward light shines through translucent muscles (cross-hatched region).

light of luminous bacteria, which are cultured within the glands of their photophores. Why there should be this contrast is evidently another of the mysteries of evolution. The association between fish and bacteria is truly symbiotic: one has a source of light; the other receives food and oxygen. Virtually nothing is known of the biological rôles of this borrowed bacterial light, but much may be made plain during the next few years.

Chapter 11

Ways of Staying Alive

TO SURVIVE in a world of enemies, fishes, like warriors, have acquired ways of concealing their presence and protecting their lives. More often than not, their enemies are other fishes, and some are just as cunning in using cover and camouflage. Now the survival value of any such adaptation ought to be expressed in statistics, but this kind of evidence is not easily got from nature. Yet recent studies of two kinds of sticklebacks in captivity have gone far to provide a model of this more rigorous kind of approach [108]. There could be no better introduction to this chapter.

The two sticklebacks, the three-spined *Gasterosteus aculeatus* and the ten-spined *Pungitius pungitius*, live in fresh waters and inshore waters of the Northern Hemisphere. Besides otters and birds, such as herons and cormorants, two of their enemies are pike (*Esox lucius*) and perch (*Perca fluviatilis*). In the experiments referred to, these two fishes were used as predators, the pike being young individuals up to a foot in length.

When seized, a three-spined stickleback will erect its spines, which lock into position, and stay quite still. The two pelvic fins emerge at right angles to the body axis and are about one-seventh of the span of the grown fish, which is 2 to 4 inches. The two large spines on the back are a little shorter, while the two smallest spines precede the dorsal and anal fins. Ten-spined sticklebacks, which have less formidable spines, react in the same way.

As a pike snaps up a stickleback, it can hardly help being jabbed by the erected array of spines (plate 2). The young pike that were tested sometimes tried to swallow the prey, but more often than not they began to cough it up, at the same time making violent body movements. On being released, the stickleback might then make for cover and on venturing out would be caught a second time, released again, and so on. 'This might go on for an hour without the stickleback being swallowed or even killed. In most cases it was finally ejected, sometimes dead, but more often alive. In other cases it was eventually eaten.' [108]

The pike were also fed a mixture of twelve sticklebacks and twelve soft-rayed fishes (minnows, roach, rudd and Crucian carp). The unarmoured species was always the first to be eaten. It was only after all the minnows, roach and so on had disappeared that the pike would turn on the sticklebacks, but these would be rejected unless the pike was very hungry. Similar results were obtained when perch were used as predators.

Young pike and perch thus suffer discomfort on seizing and trying to swallow sticklebacks. Grown pike are probably not troubled to the same extent, but young ones, which are much more numerous than the adults, would evidently take a more severe toll of sticklebacks if it were not for their spiny armour.

The tests yielded another interesting set of statistics: the more formidably armed *Gasterosteus* is better protected than the ten-spined species. Evidently in keeping with this, *Pungitius* is the more timid and secretive fish, prone to use any means of cover. Moreover, the nuptial colours of male *Pungitius* seem less conspicuous than those of male *Gasterosteus*. Outside the breeding season these two sticklebacks live in schools and both are cryptically coloured. As in many other fishes, the back is darker than the underparts (countershading), while the overall tone of the fish tends to match that of the background. Superimposed on this, the colour pattern is broken up by a series of dark vertical bars that cross the flanks, disruptive colouration.

During the breeding season, the males assume a very different colour pattern. The back of the three-spined stickleback becomes a shining bluish-white, while the underparts turn a vivid red. With a reversed countershading and bright colours, the male is now very conspicuous, as, no doubt, he must be to attract the females (p. 224). Though the males are very lively, they are loth to leave their nesting territories and there is evidence that they are readily taken by such birds as herons and cormorants. The breeding males of *Pungitius* also have a reversed countershading – the underparts are jet black – but they are not so conspicuous as the males of their bolder relative. Moreover, the nest is built in the cover of dense vegetation, whereas the three-spined stickleback usually chooses a more open site [109].

We have refined senses of touch and pain, so it is natural for us to suppose that spines can be an excellent means of defence. The experiments confirm our supposition, but their value is in revealing just how much protection is given by spines. Even though well protected, and behaving in conformity with the degree of protection, both sticklebacks are cryptically coloured. But a good many fishes lack spines and, apart from their scales, have no specially developed forms of protection. It will thus be best to turn first to cryptic colouration or camouflage.

An illusion, according to *Webster's Dictionary*, is an 'unreal or misleading image presented to the vision; a deceptive appearance; a perception which fails to give the true characters of an object perceived.' As will soon be evident, countershading and disruptive colouration, which transform the appearance of fishes, may be regarded through our eyes as optical illusions. The form and discrete relief of fishes may also be disguised by the merging of their colours with those of their environment. Deceptive resemblances, or the similarities between animals and definite objects in their immediate surroundings, also involve a kind of visual illusion. One kind of animal may also be deceptively like another kind that enjoys some protection from its enemies; this is mimicry.

All these ways of camouflage are inevitably judged through human eyes. Are the predators of these fishes also deceived? There is evidence to suggest a positive answer. The optical properties and visual-cell mosaics of squid and teleost fish eyes are such as to indicate that an accurate image is formed over much of their visual field. In addition, teleost fishes, like birds, have colour vision; though a black and white photograph of a camouflaged animal reveals just as much concealment as a colour plate.

If fish and squid eyes form accurate images, it is reasonable to assume that the animal sees a clear picture. We have good eyesight and are deceived by camouflage, but is this true of fishes and squids? For fishes, at least, there is undoubted evidence that they are misled by just the same kind of optical illusions as are human beings.

Minnows can readily be trained to take food near either one of two adjacent figures, which can be made similar in shape but different in size. In one series of experiments, Dr Konrad Herter trained minnows to associate the taking of food with the larger of two upright rectangular blocks, which were coloured black. On replacing these figures by an optical illusion, in which one of the two upright and identical rectangles seems larger than the other, the minnows showed a decided preference to feed near the apparently larger figure (figure 47). Similar tests with other pairs of figures and corresponding optical illusions gave quite similar results [110]. But, aside from such deceptions, fishes are able to distinguish between two figures showing rather subtle differences in pattern.

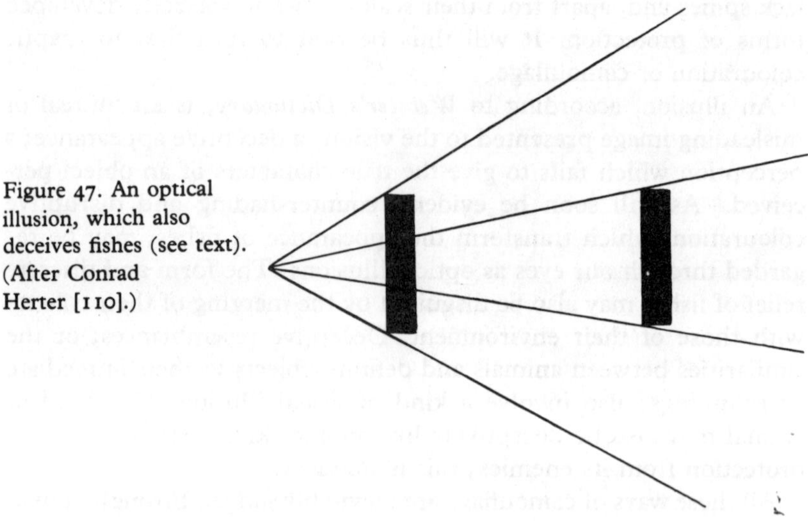

Figure 47. An optical illusion, which also deceives fishes (see text). (After Conrad Herter [110].)

Countershading

Like the two kinds of sticklebacks considered earlier, many fishes are countershaded: a dark back grades into lighter or silvery coloured underparts. By these means a fish tends to merge with the background of light in water. Looking downward, the background is dark, for sunlight is markedly absorbed and scattered as it passes through water. The dark backs of fishes thus blend with the downward and darkening view of a predator. When he is searching for fishes that are swimming above him, their silvery or light underparts mirror or match the light that streams down from the surface. Seen from the side, the light catching the back of a fish makes this part take on the same general tone as the flanks and underparts, which are more shaded from the downgoing rays of sunlight (figure 48). The fish appears flat and insubstantial. 'The diver who has seen the barracuda swim toward him like a ghost . . .', [18] has seen countershading in action.

If countershading is a protective response to the play of light both around and on the forms of fishes, we might expect it to be particularly well marked in species that live close to the surface. By and large, this is true. In the open ocean countershading is superbly displayed by mackerel sharks, blue sharks, hammerhead sharks, flying-fishes, skippers, tunnies, marlins, sword-fish and so on. Among the many coastal fishes with sharp obliterative shading are herrings, half-beaks, garfish, silversides, grey-mullet, barracuda, many sea-perches, jacks, king-fish, mackerel, various snappers and certain drum-fishes. In fresh waters, fine instances of countershading may be found in the salmon family and in many characins, cyprinids and cichlids.

Certain tropical African cat-fishes of the family Mochocidae are particularly intriguing. Three species of the genus *Synodontis* have the habit of swimming upside-down, and in these the countershading is reversed, the underparts being dark and the back pale. Yet two other upside-down species have light coloured bellies, which

Figure 48. Counter-shading in a mako-shark. Light falling on a uniformly coloured fish would throw a shadow on the underparts (top figure). A countershaded fish in all-round illumination would appear as in the middle figure. When sunlight falls on countershaded colouration, the fish becomes inconspicuous. The dark back merges with the light, but shaded, underparts. (After Ray and Ciampi [18].)

seems to spoil the story [111]. Close study of these two in their natural habitat would be especially interesting.

By no means every fish that lives in well-lit surroundings has well marked countershading. There are plenty of coral fishes* that possess gay and startling colour patterns with little or no hint of countershading. Such fishes are so conspicuous that Dr J. E. Reighard wondered whether their brilliant markings might be instances of warning colouration. Their predators might learn to avoid them because they are poisonous or dangerous to eat, qualities that are advertised in brilliant and distinctive colour patterns.

* For instance, wrasses, parrot-fishes, surgeon-fishes, butterfly-fishes, angel-fishes, trigger-fishes, etc.

In the Tortugas one outstanding predator of fishes is the gray-snapper (*Lutianus griseus*). It lives over stretches of coral sand and is well countershaded. Dr Reighard [112] found that the gray-snapper was able to discriminate between colours. In particular, he experimented with the silverside (*Atherina laticeps*) which is an important food of the snapper. The silversides were artificially coloured and injected with certain unpleasant chemicals. On feeding, for instance, red-tinted silversides treated with formalin to the snappers, the food was seized but soon rejected. In quite a short time the snappers began to avoid red and other coloured silversides that had been made unpalatable. Not only did they learn quickly, but the lesson was remembered for relatively long periods.

Since the predators can thus associate unpalatability with particular colours; living and dead coral fishes, such as butterfly-fishes, angel-fishes, damsel-fishes and wrasses, were then used as food. None of these twenty-one food species was refused by the snappers. As the latter prey on a wide variety of fishes, it is evident that many coral fishes are not displaying warning colour patterns. Indeed, the very boldness of these patterns in those that live singly or in pairs, notably butterfly-fishes, angel-fishes and certain wrasses, may well be to give other fishes a brilliant visual warning that a territory is occupied. Parts of a colour pattern may also assume a special significance in display during courtship. At all events, a great many coral fishes are well able to avoid danger by darting into holes in the reef or by slipping between the branches of coral heads. And certain brightly coloured species, such as surgeon-fishes and parrot-fishes, live in schools, which seems to be a means of ensuring greater survival (pp. 207–209).

Disruptive colouration

Numerous coral fishes and indeed many other marine and fresh-water species display something like the dazzle camouflage that has been used to make warships look less conspicuous. The natural outline of the fish seems to be broken up by bands, spots or patches of pigment that contrast strongly with the ground colour of the skin (plate 1). These disruptive markings draw the observer's gaze, diverting it from the fish as a whole. To human eyes the effect is a kind of optical illusion and it may well be that the enemies of fishes are also deceived. Tests have yet to be devised to discover how effective this camouflage may be, but the pattern usually has one significant and suggestive feature. A patch or band or colour, or even the details of a more complex pattern, extends over each eye (plate 1), masking or disrupting its conspicuous features, which are the roundness and the blackness of the pupil.

Unmasked eyes on a camouflaged body hardly seem to make sense, and here some mention of eye-spots or ocelli is relevant. Certain rays, butterfly-fishes, wrasses, damsel-fishes, cichlids, gobies, blennies and flat-fishes, to name but a few instances, have one or more eye-spots on the body or unpaired fins, each having a bull's-eye which is set off by an outer ring of contrasting colour (plate 1). At the same time, the eyes are usually masked.

When a predator attacks a fish it is often attracted to the eyes. The eye-spots are believed to divert an enemy from this vital region and there is some supporting evidence. There is an Indo-Pacific blenny (*Aspidonotus taeniatus*) that preys on other fishes, tearing off pieces of their skin. It may also make for the eyes of its victim, but when attacking certain butterfly-fishes it tends to aim for the eye-spots on the rear part of the body.

If eye-spots are marks that deceive predators, this may not be their only, or even their primary, function. In certain cichlid fishes, for instance, eye-spots are displayed when one fish is trying to intimidate a rival. Concerning disruptive colour patterns, the threat display of several cichlids involves a marked darkening of stripes that extend across the body (figure 58). During such displays it seems that fishes are making every effort to look more imposing. But it must be remembered that the intimidator is close to his rival. To be effective, disruptive patterns must be seen at a distance, and against a natural background, which may itself be broken up by the play of light and the parts of plants, corals and so on. Even so, generalizations concerning such means of camouflage must be regarded as working ideas. What is needed is close study of individual species.

Matching the background

Besides merging with the background of mid-water light through countershading, many fishes are able to match the shade of the bottom by means of sympathetic changes of colour. Before considering this aspect, something must be said concerning the colour cells or chromatophores of fishes.

The colour cells, which bear finely branching processes, are seated at various levels in the dermal layer of the skin. Some cells hold black, brown or gray melanin pigments; others are full of yellow, orange or red carotenoids. A third type of colour cell contains opaque white granules of guanin.

In these kinds of cells, colour changes are produced by an altered disposition of the pigment. Parts of a colour pattern where black pigment cells are predominant will appear darkest when the melanin granules are spread through the chromatophores and their processes. When the pigment retreats into the central body of the cells, these same parts will assume a much lighter tone (figure 49). Parallel changes occur in the colour cells holding carotenoid and white guanin pigments. But in certain cells called iridocytes a whole range of colour changes can be produced by other means. The iridocytes,

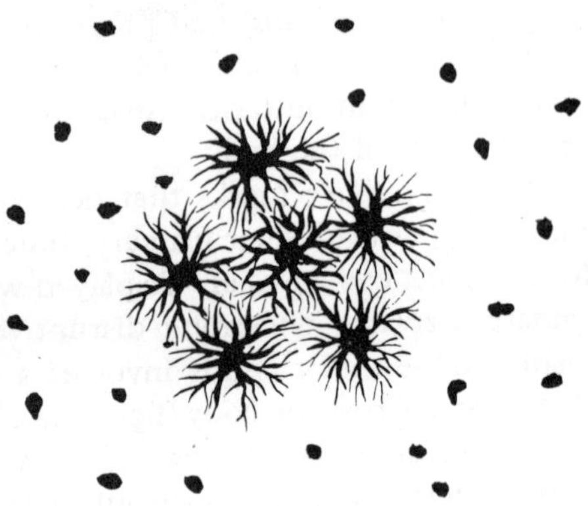

Figure 49. Black colour-cells with the pigment expanded (centre). Around this region are cells with the pigment contracted to a dot.

which are found in a variety of teleost fishes, contain elongated guanin crystals that catch the light and emit an iridescent sheen. Just how these cells change colour is not clear; for in about five seconds they can produce a spectral sequence of hues ranging from bluish-green through yellow and orange to red. The iridocytes are usually grouped with the other three kinds of pigment cells, the colours of one interacting with those of the others. Thinking now of the entire complement of colour cells, a fish may have the means not only to change its shade or colour, but also to alter its colour pattern.

Such changes may require no more than seconds or minutes, but in certain circumstances the process is continued at a much slower tempo. When a fish is exposed to a white background it may quickly become pale. If it continues to live in these surroundings it becomes even paler and this longer term change is due to an actual loss of

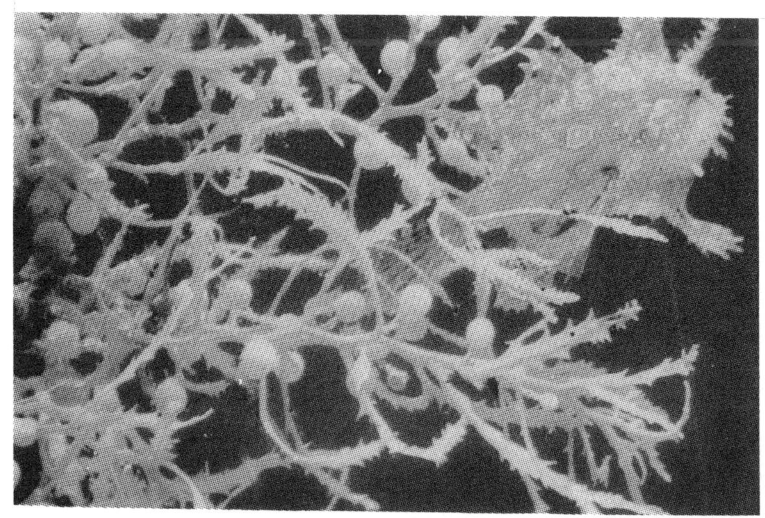

14. Sargassum-fish (*Histrio*) with Sargassum-weed.

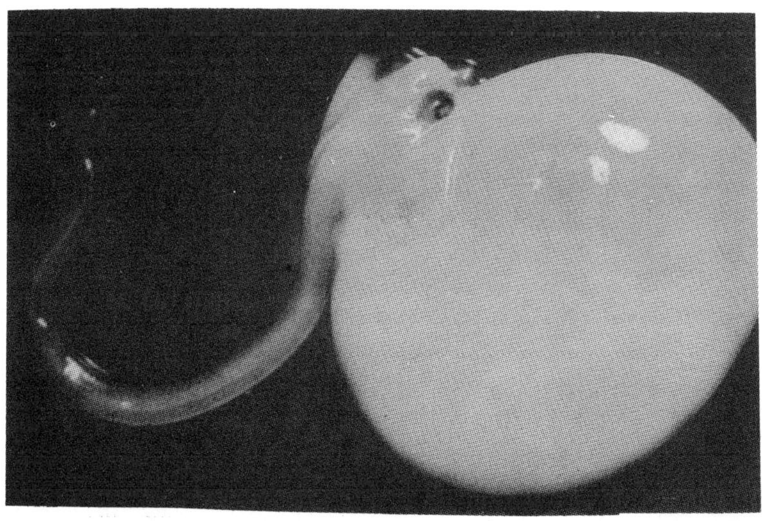

15. An embryo skate (length about 10 centimetres). Note the large yolk-sac.

16. A lantern-fish (Myctophum sp.) showing the small button-like light organs.
17. Head of a lantern-fish (*Diaphus* sp.) showing the large light organ on the snout. See the large eye with wide pupil and large lens.

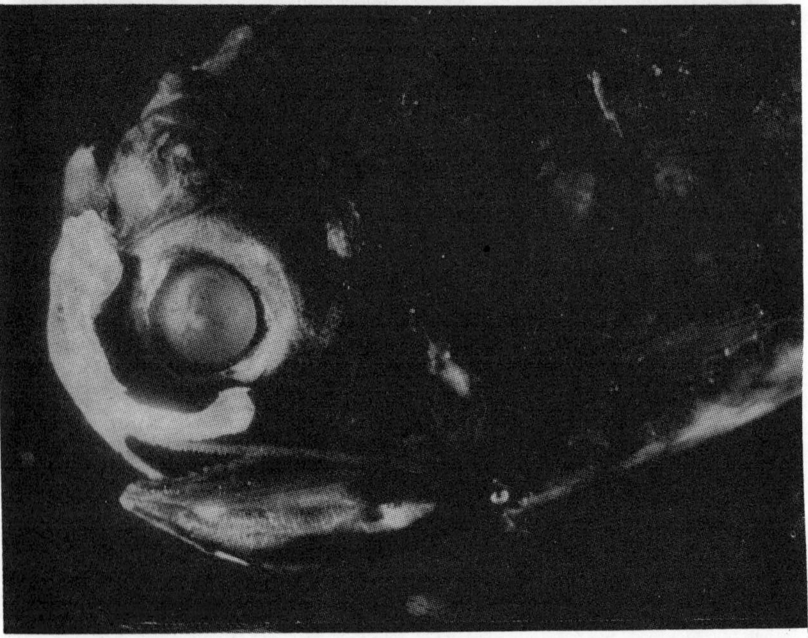

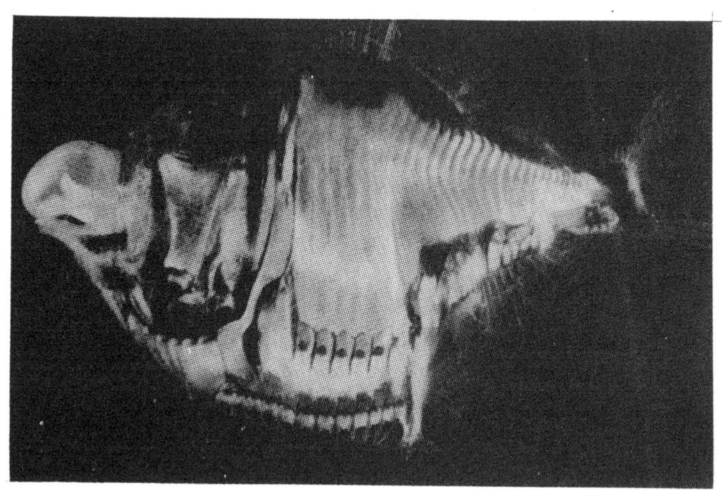

18. A hatchet-fish (*Argyropelecus* sp.). Note the tubular eye, the large light organs and the balloon of viscera from the mouth, caused by the expansion of the swimbladder as the fish was hauled to the surface.

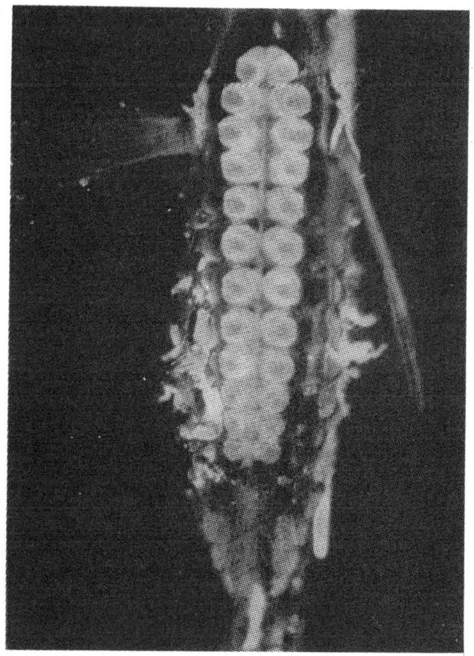

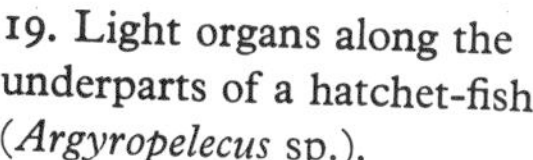

19. Light organs along the underparts of a hatchet-fish (*Argyropelecus* sp.).

20. A rainbow parrot-fish (*Scarus guacamaia*) among coral in a Bermudan reef. Note the large wing-like pectoral fins, which are used in swimming.

21. Bermudan coral-fishes. In the left foreground is an angel fish (*Holacanthus bermudensis*) and to the right a damsel-fish (*Abudefduf saxatilis*), sometimes called the sergeant-major.

some of its black pigment cells. During experiments with the killifish (*Fundulus heteroclitus*) over half the complement of black cells disappeared in sixty-nine days. There is no loss of these cells if a fish is kept on a black ground. Indeed, their number may even increase, the fish taking a very dark colour.

Marine backgrounds range from silvery coral sands to dark mud and rocks. In fresh waters there is a parallel diversity from chalk streams to sombre, muddy bottoms. The general ground colours of many fishes tend to match those of the substratum, though experiments show that the capacity to blend with a background varies in extent from one kind of fish to the next.* The time taken to change from pale to dark or vice-versa varies from ten seconds to about twenty days.

Fishes that assume the shade of their surroundings ought to enjoy some protection from their enemies. Having this in mind, Dr F. B. Sumner [113] devised a series of tests with mosquito-fish (*Gambusia*), which were kept for about two months in tanks with black and white backgrounds. After this period the dark or pale shades of these fishes were not quickly changed when they were transferred to new and different surroundings.

Equal numbers of black and white *Gambusia* were introduced into two tanks, one painted black within, the other a very pale gray. Galapagos penguins were used as predators, and on being released in the tanks they soon began to attack the fishes. At the end of several trials in the pale tank, 62 per cent of the black fish were eaten and 38 per cent of the whites. The figures for the black tank were 27 per cent of the blacks and 73 per cent of the whites. It was thus perfectly clear that fishes matching the colour of the background were not so readily seen by the penguins.

Similar tests with blue-green sunfish (*Apomotis cyanellus*) as predators yielded rather similar results. But it was evident that the mosquito-fish could often evade capture, particularly when faced with smaller sun-fishes. On being anaesthetized to a quiescent state, they were taken more readily, which is not to say that staying still is always a disadvantage in natural surroundings.

When a fish blends quickly with a background, the shade it assumes depends on the proportion of incident light reflected from

* To cite a few instances, blending responses have been found in the hag-fish (*Myxine glutinosa*), dog-fishes (*Mustelus*, *Scyliorhinus*), salmon (*Salmo salar*), minnows (*Phoxinus*), loach (*Nemachilus barbatulus*), cat-fish (*Ictalurus nebulosus*), eels (*Anguilla*), guppies (*Lebistes reticulatus*), mosquito-fish (*Gambusia*), killifish (*Fundulus*), perch (*Perca fluviatilis*), stickleback (*Gasterosteus aculeatus*) and various flatfishes.

the substratum. If it were simply reacting to the amount of illumination coming from the background, it would respond in much the same way to dimly lit, light-coloured substrata as to brightly-lit, dull ones. Experiments have shown that the fish is able to do this through its eyes. In each eye, the lower half of the retina gives a sensory measure of the incident light coming from above, while the upper half does the same for light reflected from the bottom [114].

Nerve fibres run to the colour cells and in some teleost fishes it seems that one kind of fibre controls the dispersion and another the concentration of the pigment granules. These movements of pigment are believed to be stimulated by chemicals known as neurohumors produced by nerve endings near the colour cells. In addition, hormones from the pituitary gland, particularly intermedin, can induce these changes. The nerve fibres controlling pigment concentration in the minnow were traced by Dr Karl van Frisch through the sympathetic nerve chain and spinal cord to a centre in the medulla oblongata of the brain. He was also inclined to think that another brain centre was responsible for the dispersion of pigment in the chromatophores [115].

Thus, starting with the eyes, which provide a sensory appreciation of the background, colour changes are controlled through brain centres and the pituitary. But the entire functioning system has yet to be fully appreciated. And, as we shall now see, some fishes can do more than match the shade of the substratum.

'Probably no animal, not even the far-famed chameleon, exhibits such remarkable colour changes as the flat-fish.' Some flat-fishes, as Dr G. H. Parker [116] goes on to say, are able '. . . to adapt themselves not only in colour, but also in pattern to their surroundings'. While gliding over a variegated bottom, waves of change pass over the colour patterns of the chameleon-excelling flat-fishes. Concerning colour matching alone, the flounders *Paralichthys* and *Ancylopsetta* have an extraordinary ability to blend with a whole range of colours: blue, green, yellow, orange, pink and brown. When *Paralichthys albiguttus* is placed on a series of chequered black and white backgrounds, the size of the squares ranging from large to small, the fish assumes a mottled appearance. Even though each background is equally black and white, the fish responds to the pattern; the larger the squares, the coarser the irregular patchwork of light and dark across the skin. Such powers to mimic a background are unusual in flat-fishes, but many can at least blend with the shade of the bottom. If they live on sand or mud, which many species prefer, flat-fishes have an even better means of camouflage while resting. On coming to rest they settle into the bottom, flouncing the body and fins so as they cloak themselves with a thin layer of deposit, leaving no more to view than a pair of watchful eyes (plate 27). Skates, rays and flatheads have similar habits.

Protection may not only be got through movements of pigment in pointilliste systems of colour cells, but through the actual extrusion of melanins into the sea. The ink-sacs of cephalopods such as squids and octopuses contain black, melanoid pigments, which are puffed out in a dense cloud when these animals are disturbed. R. Buchsbaum and L. J. Milne write: 'Ink may provide not only a smoke screen, but combined with colour changes enables a neat deceptive trick to be performed. Thus a small sepiolid squid is reported to

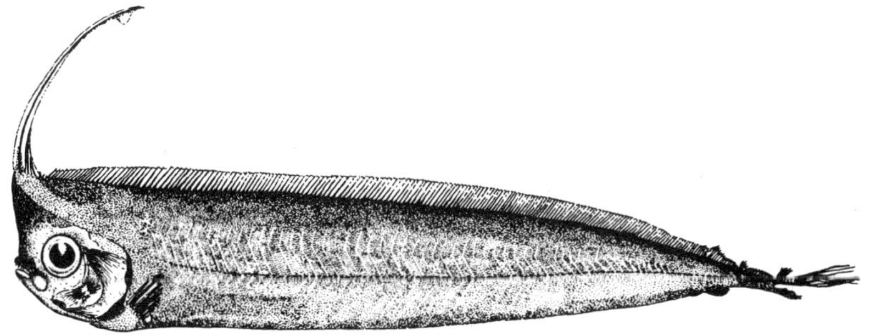

Figure 50. A unicorn-fish (*Lophotes*).

react to dangers by first quickly assuming its dark colour, next emitting a compact cloud of ink to form a roughly shaped decoy, and then rapidly blanching and darting away unnoticed [117]. Less well known is the fact that unicorn-fishes (Lophotidae) (figure 50) also have an ink-sac. This lies below and behind the swimbladder and discharges into the hind gut just before the anus.

The melanoid ink of a unicorn-fish has viscid, adhesive qualities and a musty smell. Two naturalists, one in Victoria, Australia, the other in New Zealand, have written that when *Lophotes* is captured, an inky fluid is ejected through the vent. Here, then, are fishes that respond to danger in a squid-like way, perhaps throwing out a smoke screen or even making a decoy. Here indeed is a curious instance of convergence: the development of ink-sacs in molluscan invertebrates and in one remarkable family of vertebrates. Even apart from the long and robust leading dorsal ray, which originates in front of the eyes, these vertebrates are well called unicorn-fishes.

Deceptive resemblances

A classic model of deceptive resemblance is provided by the leaf-fish (*Monocirrhus polyacanthus*), a spiny-finned nandid that lives in the Amazon and Rio Negro Basins. A grown fish is from 3 to 4 inches in length and it has a greatly compressed, leaf-shaped body, which displays a range of colours – plain or mottled brown, light grey, pale tan or black – colours to match those of dead, waterlogged leaves that drift in the water. A slender, fleshy barbel, looking like a stalk of a leaf, emerges from the chin (figure 51).

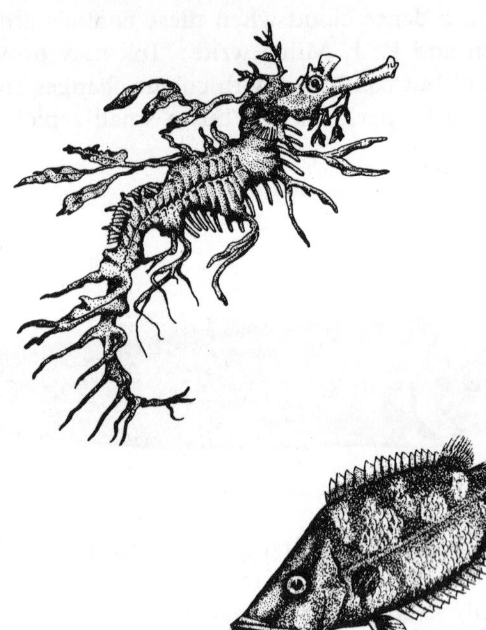

Figure 51. A leafy sea-horse (*Phyllopteryx eques*) and a leaf-fish (*Monocirrhus polyacanthus*).

The pectoral fins and upper parts of the dorsal and anal fins are quite transparent, and it is these 'invisible' members that are constantly in motion. Driven and canted by these fins, the leaf-fish hangs head down in the water, seeming to drift like a dead leaf. It will also drift up to its prey. 'Then in one motion faster than the eye can follow, the unlucky fish is virtually inhaled and disappears within the cavernous maw.' [118] If it can thus steal up to a fish, it may also deceive its enemies. And, if need be, it also has the power to move with astonishing speed, even leaping out of the water in its haste.

Leaf-fishes live near dead leaves: they both look and behave like their models. Such close and detailed resemblances can hardly be fortuitous, and most biologists see them as striking evidence for the efficacy of natural selection. Perhaps the beginnings of deceptive resemblances were due to chance genetic changes, but once these changes began to take effect, there is likely to have been a continuing selection for more cunningly adapted fishes. If this is granted, the evolution of deceptive resemblances must have been a cumulative process over long periods of time.

Using the leaf-fish as a paradigm, we may now turn more briefly to other instances of deceptive resemblance. A number of naturalists have been impressed with the likeness between certain pipe-fishes and the leaves of marine plants. Eel-grass (*Zostera marina*), which grows best in the inshore waters of north temperate regions, harbours several kinds of pipe-fishes. When poised and still among the long narrow leaves of *Zostera*, their bodies aligned in leaf-like attitudes, it is not easy to tell fish from plant (figure 52). These postures are nearly always assumed when they move, their slow and stately movements being something like the swaying of leaves. The frog-fish (*Histrio histrio*), which roosts and clambers among floating tangles of Sargassum weed, has mottled markings, appendages and tags to match the algal pattern around it (plate 14). The leafy sea-horse (*Phyllopteryx eques*) looks as much like a piece of seaweed as a fish (figure 51). Judging from a fine photograph in *Living fishes of the World*, by Earl Herald [118], another frog-fish (*Antennarius scaber*) resembles a weed-covered stone. Stone-fishes (*Synanceia*), the most poisonous of all fishes, have the inert and dead appearance of a rock.

During their early life, fishes fall prey to a great many enemies. Camouflage will not help them escape from diseases, jelly-fishes, sea-anemones and the like, but it will reduce the inroads of predators with highly developed eyes in the form of fishes, cephalopods, pistol-shrimps, mantis-shrimps, and so on. Now a good many fishes spend their early lives in inshore waters, which often contain the remains of plants. In the Tortugas area, Florida, the Virgin Islands and Hawaii, young garfishes (*Strongylura* spp.) have been found among drifting twigs and other plant debris, which they closely resemble [119]. Juvenile barracuda also look like small floating sticks. In North American fresh waters, their place is taken, as it were, by young gar-pike. Here are the observations of Dr Carl L. Hubbs:

> Soon after hatching the baby gars appear at the margins of Michigan Lakes, where they lie at or near the surface. They appear to be as stiff as the little twigs and leaflets of conifers among which they float; and their

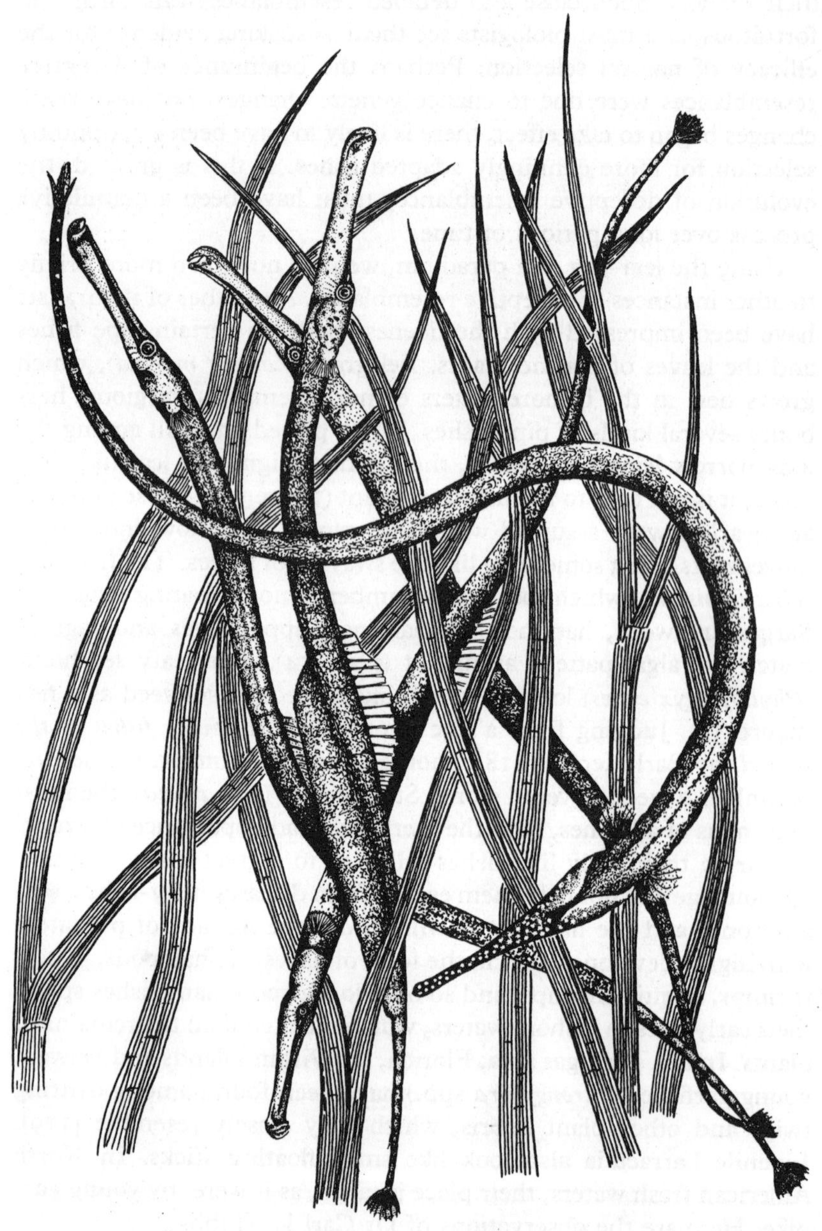

Figure 52. Pipe-fishes (*Siphostoma typhle*) among eel-grass (*Zostera*). Note the brood pouch of the most upright fish.

black sooty colour matches that of the decaying wood tissue. Until they make occasional darts the baby gars are hard to tell from the plant fragments, and even these actions simulate the movements of the twigs by wavelets [120].

Garfishes, barracuda and garpikes are very different kinds of fish, but they are basically alike in form and fin pattern. They are long-snouted, elongated fishes, with opposed dorsal and anal fins that are inserted on the downstream part of the body. To these convergences we may now add another: the stick-like appearance and behaviour of certain species when young.

Pipe-fishes are not the only 'leaf-fishes' in the sea. Young and adult bat-fishes (*Platax*) have an uncanny resemblance to the faded, yellowing leaves of red mangroves (*Rhizophora*). They adopt the attitude of drifting leaves and may even keel over and feign death if pursued. The young of the leather-jacket (*Oligoplites saurus*), a common carangid fish off the coasts of tropical America, is another aper of decaying mangrove leaves. Along the mangrove coast of Western Florida, these young fishes '. . . rest near the surface of the water, generally head down or twisted in such a manner as to somewhat resemble a half water-logged leaf and present a most non fish-like appearance' [120]. Young spade-fishes (*Chaetodipterus faber*), which grow up in the inshore waters of eastern America, mimic the black, decaying seed pods of the red mangrove. Spreading their black pelvic fins, they allow themselves to be tossed about in the waves, when they look very like their models. But there are no mangroves in North Carolina, and here the young of the spade-fish are rarely black. Are the deceptive resemblances only developed in mangrove-bearing regions?

Lastly, a number of wrasses have the appearance of marine plants. In Papetoai Bay, Moorea (Society Islands), young *Novaculichthys taeniourus*, which are translucent green with dark brown lines and white blotches, looked like floating green algae. Again, the fish were allowing themselves to be carried to and fro in the surf, their bodies limply following the movements of the sea. At the same locality, young *Hemipteronotus pentadactylus* and *H. pavo*, behaving in much the same way, closely resembled the leaves of land plants [119].

All these fishes, whether adult or young, not only look like their model, but behave in appropriate mimic ways. Perhaps it is not so surprising that there are a number of 'leaf-fishes'; for many fishes in profile have the appearance of the simpler forms of leaves. Again, the young of slim-bodied fishes already have a certain resemblance to twigs and stems of leaves. If a young fish has the not unusual habit of 'schooling' with the floating remains of plants, the way may be open to the evolution of closer, deceptive resemblances. For the most part, the steps of this process are concealed in the past, but close study of certain species, such as the spade-fish, may well reveal that deceptive resemblances, including the behavioural aspect, vary from place to place. So it need not be very long before we have a better understanding of these natural *faits accomplis*.

Mimicry

Some fishes, as we have seen, gain protection by resembling certain objects in their habitat, but these objects are of no interest to their enemies. Mimicry is a special and rarer kind of deceptive resemblance, being the protective similarity in appearance and often in behaviour of one species of animal to another. In one form of mimicry known as Batesian, one of the two animals is poisonous, distasteful, or protected in some other way from its predators. The mimic may thus gain an advantage. In the other kind of mimicry, Mullerian, both species are protected from predators, but this is not known in fishes.

A well-known instance of mimicry occurs in European seas between the common sole (*Solea solea*) and weever fishes (*Trachinus*). Young soles grow up in sandy bays, often in company with lesser weevers (*T. vipera*); the adults live in deeper waters, often together with greater weevers (*T. draco*). Weevers, which bear venomous spines on the gill covers and back, have the habit of lying half buried in the sand. On being disturbed, they raise the poisonous dorsal spines which are connected by black membranes. This is a warning to an experienced intruder. In similar circumstances the sole warns by raising its pectoral fin, which is also black. By this simple means, it is believed, common soles are sometimes protected from their enemies.

All sea-snakes are venomous and one species, *Platurus colubrinus*, appears to be mimicked by a snake-eel (*Myrichthys colubrinus*). Both have black bands round a blue body and much the same serpentine shape and way of moving. Dr T. Mortensen, who saw these two species in muddy waters off Amboina, Indonesia, was convinced that an experienced ichthyologist would not be able, except on close examination, to tell fish from snake [121].

The mimicry of an Indo-Pacific wrasse (*Labroides dimidiatus*) by a blenny (*Aspidonotus taeniatus*) is the most convincing instance of this phenomenon in fishes. Later on we shall see (p. 210) that the wrasse is visited by various kinds of fishes, who are relieved of their parasites. Through these cleaning symbioses the wrasse not only obtains food, but is tolerated by predatory fishes. The blenny gets much of its living by tearing pieces of skin from the fins of other fishes.

These two fishes, which often live in close proximity, are remarkably alike in size, form, colour pattern and mode of swimming. Both have a rather slim body and long-based dorsal and anal fins. The outstanding similarity in colour pattern, which has a blue ground, is a long black band that spans the entire body and becomes gradually broader from head to tail (figure 53). When darting to attack its victim, the blenny moves by rapid undulations of the body,

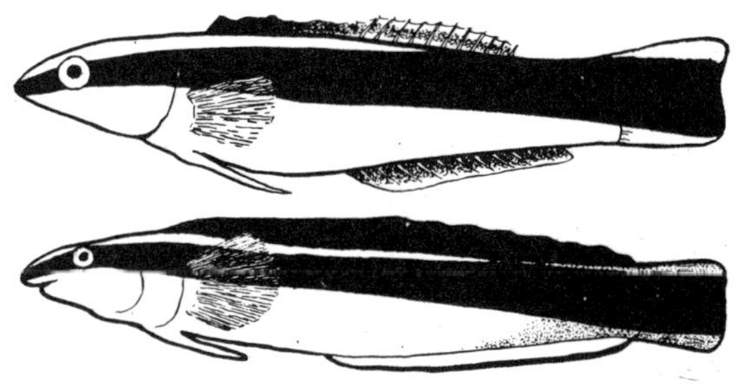

Figure 53. Mimicry: The lower fish (*Aspidonotus taeniatus*), a predatory blenny, not only has much the same form and colour pattern as the wrasse (*Labroides dimidiatus*), which cleans other fishes, but behaves in similar ways.

but the approach is made in a smooth, wrasse-like way, the fish being rowed by strokes of the pectoral fins.

By means of such mimicry – and the mimic is less common than its benign and protected model – the blenny is able to surprise its prey. Yet Dr and Mrs Randall found that: 'Adult fishes, in general, are able to distinguish the blenny from the wrasse. When *Aspidonotus taeniatus* approaches an adult fish, the latter usually swims warily away, keeping a distance greater than that which the blenny could traverse with a burst of speed.' [119] Younger and less mobile fishes, which are also likely to be less experienced, were more easily attacked. Since there are more growing fish than grown, the mimic may still make a good living.

There is a parallel instance of mimicry in the West Indies region, where lives a wrasse, *Thalassoma bifasciatum* that gets part of its food by removing the parasites of other fishes. The mimic is the blenny, *Hemiemblemaria simulus*, which resembles the wrasse in form, fin pattern, colouration and way of swimming. Again, the blenny is rarer than its model, which it has been seen to accompany, winging its way smoothly by strokes of the pectoral fins. It feeds on various crustaceans and small fishes. By mimicking a tolerated fish, it is believed to obtain some protection from predators, and may, like *Aspidonotus*, sometimes deceive its quarry [119].

There are other presumed instances of mimicry in fishes, but those just cited must suffice. Mimicry may be suspected when two kinds of fish have much the same form, fin pattern, colour pattern and behaviour. If the mimic is rarer than the model, and if the model is protected in some way from predators, suspicion may be allowed to grow. Close submarine study is still needed, which is now perfectly feasible thanks to the invention of aqualungs.

Defence and offence

If a male three-spined stickleback approaches the nest of another male, a fight is almost inevitable; 'Thrust and counterthrust follow each other so quickly that the eye of the observer can scarcely follow them.' But the two fishes rarely suffer damage. Sticklebacks have a very resilient skin, which is reinforced with keeled, bony plates along the flanks. As Konrad Lorenz goes on to say: 'Owing to the extreme toughness of the stickleback's skin, no serious wounds can be inflicted in their natural battles, which, as compared with those of the fighting fish, are absurdly harmless.' [122] Even the unprotected parts of the skin are relatively tough, which is even more true of a great many eels.

A leathery, formidably armoured skin must afford yet better protection. Among the major groups of fishes, such protection is best developed in sharks and plectognath teleosts. Concerning the second group, trigger-fishes have a leathery skin that is studded with bony scales set edge to edge. File-fishes or leather-jackets bear small, close-set scales in the skin. In the trunk-fishes, the head and trunk are invested with a strong bony box made of fused hexagonal plates. The skin of puffer-fishes and porcupine-fishes is armed with spines, which are strong and formidable in some members of the second family (plate 12).

Such heavily armoured fishes ought to enjoy some protection from the teeth of predators, provided, of course, that the attacker is not overwhelmingly large. If pursued by a human swimmer, trigger-fishes, file-fishes and trunk-fishes will seek shelter in coral or rock.

Porcupine-fishes and puffer-fishes may react to danger in other ways. When molested or threatened they rapidly inflate their bodies with water, which is gulped down and forced through the stomach into a distensible sac below. Certain species are able to swell up to three times their normal bulk, when the spines in the skin become fully erect (plate 12). More than once, Dr William Beebe and Dr J. Tee-Van saw these fishes save themselves from the jaws of larger fishes by this sea-swallowing, spine-bristling feat. But elsewhere, when writing of the porcupine-fish, *Diodon hystrix*, they remark that 'The numbers of species of fish that consume these apparently unpalatable pin cushions is quite amazing.' [123] In the Haiti area, several kinds of snappers (*Lutianus*) were among the most persistent predators of *Diodon*. Out of twenty-five schoolmasters (*Lutianus apodes*) that were opened, fifteen contained the remains of this fish, the stomach of one even holding a fully inflated individual.

Yet heavily armoured fishes are the exceptions in modern times. Like their diverse fossil relatives, the bichirs and reed-fish (*Polypteridae*) and the garpikes (*Lepisosteus*) are covered by a glassy parquetry of heavy ganoid scales. The panzer groups of teleosts include the mailed cat-fishes of tropical America (Loricariidae and Callichthyidae), which are covered with a series of bony plates. The armed bullheads (Agonidae) also have the body encased in a pattern of bony plates, while in the sea-dragons (Pegasidae) the trunk is protected by a broad, bony box, the tail by a series of bony rings. There are other instances, but the listing of these would not affect a statement made elsewhere (p. 61) that one outstanding feature of the teleosts is the lightness of their skeleton, which is yet very strong; it is a skeleton that goes with great mobility.

In more than half of the teleosts, some of the fin rays have become strong and formidable spines. In the main group (Percomorphi), these spines are found as forward rays of the dorsal and anal fins and as an outermost ray of each pelvic fin. Like the soft rays, each spiny ray is provided with special muscles (figure 10), one set of which instantly erects the spines if a fish is threatened.

A sudden bristling of the spines may often deter a predator, but not always. Pike and barracuda tend to swallow a fish head first, and during this act the spines will be pressed backwards and out of harm's way. Spines that lock into an erect position, as in sticklebacks, are found in very few fishes, well-known instances being those of the first dorsal fin of trigger-fishes and the pectoral spines of mochocid cat-fishes.

If a spine is also the injection-needle of a poison gland, we may suppose that it becomes an even greater deterrent. Most venomous fishes are sluggish, bottom-dwelling animals, the exceptions being rabbit-fishes which swim among corals, piked dogfish, eagle-rays and cow-nosed rays. Weever-fishes and star-gazers lie buried in the sand or mud, only the upper part of the head being visible; sting-rays cover themselves with sand, while stone-fishes, scorpion-fishes and toad-fishes are well camouflaged, but they may seek shelter in cavities and crevices of coral or rock. Even so, their venomous means of defence, judged by their effects on man, must be powerful deterrents to fish and other attackers. Symptoms of a severe sting from a weever are '. . . headache, fever, chills, delirium, nausea, vomiting, dizziness, cyanosis, joint aches, loss of speech, slow heart beat, palpitation, mental depression, convulsions, difficulty in breathing, and death' [124]. Yet venomous spines may not always deter. Sharks are sometimes caught with the tail-spines of sting-rays embedded in their jaws. Or do sting-rays sometimes lose their stings but not their lives? After all, a lost sting may be regenerated.

Weever-fishes, as we have seen, instantly erect their five to seven poisonous dorsal spines on being provoked. At the same time, they flare out the gill covers, each of which bears a sharp, backwardly directed, venomous spine. Each of the spines is grooved to receive the secretions of the venom glands, which are housed near the needle-sharp tip. Venomous spines are also found on stone-fishes, certain cat-fishes, scorpion-fishes, toad-fishes, rabbit-fishes and star-gazers. Like the weevers they live in coastal seas.

Stone-fishes (plate 2), which are probably the most venomous of all teleosts, live in parts of the Red Sea, Indian Ocean and Western Pacific Ocean. Their venomous equipment consists of some thirteen dorsal spines, three anal spines and two pelvic spines, all of which are stout structures with very large venom glands. The related scorpion-fishes have much the same noxious armament, but the venom glands are not usually so well developed. In toad-fishes (e.g. *Batrachoides*, *Opsanus* and *Thalassophryne*), two dorsal fin spines and two gill cover spines are provided with poison glands (plate 2). The rabbit-fishes (figure 86) bristle with venomous spines, there being thirteen along the back, two in each pelvic fin and seven in the anal fin. Each spine is deeply grooved and carries a gland.

There are also venomous sharks, rays and chimaeras. The spine leading each dorsal fin of the piked dogfish (*Squalus acanthias*) bears a glistening white poison gland that lies in a shallow groove at the back. Sting-rays, eagle-rays and cow-nosed rays (plate 3) have a serrated dagger-like spine, which is probably a modified denticle and is inserted at the base and on top of the tail. On the underside of the spine and near each edge is a dark groove containing a greyish tissue that produces most of the venom. A sting-ray will defend itself by lashing its tail, when the spine may enter the body of the attacker. The venom gland of a chimaera is found in a shallow groove at the back of the spine that heads the first dorsal fin.

Apart from venomous glands, the tissues of some fishes, notably those of certain plectognaths, are poisonous to man. Puffer-fishes (plate 12) '. . . are among the most poisonous of all marine creatures, and must be treated with respect. The liver, gonads, intestines and skin usually contain a powerful nerve poison, which may produce rapid and violent death.' [124] If such fishes are also poisonous to their aquatic predators, and if these predators learn to avoid them, their toxic qualities would be a form of protection. But we saw (p. 199) that the porcupine-fish, which is also poisonous to man, may often be eaten by barracuda.

Recalling that electric organs may also be used in defence, this chapter may be concluded. Though our appreciation is more empathetic than mathematic, it is fair to say that fishes have acquired many ways of protecting their lives. Moreover, the advantages of any life-keeping adaptation need not be so far-reaching as those given by the spines of sticklebacks. Regarding certain common misunderstandings of the theory of evolution through natural selection, Dr P. B. Medawar [125] wrote thus:

> A second misconception may be aptly called the Zenonian, because of a certain family likeness to the argument which purports to show that Achilles can never overtake the tortoise, nor an arrow reach its target. Any substantial adaptation, it is argued, can only be achieved by the adding up, over very many generations, of single all but infinitesimal adaptive changes which, being of inappreciable advantage to their owners, offer nothing for selection to get to grips upon. Luckily selection does not abide by human judgements of its efficacy; it can be shown that even so slight a selective advantage as that which allows one thousand and one of its possessors to perpetuate themselves for every thousand that lack it, must eventually prevail.

Chapter 12

Kinds of Association

THE LIFE of fishes is a striving for the means of existence and persistence. A fish not only shares its living space with members of its own species, but with all manner of organisms from bacteria and plants to seals and whales. If it manages to exist until the breeding season, its sexual products may then contribute to the future of its kind.

By living in associations, fishes may improve their chances of survival. The schooling habit, which seems to be instinctive, is found in so many species that we are bound to wonder what may be the special advantages of this group activity. During their evolutionary history, some kinds of fishes have taken to associating with other kinds, and in one order, the shark-suckers, their way of life is plainly imprinted on their design. Other fishes have discovered means of living together with various kinds of invertebrate animals. Schooling and these other forms of association, but not sexual congress, will be the subject of this chapter.

Schooling

A school is not just a group of fishes belonging to one species. One special feature is that the members of a school are of much the same size. Even so, an idling group of related, like-sized fishes might still be a gathering or aggregation and not a school. Yet there is one good but probably not infallible way of distinguishing such a group from a school, and this is by a sudden, but not violent, disturbance. When startled, a school of fishes will quickly come together and move almost as though it were a single organism. The aggregation will disperse.

Schooling fishes can swim with some precision along more or less parallel paths, and in doing so, they may be formed in close array (figure 54 and plates 42 and 43). A school, as Dr C. M. Breder says, can '. . . perform as a troupe of like-acting individuals in which independence of action is reduced to near the vanishing point' [126].

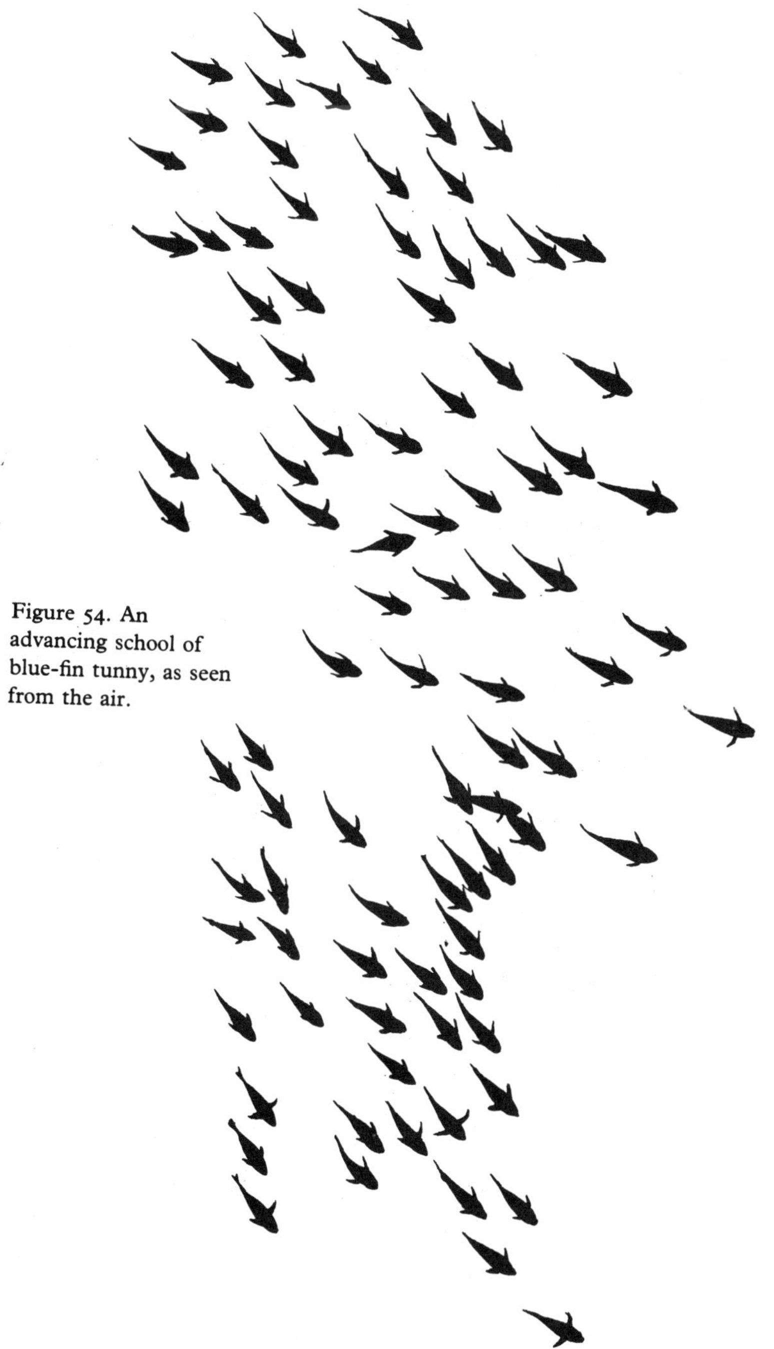

Figure 54. An advancing school of blue-fin tunny, as seen from the air.

There are, however, special circumstances when a non-schooling species might appear to be schooling.

In a strong flow, for instance, it is essential for neutrally bouyant fishes to face into the stream and swim upstream as fast as the current carries them down, if they are to hold a steady position. Holding such a position is optically mediated, and if several take an optical 'fix' on a single rock, it could easily appear that this was a social phenomenon [126].

But sooner or later, such a group of fishes would give itself away as a non-schooling kind.

There is also an intriguing photograph of rainbow trout lined up in a regular series of ranks across a stream. Again, this is not a school. This military formation is actually a reflection of the river bed, which is ridged from side to side. Each column of trout is facing into the stream and keeping station just above the hollow between two adjacent ridges. On flowing over the ridges, the stream will tend to eddy in the hollows, so forming transverse zones of relatively still water. Now salmon and trout are adept at saving energy by resting in eddies, as in those created behind any considerable obstruction to the free flow of water; hence the orderly ranks of rainbow trout.

Lastly, certain fishes, and they may be schooling kinds, are sometimes found in 'pods'. A pod is simply a closely packed school, a group in which the fishes are in physical contact as they swim. Off the Gulf coast of Florida, for instance, such groups are found in grey mullet from September to February, a period that includes the spawning season. But this would seem to be a rare form of schooling [126].

Schooling is confined, as it must be, to fishes that swim freely in aquatic space. Social groups are not common in elasmobranch fishes, although spiny dog-fish, little devil-rays (*Mobula*), eagle-rays, cow-nosed rays and even medium sized whale sharks form groups that may be schools. But these are not usually so closely knit as are the schools of so many teleosts.* Schooling is found in fresh water as well as marine fishes and in predatory as well as plankton-feeding species. The size range for schooling teleosts is from 1 inch or so to about 12 feet.

* Herring-like fishes (Clupeidae), ten-pounders (*Elops*), milk-fish (*Chanos*), salmon, trout, various cyprinids and characins, skippers, flying-fishes, cod-like fishes, silversides (Atherinidae), grey-mullet (Mugilidae), barracuda (small species), red-mullet (Mullidae), mackerel, tunnies, carangids (especially when young), blue-fish (*Pomatomus*), sea bass (*Morone*), striped bass (*Roccus*), snappers (Lutianidae), grunts (Haemulidae), some drum-fishes (Sciaenidae), sticklebacks, etc.

Teleost fishes have definite spawning grounds, which may be a nest or part of the ocean and a single female, depending on the species, will produce anything from tens to millions of eggs. In any one place there may thus be many young fishes of the same kind, which begin to move about in groups as they develop their swimming powers. It is not long before they have acquired the schooling habit, which may continue into adult life or be lost. So far as we know, the herrings, many cyprinids, mackerel and tunny fishes are life-long schoolers. Sticklebacks and certain cichlids live in schools except during the breeding season, when the male fishes take up territories for future sexual activities. Rockling and two Antarctic fishes (*Notothenia rossii* and *N. coriiceps*) have silvery, schooling young, but for part or the whole of their adult life they are dull-coloured, bottom-dwelling fishes.

When a school is feeding it may be more closely knit than at some other times. Mackerel that are taking planktonic food describe loops in the sea and as the school ascends, each fish opens its mouth and gill chambers to the fullest extent, thus allowing the largest possible volume of sea water to filter through its gill-rakers. Menhaden (figure 30), which are able to screen much finer planktonic food than mackerel, behave in much the same way. Dr William Schroeder wrote as follows:

> The fish swam swiftly in circles, like the dust driven by a whirlwind; then suddenly formed into a straight line, continually rising and falling at various depths. Each time they rose their mouths were wide open but it was not possible to see whether or not their mouths were open when they swam downward. The fish near the shore seldom broke water, but those observed in the open swam in compact schools, causing ripples at the surface; at times hundreds of them swiftly darted a few inches out of the water, causing a noise that could be heard at a distance of 300 feet. One large school was seen to divide into two parts. Some schools swam against the tide and then suddenly turned back with the tide. No general direction seemed to be maintained [127].

In shallow waters, schools of alewife (*Pomolobus pseudoharengus*) have been seen to swim along an undulating course from the sea bed to the surface. When the leading fishes broke surface, the rest followed.

Do schools of fish, like flights of birds, have leaders? In general, the answer seems to be 'no', though the leaders of goldfish schools may have distinctive colours or markings. 'Thus, in a group of plain yellow goldfish in which a few pure white individuals are included the latter may usually be found at or near the head of such an aggregation when it moves forward more as a school. This feature disappears when the group stops and fans out randomly.' [126]

Schools of blue-fin tunny and large amber-jacks may contain less than twenty-five individuals, but herring-like fishes school in thousands. All kinds of schools are alike, however, in consisting of fishes of much the same size. Seen from above some fish schools have an elliptical or drawn-out shape, but they occupy little depth of water. Measured in numbers of fishes (and because most fishes are elongated in form), schools are not much 'wider' than 'deep'. When swimming in moving water, the long axis of the school, as in herring and some cyprinids, is parallel to the direction of a current or tidal stream.

A startled school of fishes will quickly bunch together and behave in closely concerted ways. When attacked by diving loons, a school of Californian sardine (*Sardinops caeruleus*) was seen to rush into a compact ball. Spherical schools have been observed in another herring-like fish, *Jenkinsia*, and in a rock-fish, *Sebastodes paucispinis*. This way of crowding, when each fish seems to be trying to hide behind its fellows, may be rather rare. But predators are not necessarily deterred by the rushing together and aquabatics of a school. Amber-jacks, barracuda and tunny will cut swathes through schools of small fishes and they usually strike with precision. On the other hand, the dart of predators may be so sudden that a school hardly has time to be alarmed. A young pike, as I have seen, can swiftly and neatly snap up a member of a minnow school. As the pike struck, the school was no more than slightly alarmed. Garfishes, which are marine counterparts of pike, also strike with quickness and precision. On putting a young garfish (*Tylosurus acus*) into a tank containing a school of *Jenkinsia*: 'The school tended to avoid the rather quiescent hound-fish, but then they tended to avoid practically any object. When the fish struck into the school and took a member, a minor tremor ran through the group, but no great rushing about.' [128]

Barracuda, which also live and look like pike, may even appropriate small schools of fish, watching and herding their prey like a wolfish kind of sheepdog. Still more remarkable, three black skip-jack or little tunny (*Euthynnus yaito*), which is an inshore, Indo-Pacific species, were seen herding a densely packed school consisting of several hundred mackerel scad (*Decapterus*). R. W. Hiatt and V. E. Brock observed that 'The three tuna usually followed the school of scads rather closely, with one tuna at each rear flank of the school and the third lagging behind them. Now and then the scads would turn off to one side, at which time the tuna on that side would move swiftly forward and herd them back into line. It became obvious that the school of scads was prevented from leaving the area over the top of the coral head, being herded back whenever it moved over deep water. On one occasion a laggard scad was swiftly picked off by the rearmost black skip-jack; however, except for this incident the tuna made no attempt to prey on the scads during our period of observation.' [129] There could be no better instance of how predatory fishes have learned to turn the schooling reactions of their prey to good advantage.

Fishes use their eyes to school, but vision may not be the only sense involved. At night the members of a school tend to drift apart, though some kind of contact seems to be kept, and this not simply because fishes are apt to swim more slowly in the dark. Herring schools can be well organized, even on the darkest nights, and many schooling fishes are most active at twilight. But blinded rudd (*Scardinius erythrophthalmus*) fail to school, although they will turn towards other fishes that are passing close by. There is also the observation of a blinded anchovy schooling with normal individuals when they were startled. Again, a single fish will approach others of its kind that are enclosed in a glass container: at first it is strongly attracted but gradually it loses interest. The cohesion of a school would seem to involve more than the mutual, visual 'satisfaction' of its members.

During its life, a school is likely to meet other groups of its own and other kinds consisting of fishes that are different in size. How does a school maintain a certain integrity of size of its members? One suggestion is that a school of small fishes will not be able to keep pace with larger and faster moving individuals. Yet in the confined space of an aquarium, two merging schools of sea cat-fish (*Plotosus anguillaris*) will sort themselves out provided that the mean length of the individuals of one school is one and a half or more times the mean length of those in the other school. It is possible that like-sized fishes know each other because they are throwing off swimming disturbances of much the same frequency. Larger fish tend to produce lower frequencies than smaller ones. Do the fishes of a school scrutinize each other through the lateral-line sense as well as by sight? It will be interesting to devise experiments to test this conjecture.

Finally, what may be the advantages of the schooling habit? A seemingly obvious one is in the propagation of the species: even if a school never meets others of its own stage of maturity it is a potential stock of breeding fishes. There is another aspect. Quite recently, V. E. Brock and R. H. Riffenburgh [130] have suggested that schooling may help to reduce the inroads of predators. They begin by contrasting the large fish predators at the end of the food chain, which are usually non-schooling, solitary individuals, with schooling fishes, those that are prey species for at least part of their lives. There would then seem to be two opposed aspects of schooling. While scattered individuals of a prey species are more likely to meet predators than would the same number of fish if formed into a school, the group runs a greater risk of being detected by sight and other senses than a solitary individual. After treating such ideas mathematically, the authors resolve them as follows:

> The quantity of fish that a predator can consume on any single encounter with a school of prey has some average limit, and once school size exceeds this quantity, further increases in school size reduce the frequency of predator-prey encounters without necessarily changing the quantity of prey consumed on the occasion of each encounter, which in turn may reduce the rate of consumption of a prey species by a predator [130].

There is a third aspect. Groups of fishes appear to feed more readily, learn quicker and possess a longer memory than isolated individuals of the same kind. There is no way of telling the part that such factors may have played in the evolution of the schooling habit, but it is possible to regard a school as a kind of super-organism, one with many sets of sense organs, brains, nerves and so on. If food supplies are patchy, and they usually are, wide scanning of the environment by a ranging school is more likely to reveal such supplies. Once a source of food is found, the suddenly altered behaviour of the discoverers will arouse the interest of near-by individuals. Starting in this way, the discovery may run through part, or – if the field of fishes is not too extensive – the whole of a school. In any event, some individuals will share in the discovery.

The food of schooling species ranges in size from small planktonic organisms to active, nektonic animals, such as prawns, squids and fishes. The searchers may be small, such as minnows and sticklebacks; medium-sized, for instance, herrings, menhaden, pilchards, grey-mullet and mackerel; or large, as in the case of tunny-like fishes. But the salient feature is that these fishes take all or much of their food in aquatic space, where the finding of prey moving in a mobile medium is likely to involve more energy and exploration than does the detection of food organisms that live on the bottom of an ocean or river. In aquatic space, the continual scanning of many sets of sense organs can cover a large volume of water.

Symbiotic associations

Symbiosis, a living together, is a close association between dissimilar organisms to their mutual advantage. Cleaning symbiosis, which is a ritual relationship between numerous kinds of marine shrimps and fishes – the cleaners – and the fishes that visit them to be cleaned, appears to be widespread in the ocean. Certain birds have long been known to groom their hosts, for example the Egyptian plover and the Nile crocodile; tickbirds and rhinoceroses; egrets and cattle; but it is only with the spread of skin-diving that the significance of aquatic cleaning symbioses has begun to be appreciated. Some thirty species of fish and six species of shrimps are now known to act as cleaners.

The Pederson shrimp (*Periclimenes pedersoni*), which lives in association with a sea-anemone (*Bartholomea annulata*), either by clinging to the anemone or living in the same hole, cleans a number of fishes such as groupers, squirrel-fishes and moray-eels that live around the Bahamas.

> When a fish approaches, the shrimp will whip its long antennae and sway its body back and forth. If the fish is interested it will swim directly to the shrimp and stop an inch or so away. The fish usually presents its head or a gill cover for cleaning, but if it is bothered by something out of the ordinary, such as an injury near its tail, it presents itself tail first. The shrimp swims or crawls forward, climbs aboard and walks rapidly over the fish, checking irregularities, tugging at parasites with its claws and cleaning injured areas. The fish remains almost motionless during this inspection and allows the shrimp to make minor incisions in order to get at subcutaneous parasites. As the shrimp approaches the gill covers, the fish opens each one in turn and allows the shrimp to enter and forage among the gills. The shrimp is even permitted to enter and leave the fish's mouth cavity. Local fishes quickly learn the location of these shrimps. They line or crowd round for their turn and often wait to be cleaned when the shrimp has retired into the hole beside the anemone [131].

This quotation, which is from an article based on the fine work of the late Conrad Limbaugh, intimately reveals the partnership between the cleaner and its customers. The cleaner establishes itself in a definite locality and signals to its hosts that it is ready to receive them. Fishes soon discover the cleaning stations, and after visiting them are relieved of their parasites, dead tissues, food remains in the mouth and so on. The cleaners have a ready and regular source of food and, almost certainly, they are less liable to the attacks of predators.

I first saw the signalling of cleaning fishes when collecting in the Red Sea, but I was not then aware of the underlying meaning. One day I caught a small black and blue wrasse (*Labroides dimidiatus*) (figure 53), which attracted my attention because it was hovering over a coral patch and appeared to be dancing on its tail. More recently, John Randall has been able to observe these wrasses at work and even feel them plucking the hairs on his legs.

Considering *Labroides dimidiatus* alone (there are three other known species of *Labroides* and they are all cleaners), Randall found crustacean copepod and isopod parasites among the contents of the stomach. This species, which is found in the Red Sea, the Indian Ocean and the Western Pacific Ocean, cleans other wrasses, amber-jacks, groupers, parrot-fishes, surgeon-fishes, red mullets, moray-eels and damsel-fishes. When being cleaned, the hosts generally hover quietly in the water, all their fins being fully displayed. Besides being relieved of skin parasites, the mouth and gill centres may also be cleansed. Here, for instance, are Randall's observations of a 2-inch *Labroides*, which was picking over a 4-foot moray-eel. 'After about thirty seconds the wrasse devoted its attentions to the eel's head, whereupon the latter opened its mouth widely, maintaining it in this position while the labrid picked inside the upper and lower jaws and then disappeared into the pharynx. A sharp lateral jerk of the eel's head preceded the departure of the wrasse.' [132] (Plate 11)

Conrad Limbaugh [131] found that a golden-brown wrasse (*Oxyjulis californica*), called the señorita was visited by several kinds of fish hosts in Californian waters. Opal-eye (*Girella nigricans*), top-smelt (*Atherinops affinis*), and blacksmith (*Chromis punctipinnis*) would so mill round a señorita that it was often not possible to see this cleaner at work. But he discovered small crustacean parasites (copepods and isopods) in the stomach of señoritas, and also pieces of a white, fluffy growth, caused by a bacterial infection of the host fishes. Señoritas, which live in kelp beds, are also visited by black sea-bass (*Stereolepis gigas*), ocean sun-fish (*Mola mola*) and bat-rays (*Holorhinus californicus*). Other cleaning fishes include a wrasse called the Spanish hog-fish (*Bodianus rufus*); gobies (*Elecatinus*); butterfly-fishes (*Chaetodon*), and juvenile gray angel-fishes (*Pomacanthus aureus*) (plates 11 and 34).

Cleaning symbioses seem to be better developed in clear tropical seas than in more temperate waters. The tropical cleaning fishes are brightly and distinctively marked and stand out against their backgrounds, particularly when they display their readiness to receive their hosts. Most kinds are solitary or live together in pairs. The temperate species have less conspicuous colour patterns and tend to be more gregarious. There is no display before they begin cleaning a fish.

After removing all the known cleaners from two small, isolated reefs in the Bahamas, Limbaugh found after a fortnight that all but the territorial fishes had deserted the area. There is thus much substance in his view that '. . . cleaners must be regarded as key organisms in the assembly of the species that compose the populations of various marine habitats.' [131] This, indeed, is new insight to an old and complex problem – the interplay of lives in the world.

Certain kinds of sucker-fishes (*Remora remora*, *Remoropsis pallidus* and *Remoropsis brachypterus*) are known to feed on copepod parasites, which presumably come from their hosts. The other sucker-fishes might also be cleaners, for the outer row of teeth in the upper jaw form a sharp and narrow blade, which is inclined forward and seems perfectly fitted for cutting off the parasites of a host. It may also be significant that sucker-fishes, like the cleaning fishes already considered, are often found in the mouth and gill cavities of their partners. Besides making fine shelters, these regions of the host may contain many parasites.

The outstanding feature of sucker-fishes (Discocephali) is a large oval suction-disc, which is set on top of the flat head (plate 35) and is evidently a highly modified spinous dorsal fin. The main part of the disc is formed by a series of narrow, transverse plates, each being a transmuted fin-ray and moved by special muscles, the whole serving as a powerful suction disc. Using its disc, a sucker-fish will cling with great tenacity to its host or to any other suitable object such as a boat or ship.

Any sizeable shark or ray may be a host to many sucker-fishes. *Echeneis naucrates* lives with sharks, the giant devil-ray, sting-rays and an ocean sun-fish (*Masturus*). *Rhombochirus osteochir* is most frequently taken on marlins and sail-fish; *Remoropsis brachypterus* on marlin, tunny and sword-fishes and it has also been discovered in the gill cavity of an ocean sun-fish. The louse-fish, *Phtheirichthys lineatus*, has been recorded on sharks, tarpon, barracudas and spear-fishes. Lastly, but not to complete a full review, the whale-sucker (*Remiligia australis*) usually associates with various kinds of cetaceans.

Even if all kinds of sucker-fish do groom their hosts, this activity, as in other cleaning fishes, may not give them all the food they need. *Echeneis* and *Remora*, which can be taken on hook and line, feed on small fishes and crustaceans. Besides giving them shelter and protection from predators, their mobile cleaning stations will carry them to new feeding grounds. There is also the idea that sucker-fishes may sometimes be able to share in the meals of their hosts. That this can happen is shown by the following observation. 'Several times when sharks with remoras were feeding by our boat we had an opportunity to observe the manner in which remoras feed. As the shark is making a meal, the remoras release their suction on the shark and dash forward to snatch a portion of the food. When the food is finished they immediately attach themselves to the shark again.' [133] Perhaps pilot-fishes, which dog sharks and swim before the mouths of giant manta-rays, also obtain a small share of their host's food.

Damsel-fishes of the genera *Amphiprion* and *Premnas* are known as anemone-fishes. They live with certain kinds of large sea anemones that flourish in the coral reefs and atolls of the Indian and Western Pacific Oceans. A large anemone, such as *Stoichactis*, may hold four or five fishes, but more often the male and female fish will share one anemone or live apart on two adjacent hosts. Anemone-fishes are small, 1 to 3 inches long, and strikingly marked, most species bearing one or more vivid white bands across the deep and compressed body. On being approached, an anemone-fish may dart for its host, where it will nestle among the tentacles in a fussy kind of way. But any anemone-fish, like other damsel-fishes, has a decided territorial sense and it will fiercely defend its home against other members of its species or against other fishes. It will butt and bite, spread its fins, and make audible threatening sounds as it sees off an intruder [134].

If an anemone is put into an aquarium containing an *Amphiprion*, the fish will immediately thrust its head among the tentacles and rub its flanks against the column of the actinian. It will continue to fuss round the anemone until it is in good shape to be a home. Now many anemones are perfectly able to paralyse fishes and swallow them, but the immunity of anemone-fishes remained an enigma until a few years ago. It appears that a substance in the mucous covering of the skin inhibits the discharge of the batteries of stinging cells or nematocysts in the tentacles of the anemone. The fishes' immunity seems to be gradually acquired [135].

Anemone-fishes have a deadly refuge from predators, but is the partnership a true symbiosis? There is some evidence that these fishes are scavengers, eating or removing the digestive wastes extruded by the anemone. They are believed to nibble away dead tissues from the tips of the anemone's tentacles, though certain observers think that they bite off living parts of these organs. In an aquarium, one species, *Amphiprion polymnus*, has been observed to feed its anemone, bringing and placing in its tentacles prawns and pieces of fish [136]. But whether this happens in nature, and whether these fishes entice other fishes to the anemone, remains to be seen. Even so, the evidence does seem to favour a give-and-take association.

The association between the stromateid fish, *Nomeus gronovii* and the Portuguese man-of-war (*Physalia*) is more obscure. As E. W. Gudger observes, on lifting a *Physalia* from the sea

> ... a number of beautiful little fishes are likely to dart wildly about with panic-striken, erratic movements. Cross-barred with the same purplish colour that one sees on the tentacles of *Physalia*, they seem to harmonize in colour with the much larger jelly fish, but their association is much more than aesthetic. These fishes, *Nomeus gronovii* by name, live under the floating pneumatophore and swim unharmed among the deadly tentacles. They never voluntarily go any distance from their host and home. They have no other home [137].

Nomeus will feed on the zooids and tentacles of its host, which suggests that it may have the same kind of immunity as an anemone-fish. Yet Portuguese men-of-war have been found clutching partly digested *Nomeus*. Clearly, there is more to be learned and in other ways as well. *Nomeus* is said to be attacked by other fishes, which may come to an abrupt end in the long, drifting tentacles of *Physalia*; but definite evidence is lacking.

Other associations

To introduce commensalism, a living together that is to the advantage of one species without serious harm to the other, we may turn to another fish and jelly-fish association. The fry of cod, haddock, whiting, horse-mackerel, amber-jacks, yellow-tails and various stromateid fishes have the habit of living under large jelly-fish, such as *Rhizostoma* and *Cyanea*. As Dr Lionel A. Walford says:

> These little fish travel with their host in the plankton, feeding around it within a few feet, darting to safe shelter beneath its umbrella when threatened by enemies. They continue this mode of life as long as it is advantageous to them, until they are ready to be independent. This association may be an essential stage in the life cycle of some fishes; that is to say, if they fail to find a jelly-fish within a certain time, they probably perish [67].

At the same time, hordes of young fishes end their lives in the tentacles of jelly-fish. Yet there is evidence that young whiting (*Odontogadus merlangus*) have some immunity to the stinging cells of the jelly-fish *Cyanea*.

A number of fishes live among the long spines of hat-pin sea-urchins (*Diadema*, etc.), echinoderms that are quite common in tropical coral reefs. The spines of these animals are nearly always on the move, and should a shadow fall on an urchin, the tips of the spines quickly converge towards the source. On one occasion the source happened to be one of my hands and I then learned how lastingly painful can be a number of jabs from the hat-pins, which penetrate human flesh with extraordinary ease.

Certain cardinal-fishes (e.g. *Apogon endekataenia*), a shrimp-fish (*Aeoliscus strigatus*) and a cling-fish (*Diademichthys deversor*) are among the other animals, such as shrimps, that live with hat-pin sea-urchins. The shrimp-fish and the cling-fish are particularly interesting. Though they belong to two quite distantly related orders of teleosts (*Solenichthyes* and *Xenopterygii*), they are remarkably alike or convergent in a number of respects. Both hover head downwards among the spines of *Diadema*, and both are elongated fishes with a long, tubular snout. Again, both fishes have a striking black and white colour pattern. The ground colour is black in the cling-fish, while four vivid white bands run along the upper and lower surfaces and sides of the body. The shrimp-fish has a light background colour and three black bands, two along the flanks and one along the back. As may be seen in figure 55, these colour patterns blend with the black or black and white spine thickets of the urchins.

Both fishes have a fine refuge and the cling-fish gets some food as well. Its small jaws bear cusped, shearing teeth, which are used to snip off the tube feet of the urchin. But the damage done to the echinoderm may not be serious, for it has hundreds of tube feet and they are easily regenerated [138].

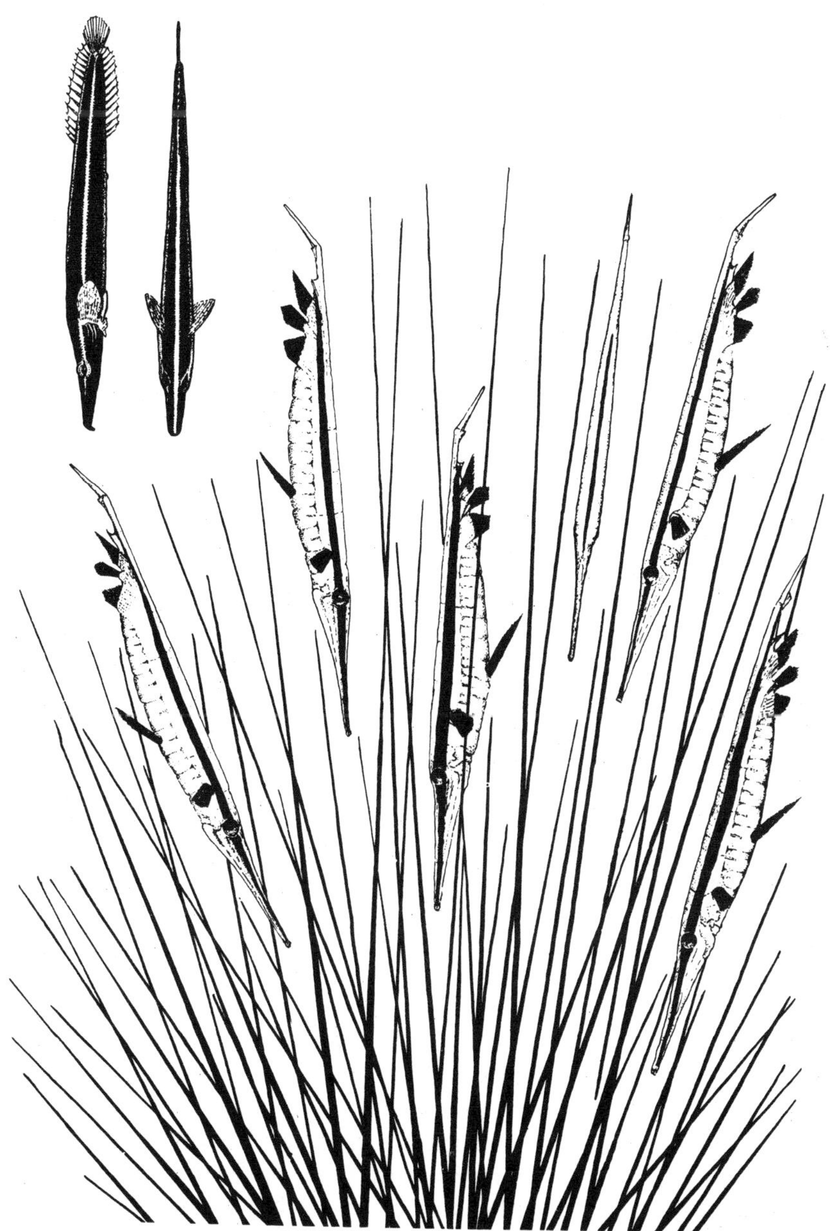

Figure 55. Two fishes that hover head-downwards among the spines of hat-pin sea-urchins (*Diadema*). The shrimp-fishes (*Aeoliscus*), which are figured among the spines, are like the cling-fish (*Diademichthys*), shown upper left, in form and basic colour pattern.

Numerous kinds of gobies are small enough to find shelter among the branches of living coral heads. Many species make their own shelters in the form of burrows, but a few have taken to living in the burrows of crustaceans. The blind Californian goby (*Typhlogobius californiensis*) lives in pairs with ghost-shrimps (*Callianassa*), feeding on bits of seaweed and other pieces of food that are washed into the burrow. The small, sand-coloured, arrow-goby (*Clevelandia ios*), which also lives along the Pacific Coast of North America, lodges in the burrows of an echiuroid worm (*Urechis*). It may have to share its refuge with a pea-crab and a polynoid worm. But it will make use of crustaceans in a remarkable way. If a *Clevelandia* finds a piece of food that is too large to be swallowed, it will take it to a shrimp or crab, and as the latter tears the food to pieces, the fish will snatch particles to eat, even trying from time to time to swallow the larger piece. When living with a pea-crab, the fish may use this crustacean as a food-shredder [133].

Lastly, Indo-Pacific gobies (*Smilogobius*) live in the burrows of snapping-shrimps.

> The shrimps maintain the burrow, constantly digging night and day to keep the hole from collapsing, and the gobies, usually a pair, stand guard, like sentinels, at the entrance. At the slightest sign of danger, the gobies dive into the hole, thereby warning the shrimps. Usually the shrimps will not emerge until the gobies are again posted at the entrance [118].

Organisms that live in some bodily space of another organism but do no harm are called inquilines or lodgers. The sponge-blenny (*Paraclinus marmoratus*), which is common along the coast of Florida, resides in the cavities of sponges, sometimes in association with toad-fishes (*Opsanus beta*), gobies (*Bathygobius soporator*) and various invertebrates. The blennies are not only lodgers but will often lay their adhesive eggs inside the sponge, where the currents created by the sponge help in aeration. However, the egg masses are always guarded and fanned by the male fish.

Certain cardinal-fishes (*Apogonichthys stellatus* and *A. puncticulatus*) are often found in the mantle cavities of giant conch-snails (*Strombus*), but an even more remarkable association is between pearl-fishes (*Carapidae*) and various invertebrates, particularly sea cucumbers. Pearl-fishes, which belong to a suborder (*Ophidoidea*) of teleost fishes, are slender, scaleless creatures with a compressed or rounded body; and some species, like certain snake-eels, have the habit of entering a refuge tail first. This shelter may be a crevice, but it may also be part of the body cavity of a sea cucumber, a clam, a star-fish, a sea-urchin or a sea-squirt.

The behaviour of the Mediterranean pearl-fish, *Carapus acus*, which lives in the body cavity of sea cucumbers (*Holothuria* spp.), depends on its age. After carefully exploring a prospective host, juvenile fishes usually enter the anus head first. Adults always enter tail first by means of a pronounced corkscrew motion of the body. Both adults and juveniles will explore a model provided it is long relative to its depth, and provided that mucous from a sea-cucumber is present in the water. Entry will not occur unless a water current is emerging from the anus of the sea cucumber [139].

The Mediterranean pearl-fish lives well within the body of the host, usually at the fore end near the branches of the gonads, which seem to be used as food. After entering the sea cucumber, the fish moves down the branchial trees and on rupturing the thin-walled duct, passes into the body cavity. Both juvenile and adult fishes can live apart from the host and small crustaceans have been found in the gut. The first, or *vexillifer* larval stage is planktonic; but the second or *tenuis* stage is entirely dependent on a host, both for food and for bringing about the metamorphosis to the juvenile condition. It is thus a parasite, but being small, does relatively little damage. Perhaps it is the older fishes that do most harm by interfering in the reproductive life of their host.

Thinking of the kinds of association that have been the concern of this chapter, there is increasing evidence from varied living spaces that there is co-operation as well as competition in the striving for life. The complex processes of natural selection are not easily discovered, but it is evident that they favour close association between species as well as individual species. Predatory activities themselves, as Anatole Rapoport [140] has written, may, in certain circumstances, be regarded as a kind of symbiosis for '. . . in the absence of predators, the prey may quickly exhaust their food supply through overpopulation and perish'. We are beginning to see that entire ecological systems of organisms have some of the qualities of a simple organism. The more elaborate these systems, the more stable they seem to be. There is a continual interplay of evolving lives, which may sometimes crystallize in successful forms of association within species and between species. Nature is much less grim than we were once inclined to believe.

Chapter 13

Aspects of Behaviour

A FISH in full motion displays, *inter alia*, the capacities of its form, fin pattern and muscles. In breathing and feeding it reveals the apt and elaborate design of its head. If a fish is unseen by a hungry enemy, it may owe its life to the concealing effect of its colour pattern. Knowledge of such aspects of fish life is far from perfect, but even if we were all-knowing we should not have discovered the entire functional expression of body, head, fins, muscles and colour pattern. These parts of a fish are used also in gestures, which are employed when they are defending their territories, threatening, fighting, submitting, courting, mating, and so forth. Just as each species of fish has its own way of moving, so does it also have an individual 'language' of gestures. These aspects of behaviour will now be considered in two distantly related fishes, cod and stickleback. Study of these two species will also serve to introduce later parts of this chapter.

Recent studies [141] of captive cod (*Gadus morhua*) have shown that its behaviour is more elaborate than one might have imagined. Ethologists, who try to make the study of animal behaviour a more exact discipline, have paid more attention to sticklebacks than to any other kind of fish. Three-spined sticklebacks (*Gasterosteus aculeatus*) are easily obtained and observed. They are also hardy and very rewarding to study.

The cod were trawled off the coast of Northumberland. For the purposes of study they were divided into three size groups; small cod (5 to 18 centimetres long), medium cod (20 to 45 centimetres) and large cod (45 to 90 centimetres). At certain times all the cod were aggressive. when one fish would move quickly and fixedly towards another. Small cod did no more than this, but the medium and large fish threatened in more complex ways. Hovering so as to display a side view of its body to a rival, the aggressor arched its back, lowered the first dorsal fin and swung forward the pelvic fins (figure 56). At the same time the head assumed a bulging, formidable aspect, which was accomplished by the flaring of gill covers and gill membranes and by lowering the floor of the mouth. There was no change in the colour pattern, but the display was often accompanied by several grunting sounds made through the swimbladder. Sometimes the aggressor thrust its snout against the head or flanks of a rival, and even made as though to bite its tail or pectoral fins.

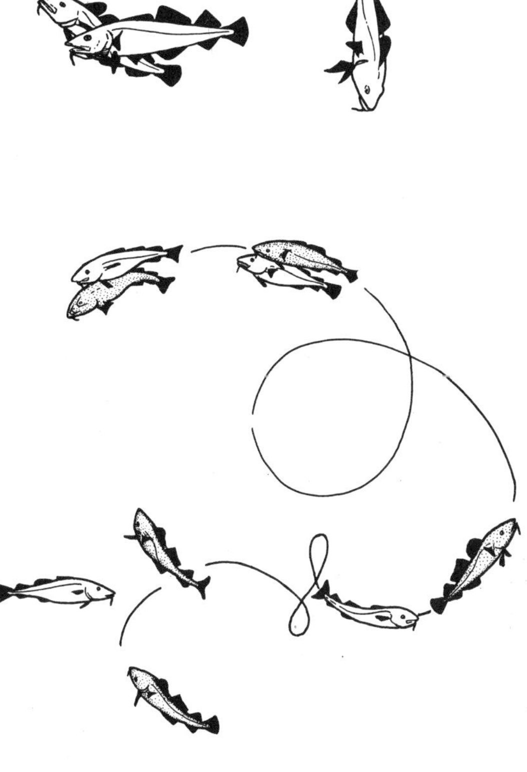

Figure 56. Behaviour of cod. Above left, a threat-display. The nearer fish, with lowered first dorsal fin and forwardly swung pelvic fins, is threatening the other fish. Above right, the attitude of a much bullied individual. Below, a courtship display. The male who is darker than the female, is enticing her to swim upwards to a mating position. (After Brawn [141].)

Threatened fish reacted in a number of ways, the rarest being indifference. Flight was the commonest response, which was usually signalled by a sudden raising of the dorsal fins. Much bullied fish sometimes assumed a head-down position close to the walls of the tank (figure 56). When threatened they did not flee, but simply spread their dorsal fins to their fullest extent. Occasionally a fish would resist a threat and keel over towards the aggressor, pressing the paired fins close against its partly exposed side. More rarely the threat was returned, when the two fish circled, pushed against each other and grunted, both in full threat display [141]. Out of these various encounters emerged a graded scale of dominance. In a group of seven large cod, for instance, three aggressive individuals established a bullying order among themselves and all three threatened the four less forceful fishes.

Such displays among large cod were confined to the autumn, when both sexes were aggressive, and a few weeks preceding spawning, which was in March. During the pre-spawning period, territories were established. Now the dominant and largest male would not allow other males and unripe females to enter his beat, but a smaller and less dominant male was able to claim a smaller holding next to him. This male threatened all but the dominant fish.

This new insight to the behaviour of cod is an outline of recent research by Miss Vivien Brawn [141]. Being well aware that cod in tanks might not behave quite like those in the sea, she has tried to relate her observations to the findings of fishery biologists. In the North Sea, maturing cod move to their spawning grounds from February to April, when they gather in rather dense aggregations. Before spawning they swim well away from the sea floor, well above fixed features which might serve to mark a territory. 'Yet many of the observed characteristics of aggressive behaviour could still occur if there was a zone around each aggressive fish within which it would not tolerate the presence of other cod. This invisible territory would move as the cod moved, changing in size according to the aggressiveness of the cod.' [141] After spawning, the spent cod make for feeding grounds, which are left in September, when the fish disperse until the return of the next spawning season. Now, captive cod, as we saw, became aggressive in autumn. If, as may well be, this happens in nature, the continuing bouts of hostility would gradually lead to the dispersal of the stock. Each vigorous fish would thus gain a larger share of the sea floor for the finding of food.

Before going on to consider courting and mating behaviour of cod, we shall first take a look at the pre-spawning activities of sticklebacks. Outside the breeding season, sticklebacks live in small schools, when they may vie for food but are not really aggressive. In spring they migrate from inshore waters or the deeper parts of rivers and streams to shallow freshwater reaches. The males now leave the schools, begin to claim territories, and acquire brilliant nuptial colours. Each resident male defends his keep and his hostility is most aroused by another male stickleback in breeding dress. Fights do occur, but more often than not the defender delivers only a threat. With his dorsal spines erect and jaws open as though ready to bite, he darts towards the trespasser. If the latter stands his ground, the threat may be changed. The defender now hovers head downwards with one or both pelvic spines erected, moving jerkily as though to bore into the sand. The new posture may lead to a fight, which most often occurs near the boundary of a territory.

When a male has won a territory, he builds a nest. This is founded in a small pit made by his removing mouthfuls of sand from the bank or bed of the river. The nest is a round affair, formed of fine pieces of grass and water weed, which are glued together by a sticky secretion produced by the kidneys. During and after the building, the male continues to defend his territory, which is usually near weeds but in an open position [109].

Returning now to the courting and mating of cod, we saw that mature males did not allow other males and unripe females to encroach on their territory. On entering male territory, a ripe female moved slowly and calmly, not quickly and warily as did male fish. Perhaps this is how a male recognizes a potential mate, though ripe females are also lighter in colour than the males.

> The male fish, on seeing the female, raised his dorsal fins momentarily, or paused, and then approached the female slowly. Positioning himself in front of the female and about a foot away, the male cod began the courtship flaunting display [figure 56]. In this display all the median fins were fully erected and the male made many exaggerated lateral bends of the body. The fins, although erected, were held laxly at their base, so that they waved from side to side at each undulation of the body. The effectiveness of the display was increased by lowering the first dorsal fin, or the second dorsal fin with the first ventral fin and then suddenly erecting them again [141].

If ready to spawn, the female followed the male, who continued his display. His aim was now to induce his partner to swim up to the top of the tank (figure 56). Displaying and grunting, even prodding his mate, the male finally achieved his end, though there was much swimming up and down before mating occurred. Once the female was moving near the surface – and unlike the male, her movements were always slow and deliberate – the male quickly mounted his partner. Then '. . . he immediately slipped down one side of the female, still with his ventral surface closely pressed against her body and clasping her with the pelvic fins. The male came to be in an

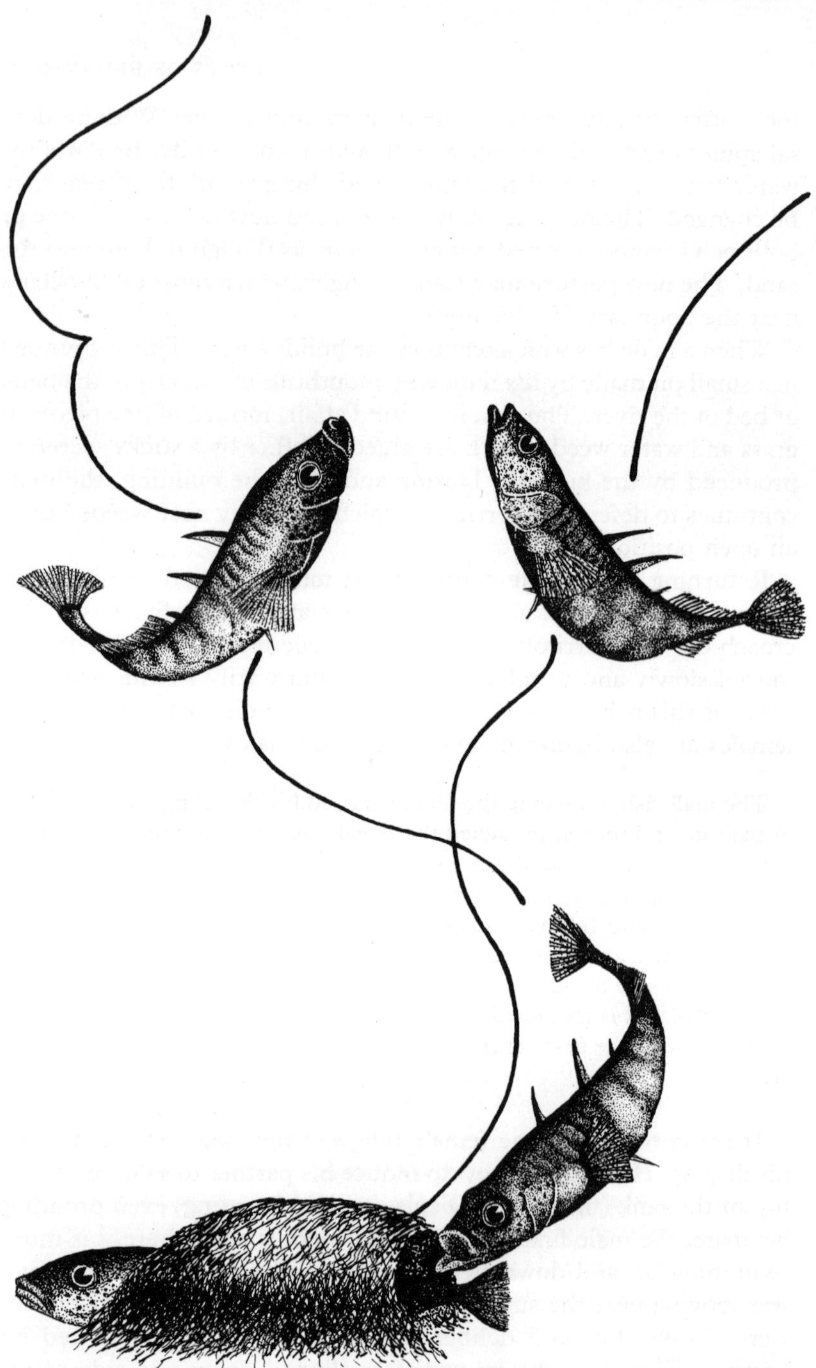

Figure 57. The zig-zag courtship dance of a male stickleback (upper left) and the entry of the female into the nest. (Based on Tinbergen [142].)

inverted position below the female and with the ventral surfaces of both fish and their genital apertures closely pressed together.' [141] The female now spawned and her eggs were inseminated by the male. After spawning the female rejoined the group of fish outside the male's territory. The male, on the other hand, sometimes became very aggressive and charged at other fish, uttering grunt after grunt.

The courtship and mating of sticklebacks is quite different. Once the nest is complete, the male assumes an even brighter breeding dress. 'The red becomes even more intense, and all the black colour cells which are found on the back contract to minute dots. Thereby the underlying glittering bluish crystals of guanin, which are situated in a deeper stratum of the skin are exposed, and the back now becomes a shining bluish white. The light back and the dark red underside, together with the brilliant eye, now make the male extremely conspicuous.' [142]

While the males are preparing, the females become silvery and swollen with eggs. If such a female is passing close by in a school, the male tries to entice her into his territory. He swims a leaping, zig-zag figure round the female (figure 57), yet all he may do is to drive the school away. But if the female is ready, she will follow him to the nest. Here he thrusts his head into the side of the nest and turns on his flank. Now the female pushes her way into the nest, so that her head and tail emerge from each end (figure 57). Urged by the male, who repeatedly butts the base of her tail with the end of his snout, she spawns and leaves the nest. The male now enters and fertilizes the eggs.

Having got two or three, or even more females to lay eggs in his nest, the male loses his sexual drive. He now guards the nest and begins to fan water over the eggs, which is done by alternate forward strokes of his pectoral fins. The eggs take seven or eight days to hatch, and during this period the male spends more and more time in fanning – ensuring, so far as he can, that the eggs are properly aerated. When the young are beginning to leave the nest, the male stops fanning and guards the brood. One of his concerns is to keep the young together: he retrieves stragglers and wanderers and spits them back into the swarm. After a fortnight or so, the young acquire schooling habits. Now the male begins to lose interest in them as well as his brilliant dress. It is not long before he leaves his brood and seeks the company of grown fish [109] [142].

Aside from an intrinsic appeal, these outlines of the behaviour of cod and three-spined stickleback are meant to reveal the dissimilar 'language' of gestures in two quite different kinds of fish. Perhaps this divergence is to be expected, though we can never be sure. Brown trout, char, and certain cichlid fishes, which are no more closely related to the cod than is the stickleback, make aggressive signals that are rather like those of the cod. Fins are raised and lowered; gill covers and gill membranes are flared; jaws are opened, and floor of the mouth is depressed. Even so, one has to remember that the number of structures that a fish can use in gesture is limited. Convergent kinds of signals thus become a probability.

More certainly, we can see that the breeding behaviour of cod and stickleback is linked to their different means of reproduction. Male and female cod have played their part when they have left many millions of eggs to drift and develop in the sea. Aggressive activities before spawning will tend to ensure that the eggs are fertilized by the most vigorous males. By displaying and urging their mates to swim higher in the sea, the eggs will be given a better upward start in life. Cod eggs are a little lighter than sea water; as they drift they will rise, which means that the larvae will emerge in the upper waters, where there is likely to be suitable planktonic food to sustain them.

It is only when the brood begins to range afield that the behaviour of sticklebacks ceases to be centred on territories and nests. As in the cod, aggressive activities before spawning will tend to select the most vigorous males and full vigour is certainly needed if a male is to perform his varied and strenuous rôle. In each nest is a limited number of relatively large eggs. They must be aerated and guarded, while the young, unlike the myriads of larval cod, are not left to time and chance. Here there is no safety in numbers. Young sticklebacks are guarded until they can school and are well able to fend for themselves.

A threat display by a cod seems a rather elaborate activity. The quick approach, arched back, lowered dorsal fin, jutting pelvic fins, bulging head, grunting, butting and feigned or actual biting – all these seem to be part of the threat language. But what is the relative significance of each part? More is known of such matters in the stickleback. Crude model fish with an eye and red undersides are more vigorously attacked than accurate models or even dead sticklebacks that are not red. The dummy will be harried more fiercely if it is also placed in a head-down position, mimicking the threat posture of a male stickleback near the boundary of his territory. Again a male will court a crude model 'female' with a swollen belly, but it responds less actively to a stickleback with an elegant figure. If the model is presented in the curved head-upward attitude of a willing female, it is courted very ardently. Similar tests have revealed that two aspects of a male are most attractive to a female: his zig-zag dance and his red colour. Moreover, there are good indications that sticklebacks have a keen discrimination of colour pattern and form, which means that these responses to certain outstanding features are not to be attributed to limited visual powers [109].

These outstanding features: gestures, sounds, or displays of colour, which have a particular power of arousing apt behaviour in animals, are known as sign stimuli or releasers. According to some ethologists, there is no evidence that the set of activities released by a sign stimulus has to be learned. When the time comes – when, for instance, a fish first reaches sexual maturity – it has the capacity to respond in the right way to rivals and mates. This capacity to acquire apt patterns of behaviour is believed to be part of an animals' inheritance. As the wards of a lock are moved by one kind of key, so each set of activities is released by relevant key sign stimuli. The 'wards' of an animal are presumably formed by certain connexions between its sense organs and nervous system and it seems that the animal has an inborn potentiality to acquire these connexions. In the terms used by ethologists, the animal develops innate releasing mechanisms (IRMs). There is also an unfolding of IRMs during the life of an animal. In male sticklebacks, for instance, the IRMs brought into play by the claiming of territory are followed by those involved in nest-building, courting, mating, fanning the eggs and parental care. Each set of activities is released by particular sign stimuli. Perhaps the waning of one set of IRMs unleashes the next, and so on. At all events, these activities require more than appropriate nexuses of sense organs and nerve cells for their release. There must be a proper balance of hormones in the blood, which is vital for successful breeding activities. To go back to the lock and key analogy, the wards will stick unless they are oiled.

More will emerge of such aspects at later points in this chapter. Meantime, we may remember that the study of behaviour in cod and sticklebacks has revealed the existence of territories that are known and defended. How well, we shall now try and discover, do fishes know their living space.

Home range, territory and homing

The home range of an animal is the region covered by its normal, day to day movements. A territory is a particular area that an animal is prepared to defend. Fishes, as we are becoming aware, may stay in the same place for a long time. Indeed, a lengthy survey of two streams in Indiana, USA, where fishes were caught and marked, revealed that four-fifths of the individuals stayed in one pool from one year to the next. Some even remained in the same pool for four consecutive years. The survey showed that fishes certainly have a home range and territory, but it also raised questions which are best given in Dr Gerking's words: 'How can we account for the fact that a sun-fish is found in the same pool for as long as four years, when during that time it has surely been chased by a variety of predators, been disturbed by changes in a fluctuating environment, and has searched for food beyond the home range?' [143] But if displaced fish have means of returning to their home range, such problems are resolved. For instance, when thirty-five long-ear sun-fish (*Lepomis megalotis*) were moved away from their pool, twenty-six found their way home.

Fishes soon come to know the fixed features of their living place. This kind of learning was subtly shown in certain tests with minnows. They were trained to feed at one side of a circular tank, where a light hung over the water. The tank was then turned through 180°, the position of the light being unaltered. Now the fish searched for food at the side opposite the light, which showed they were not associating it with food. What they were using for guides, as subsequent tests revealed, were fine solder marks and irregular oxidation patterns on the sides and bottom of the tank. They had come to associate these marks with the taking of food.

Training tests with minnows, plaice, sticklebacks, gobies, pipe-fish (*Syngnathus*) and jewel-fish (*Hemichromis*) have amply shown that fishes should be able to recognize the size and form of fixed features in their natural surroundings. Moreover, many stream fishes swim into a current and it is now well known that they keep station by maintaining in view the fixed objects in their immediate environment. It is likely, too, that fishes make subtle use of their olfactory sense to recognize their home range. Not only can they discriminate between the odours of other species of fish, but also between those associated with different kinds of water plants (p. 148).

Territorial activities lead to stability in fish populations because they separate '. . . individuals from one another in a regular and orderly fashion in addition to making the fish intimately aware of its surroundings' [143]. Schooling fishes, which move closely together, are not aggressive, whereas looser associations are liable to contain aggressive members. This implied relation between aggressive behaviour and the spacing between fishes is well shown by certain Pacific salmon. Soon after hatching, the young of pink and chum salmon leave the redds, or spawning beds, and move downstream for the sea. Coho salmon spend a year or more in the rivers before migrating to the ocean. Observations in aquaria showed that the schooling fry of pink and chum salmon are not aggressive; but coho fry nip one another and establish territories, which are marked by certain outstanding features. In nature, the antagonism must lead to a partition of the feeding grounds, which will tend to fall to the more vigorous individuals. After a year or so, lusty, well-nourished smolts will move down the rivers. Indeed, when they become smolts, coho salmon stop being aggressive and band together for their seaward migration. The fry of pink and chum salmon run the gauntlet as they migrate; but coho salmon, partly by competition among themselves, pass this critical stage in particular reaches of rivers.

There is competition among species as well as among individuals of the same species. Here, for instance, are some interesting observations, quoted by Dr Gerking [143].

> Four rainbow (*Salmo gairdneri*) and two brook trout (*Salvelinus fontinalis*) were observed by Newman from an observation tank submerged in Sagehen Creek, California. Essentially the same behaviour prevailed in the stream as in the aquaria. A complete nip-order was established in relation to the size of the fish, and territories were taken. Two large rainbows were dominant over the brook trout and small rainbows. The brook trout dominated the smaller rainbows and competed for dominance among themselves during the fifteen-day period of observation. The reward for dominance was territory selection within the home range. The larger fish occupied the deepest and darkest portion of the stream and the smaller ones were relegated to apparently less favourable territories.

Darters, which belong to the perch family (Percidae) and occupy particular home ranges in North American streams, were studied in aquaria, where they established territories. Ten species confined their attacks to members of their own kind. Seven other species, however, were hostile to any kind of intruder.

Salmon and trout return to their birthplace, which is recognized, by salmon at least, through its characteristic odours. Outside the breeding season, more limited homing powers are displayed by other freshwater fishes. Thinking first of this short-range kind of homing, over twenty species of marine and freshwater fishes have shown that they can return to a certain area after being displaced by misfortunes, by natural hazards, or by curious biologists. Coral reef fishes have long been suspected of having a home range and some are able to home. When tagged fish belonging to two species of grouper, the redhind (*Epinephelus guttatus*) and the Nassau grouper (*E. striatus*) were taken from one Bermudan reef to another several hundred feet away, they repeatedly returned to their home reef. But the green angel-fish (*Holacanthus bermudensis*) and a butterfly-fish (*Chaetodon melapterus*), which, like many of their kind live together in pairs, did not return after being transferred. They are evidently tied to a more restricted home range. Squirrel-fishes (*Holocentrus*), which hide by day and come out to hunt at night, were also unable to home [143].

Tide-pool fishes also occupy definite territories. One species called the mapo (*Bathygobius soporator*), will home to its tide-pool holding in an uncanny way. When the tide comes in, these gobies leave their pool and range afield in search of food. At low tide, when individuals may come to be in pools that are not their own, they move unerringly from pool to pool until they arrive home. Where a sand bar separates the pools, the gobies leap over the barrier, and they know which direction to take even though the next pool is out of sight. The most reasonable, but tentative, conclusion is that when the gobies range round at high tide they also learn the general features and topography of the bottom around their home pool. At low tide, this learning is put to the test [25].

Concerning homing in fresh waters, thorough tests have been made on brown trout living near Pitlochry in Scotland [144]. Here there are concerted movements of immature trout from the streams to the lochs in autumn. In spring, comparable return movements repopulate the streams. The homing tests concerned Dunalastair reservoir, which has five distinct spawning territories and tributaries. During the summer of 1952, over four hundred immature fish were caught and marked in these streams. The unerring way these trout returned to their home ground was particularly well shown in two spawning streams which join before reaching the reservoir. Over a period of five years not a single marked fish was recaptured in the wrong stream, this despite the fact that the access to one stream is much easier than that to the other, where the trout must negotiate a small waterfall.

This being so, tests were then made with ripe males and females, which were caught and marked in two streams on opposite sides of the reservoir. Twenty-two marked trout from one stream (A) were conveyed by road to the other (B), while from this stream, twenty-two other fish made the reverse journey B to A. Half the fish from the first exchange were found again in their home spawning grounds. Four fish only from the second exchange (B to A) were recaptured in their home stream, but the missing trout were not found spawning in the stream where they were carried, nor in any other spawning stream connected to the reservoir. One fish returned home in less than twenty-four hours, which is a striking indication that a trout has very sensitive means of knowing the way. Perhaps, like salmon, trout have a keen olfactory awareness of their home stream, though this has still to be proved [144].

It would be fascinating to have followed these displaced fish, which faced a situation that can rarely, if ever, be posed by nature. Did any swim upstream away from the reservoir ? After finding the reservoir, did they swim in a random way until sensing the inflow of their home stream ?

More is known of homing paths in the white bass (*Roccus chrysops*) of Lake Mendota, Wisconsin, which covers an area of 39·4 square kilometres and has a shore line of 32·4 kilometres, Here, over two stretches along the northern shore, the fish gather to spawn in late May or early June. During the spawning season, white bass were caught in fyke nets marked with numbered tags and then carried in tanks to different parts of the lake, where they were released. Most of these fish were recaptured on the spawning grounds. One group, for instance, returned to their grounds from a release point 2·4 kilometres away. By attaching a nylon thread and float to some fishes, their take-off direction could be followed. After many tests, Dr Arthur Hasler and his colleagues [145] found that

> . . . the course taken upon release is generally north, towards the spawning grounds, on sunny days. On cloudy days, however, the fish swim randomly. It would appear that this tendency to take off towards the north serves the purpose of bringing the fish promptly to shore in the general vicinity of the spawning grounds. Once there, they appear to locate their specific spawning areas by other cues.

Do white bass home by the sun ? Tests have been made in a round tank containing a central merry-go-round of sixteen boxes. When released from a central chamber, a fish will move to shelter in one of these boxes. But first, all except one box is closed and the fish is habituated to seeking cover in a particular direction. On opening all the boxes, a well trained fish will still go to the original box or to a nearby one in the same compass direction chosen for training. When the sky was completely overcast the fish were confused: they sought cover in all directions. Tests at different times of day not only showed that fish were using the sun as a guide, but that they were allowing for its changing position in the sky.

The crucial and definitive test was then conducted – namely, substitution of an artificial 'sun', indoors, for the actual sun. A sun-compass fish responded as though it was responding to a real sun out-of-doors at that time of day, choosing a hiding box at the appropriate angle to the artificial sun. Hence, the existence of an orientation rhythm which is associated with the so-called 'biological clock' has been established [145].

Lake fishes and surface-dwelling oceanic fishes, have room to navigate by the sun. In a meandering river, conditions are more complicated. If, for instance, a fish is moving sunwards in one arm of a U-bend, it will have its back to the sun in the other arm. Recent tests with long-eared sun-fish (*Lepomis megalotis*), which is able to return from foreign parts to its home range, have given some interesting results [146]. Blinded fish homed as quickly and accurately as those with normal vision. But when the sensory lining of their nasal organs was cauterized, they were not so successful. Out of one hundred and thirty-six fishes so deprived of their sense of smell but not of their vision, thirty-one were recaptured after being released. Seventeen returned to their home ranges, while fourteen remained in the area to which they were taken, suggesting that they were moving more or less at random. At all events, fishes deprived of both smell and sight were thoroughly confused, only one individual out of the forty being retaken. Displaced sun-fish thus need their olfactory sense but not their vision if they are to return home in significant numbers. On perceiving their home odours they knew of their arrival. The tests also revealed that fish homing in the upstream direction were more adept at returning than those which had to move with the stream. This seems understandable, for odours from the home range should be stronger in the downstream direction.

After spending a year or more at sea, where they feed and grow fat, salmon return to spawn in their natal stream. Once in coastal waters, they search for their river, which is probably recognized by its smell. Now they run, moving upstream from one lying place to the next, and eventually reach their spawning beds. Here they sense their home-coming by remembering the characteristic odours of their natal stretch of stream.

How do they find their way in the sea? While we know now that Atlantic salmon swim to South Greenland waters, more has been learned of the migration of Pacific species. Salmon and steelhead trout can be netted near the surface over central and easterly parts of the subarctic Pacific Ocean, where many have been tagged. Certain of these marked fish must have swum well over 1,000 or even 2,000 miles before being recaptured in coastal waters. If a salmon takes a direct route home and swims at the mean speed of 2½ miles per hour, it will take twenty-five days for a 1,500 mile journey. But marking tests indicate that salmon are not able to make this journey in such good time.

Recent indoor trials have shown that salmon have a sun-compass sense of direction rather like that of the white bass and certain observations indicate that salmon may be able to keep direction by night as well as by day. Professor Saul Saila and Dr Raymond Shappy have approached the problem thus: 'How precisely do fish need to determine their course in order to have a good chance of finding the vicinity of the natal stream?' [147] To make a mathematical model, they first assumed the following: (1) searching by salmon is a random activity (at any turning point fish may swim in any direction with equal probability); (2) the distance a fish moves without changing course is randomly spread along a range of 0 to 20 miles; (3) salmon swim at a mean speed of 2½ miles per hour in the open ocean and at 1¼ miles per hour when searching for their home river along the coast; (4) salmon will search for one hundred and seventy-five days before giving up trying to find their river. The model, in the form of an equation, was presented for solution to a digital computer.

Concerning movements at sea, the computer gave a most interesting answer if fishes were assumed to have a slight tendency to move, say, in an easterly direction. Out of a hundred hypothetical fish, rather more than a third would have reached coastal waters, which exceeds the observed percentage return (10 to 22 per cent) of mature salmon tagged at sea. A fixed sense of direction is thus unnecessary. The two authors argue that '. . . a very high degree of orientation or precise navigation is undesirable if these are to be the only operative mechanisms. This is true because, even if the salmon could maintain a precise course at sea, they would be displaced by currents to such an extent that they would be driven to shore at considerable distances from their natal streams.' [147] When more has been learned of salmon movements, the computer can be asked more definite questions. Meantime, we have been given a meaningful hint towards comprehending a natural mystery.

Leaving these new insights to the life of fishes, we must end this review of territory, home range and homing. We have been largely concerned with aspects of behaviour outside the breeding season. It is now time to turn to reproductive behaviour.

Aspects of behaviour in the breeding season

Having seen something of breeding behaviour in the cod and three-spined stickleback, a more general review is now in order. Threat display in these two fishes rarely leads to a damaging fight, which is probably true of a great many species. An exception is the savage aggression which may occur in *Betta splendens*, the Siamese fighting fish. When two breeding males meet, they fan out their gill membranes and fins, which Konrad Lorenz [122] likens to '. . . a swaggering act of self-display in which every luminous spot and every iridescent ray of the wonderful fins is brought into maximum play.' This spectral rivalry may go on for hours, but '. . . should it develop into action, it is only a matter of minutes before one of the combatants lies mortally wounded on the bottom'.

Male rivalry has been well studied in certain cichlid fishes. At the beginning of the breeding season, the males claim territories, but the way they stake their claim varies from one kind to the next. According to I. Eibl-Eibesfeldt and S. Kramer:

> In genera like *Haplochromis*, a furious and damaging fight follows a very short intimidation display; in genera such as *Hemichromis*, a very elaborate intimidation display is followed by a serious damaging fight only if the males are well balanced, so that neither is frightened off; finally, in *Herichthys* and some *Cichlasoma* species, the damaging fight has disappeared and a highly ritualized intimidation display is performed until complete exhaustion of one of the contestants decides the battle [148].

These ritualized contests begin with a broadside-on display, after which both males erect their vertical fins. Next, moving closer together, they beat their tails, so driving water against the flanks of their rival. These blows of water are most probably perceived through the lateral-line sense organs and the larger the fish the more impressively should he be able to shock his opponent. The two now move to a head-on clinch, which is followed either by a joust of mouth-pulling or mouth-pushing, depending on the species. Eventu-

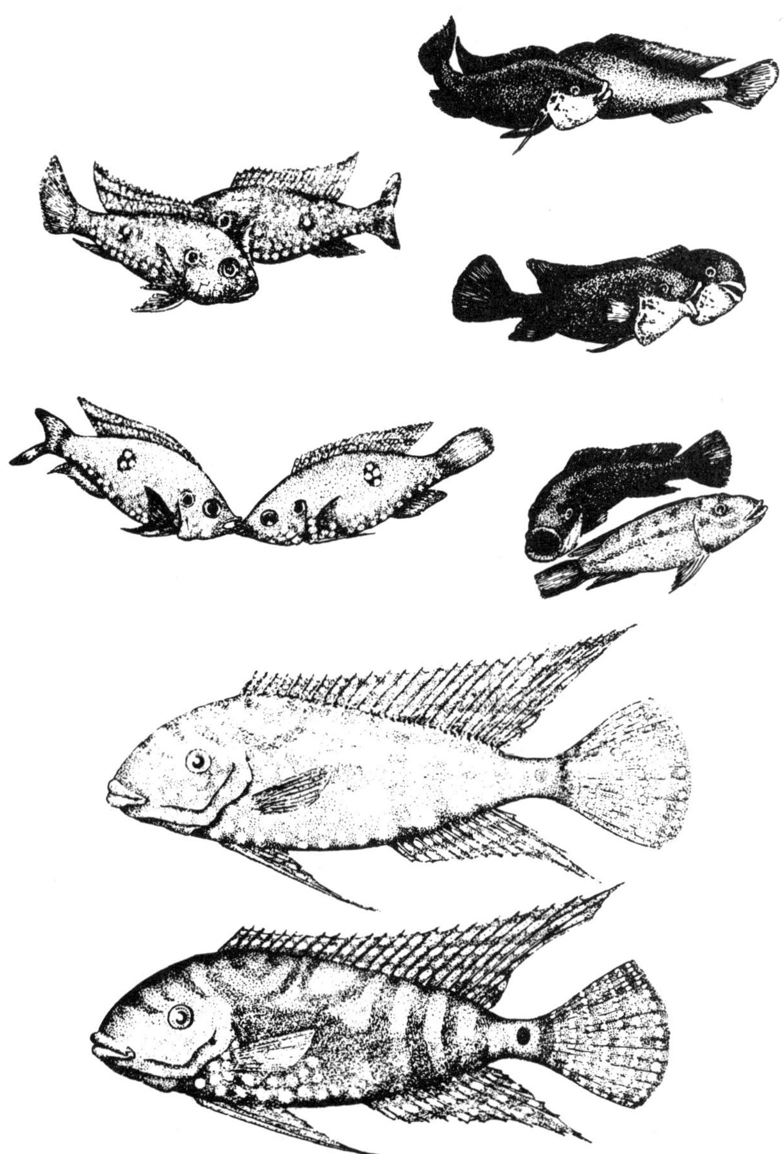

Figure 58. Behaviour in cichlid fishes. The two top-left figures represent lateral display and 'mouth-fighting' in the jewel-fish (*Hemichromis bimaculatus*). The three top-right figures show rivalry in male *Tilapia mossambica*. The topmost figure shows a lateral display, the middle figure 'butting' and the bottom figure the paling and snout-upward posture of a beaten fish.
The two lowermost figures are of *Cichlasoma biocellatum*, the upper showing the rather pale colour pattern of a 'quiescent' fish, the lower the darker and more definite colour pattern of an aggressive individual. (After Baerends and Baerends [190].)

ally, one of the fish begins to tire, grows paler, and makes his escape (see also figure 58).

Ritual fighting also occurs in anemone-fishes, such as *Amphiprion percula* and *A. akallopisus*. The territory of these fishes is the anemone, which is vigorously defended. Two rival males will meet each other head on, signalling and uttering threats. Each will jerk his head, curve the body into an S; spread his fins; ram the other and grunt [134]. Their repertoire is rather like that of certain cichlid fishes, which is another instance of convergent behaviour. Male pike-blennies (*Chaenopsis ocellata*), which live off Florida, also engage in mouth-to-mouth contests. If a male encroaches on another's territory, the defender meets him with open mouth, pouting gorge and erect dorsal fin, which is a long brilliantly-flecked sail. The two stay mouth to mouth until one bites the other, which usually signals the end of the contest.

Reverting to freshwater fishes, it will be evident that the males of many species hold and defend territories, which may contain a nest. Just as an anemone is the marine territory of an *Amphiprion*, so is a freshwater mussel the territory of a ripe male bitterling (*Rhodeus amarus*), a small cyprinid fish found in Central Europe. During the breeding season the female develops a long tube through which she lays eggs in the gills of a mussel. The male, who is close at hand, exudes his milt near the siphon bearing the ingoing respiratory stream of the mussel. Near spawning time the male acquires reddish fins, orange-yellow underparts and a darker back. Displaying this brilliant dress, he will drive another male from a chosen mussel, but the female is not aroused by this activity. 'Left alone with a ripe male her tube does not lengthen, while alone with a mussel she will spawn.' [149] Both the sight of the shell and the water currents created by the gills of the mussel are stimulating to both sexes. As well as numerous cyprinids, cichlids, sticklebacks and darters, the North American sunfishes (*Centrarchidae*), also hold territories. Male sun-fishes build the nest, defend it and guard the eggs. The female has played her part after spawning in the nest.

Territorial activities are well marked in many intertidal and coral-reef fishes. Gobies and blennies, to take but two groups, live in both habitats. More study has been devoted to those that live between the tides, particularly to the mapo, *Bathybogius soporator*. Besides having unusual homing powers these gobies hold territories. When claiming their ground, male gobies continually chase and nip each other, which leads to a bullying-order ranked according to size. The fiercest rivalry is between mature males of much the same length. As they meet, moving with lazy body waves, the two grow darker and stiffly raise their fins. Soon they lie head to tail, signalling a series of threats either in turn or together. The threats take the form of body-quivering, head-raising, throat-puffing, jaw-gaping and gill cover-flaring, which preface a bout of biting and butting. If the rivals are evenly matched the contest may last an hour or more before one gives way to the other. Such fighting also occurs among the females. Darkening, throat-puffing, gaping and fin-stiffening also occur in a European goby, *Gobius minutus*. Two European blennies (*Blennius sphynx* and *B. ocellaris*) bristle their fins and gape [150].

More might be said, but it is now time to try and see what are the biological ends of territorial behaviour. Outside the breeding season, aggressive activities space out fishes over the feeding grounds, the best areas tending to fall to the most vigorous individuals. Weaker fish accept defeat after a struggle based on conventional signs and rules of behaviour: there is little or no mortal combat. Here we should note that the signals used to intimidate are not only those made by the head, fins and body. Sound signals may be uttered by cod, toad-fishes, drum-fishes and singing midshipmen (*Porichthys*). One scientist has even likened the discharges of electrical fishes to the singing of birds. The trains of pulses are believed to be warnings to rivals that a territory is occupied. As different kinds of mormyrids and knife-fishes produce their own pattern of pulses, electrical 'singing' may well be one function of electrical activity.

During the mating season, it is generally the male who defends the spawning territory and he is usually much more strikingly marked than his mate. Again, there is little savage fighting. The contenders generally produce a conventional repertoire of threats, which may culminate in some relatively harmless kind of bout. Then the loser retires and in sticklebacks at least he may never develop his nuptial dress and the urge to mate. The females are thus likely to be won by the most lusty males, which must help to improve a local stock. Moreover, where there are fixed and defended territories, males can meet, pair and spawn without undue disturbance.

We may turn now to a brief review of some aspects of mating and breeding behaviour. In schooling species the procedure is relatively simple; there is no elaborate display such as occurs in territorial fishes. Male gold-fish, like other male cyprinid fishes, develop nodular pearl organs on the head and body. These organs may serve to buffet and bruise another male, but they are also helpful in gripping a mate. During the breeding season both sexes become a little more brilliant and among the schools one or more males will keep close track of a gravid female. When ready to mate, she signals by vibrating her body. In general, as Dr G. K. Noble [149] wrote, when '. . . cyprinids maintain their schools throughout the breeding season, no sexual differences of colour usually appear, but the retirement of breeding males to territories is almost invariably correlated with a change of colour or at least an increase of brilliancy in their nuptial dress'. The same is true of a good many characins.

Schooling marine fishes also mate in a simple way. Naturalists once believed that a mass spawning took place, but it now looks as though such activities are more discrete. Mating surgeon-fishes and parrot-fishes leave the school, swim swiftly upward, release the eggs and sperm, and then rejoin the school. Here are Dr John Randall's observations of spawning in a common surgeon-fish, *Ctenochaetus striatus* [151].

> The fish were more active than usual and milled constantly. Actual spawning took place when about three to five fish swam very rapidly upward about 4 or 5 feet above the main school and almost as rapidly down again. The fish remained remarkably close together during the ascent. There was no pause at the upper terminus of the vertical movement; the fish turned abruptly and dispersed as they returned to the school. A small cloud of white, presumably milt, could be seen at the peak of the movement.

Now during a rapid ascent the swimbladder will be subject to a certain expansion, and it is this, Dr Randall believes, that helps to release the eggs and sperm.

The blue-fin tunny (*Thunnus thynnus*) evidently spawns in a similar way. Two tunnies, as one account goes, '. . . would rise from deeper water to a depth of 4 or 5 fathoms and roll around touching their vertical surfaces together. At this moment the eggs and milt would be released. Then the fish would descend again and the whole act was seen to be repeated several times.' [152] In a school of five amber-jacks (*Caranx sexfasciatus*), two that were trying to spawn were ahead of the others. 'The whole group was moving at higher than usual speed, giving the impression that the paired fish were attempting to get away from the rest. The two fish that were spawning or attempting to spawn speeded up their swimming as they drew together in a side to side position with their vertical surfaces in contact.'

Turning from schooling to intertidal fishes, we enter a very different world. Again, the mapo (*Bathygobius soporator*) will serve as a model. The male first clears a nesting site. In breeding dress he is light coloured and has a blackened chin and throat. When meeting a gravid female he vibrates his body and tail. He also courts by emitting loud-pitched grunts, which are evidently made by quick downward jerks of the head. Eventually the female lays some fifteen to eighteen thousand eggs in the nest, and she is urged by occasional nips and butts from her mate. Her part being played, the male begins to fan the eggs.

Short jerky motions, stroking, chasing, butting and trembling are the gestures used by the brightly dressed males of a clinid blenny (*Clinus argentatus*) to court their mates. In turn the female attracts the male by quivering and supine posturing. The males of a European goby (*Gobius minutus*) court by body tremors and quick breathing movements. His colour cells brilliantly expanded, a male will move towards a mate in a series of short hops. His fins bristle, his head is raised, his throat is puffed out and his mouth is open.

In these territory-holding, intertidal fishes, the female responds to a certain conventional code of sign stimuli flashed by the male. In turn the courting male recognizes his mate by certain gestures made when she is ready to spawn. Is there any female choice? More is known of sexual selection in certain freshwater fishes. Tests made on the jewel-fish (*Hemichromis bimaculatus*) are particularly interesting [149]. Gravid females were given the choice of laying their eggs on one of three flower pots, which were placed close to one glass side of an aquarium. On the other side of each pot, in another aquarium, was a male or a female jewel-fish. Experienced females always laid their eggs opposite male fish and when one male was injected with yohimbine so as to brighten the colours of his underparts, the females nearly always chose to spawn near him. As he was not making courtship movements it was clear that the females were choosing by colour alone. When ready to breed a female is thus likely to select the brightest male at hand.

But the males of most live-bearing cyprinodont fishes (guppies, sword-tails and so on) are so persistent and active that the females have little chance to choose. The male bears a gonopodium, formed from the anal fin (figure 62). This mobile member not only serves to inseminate the female, but it endows the male with a decidedly leading rôle in mating. One set of muscles pulls the gonopodium forwards, other muscles move it sideways, while a third set pulls it back against the underparts. Mating behaviour before copulation is related to the length of the gonopodium. In species with a long gonopodium, such as *Tomeurus*, *Heterandria* and so on, there is little or no physical contact between the sexes and the male is not given to display. Males with short gonopodia, for instance *Gambusia*, *Xiphophorus* and others, display ardently and make frequent contact with their mates. Before displaying, the male moves into a position near the head of the female. His gestures may involve fin flourishing, curvature and vibrations of the body, and colour changes. Eventually, the female slows down or stops, when she is inseminated. But the males of species with long gonopodia are generally able to make copulatory contact with a freely moving mate [153].

The end of amatory display is the removal of any initial reluctance to mate. Whether a species schools, gathers in breeding grounds or is dispersed over territories, the pairs must mutually reach a certain excited state before the eggs can be fertilized. By no means every mature fish is able to cross the barrier. As Professor V. C. Wynne-Edwards contends, restrictions are often imposed on the size of the breeding stock [154]. Indeed, he has gathered impressive evidence that species are not likely to over-exploit their resources. Restriction of breeding and feeding rights, by way of conventional competition and barriers, are two of the main means to this end. So far as the local populations of species are concerned, nature is seldom red in tooth and claw, but tends to be governed by rules of competitive behaviour.

Study of the behaviour of fishes is gaining enough ground to make us realize that species do have their own 'language' of gestures. It is likely, as we can well imagine, that the individual language of a species preserves its integrity during the breeding season. Related species do sometimes hybridize, but this is rather rare in nature. One of the most careful studies has concerned the platyfish (*Xiphophorus maculatus*) and the sword-tail (*X. helleri*), two cyprinodont fishes that cross in captivity, but never, seemingly, in their natural surroundings, where they may live together. Courting, mating and copulatory activities differ in the two kinds and the divergences are evidently governed by genetic factors. The researchers concluded that the two species are sexually isolated by a variety of factors, each not effective on its own, but quite sufficient in concert under natural conditions [155].

Learning

Underwater naturalists have discovered that fishes are not only curious of man, but will soon learn to avoid him in the guise of a hunter. Concerning the European sea bass (*Morone labrax*), Pierre de Latil wrote thus: 'Aristophanes once described the bass as the most cunning of fish and all underwater hunters agree with him. The bass is also the most intelligent of fish. Since the beginning of underwater fishing, hunters are unanimously in agreement that the reaction of bass has changed in face of an unexpected visitor, man. The bass, they insist, has adapted itself to the new danger and has become definitely more circumspect. Thus it is more difficult to take a bass underwater than it was, and it is becoming steadily more and more so.' [156] Elsewhere in this book, I have written or implied that fish are well able to learn. In defending and exploring, fishes learn the topography and leading marks of their holdings. Tortugas grey snapper (*Lutianus griseus*) learned to avoid unpalatable fish that had been artificially coloured and they remembered the association of taste and colour for fairly long periods (p. 186). Salmon, we suppose, begin learning the odours of their birthplace soon after they hatch. At all events, this must be true of Pacific pink and chum salmon, which begin to make for the sea when they are small fry. Again, the learning is retained for periods up to seven years, which is surely one of the most astonishing instances of olfactory memory among the lower vertebrates. We have also seen that older fishes learn to keep a safe distance between themselves and a predatory blenny (*Aspidonotus taeniatus*), which mimicks a benign, parasite-removing wrasse (*Labroides*) (p. 197).

By exploration and experience fishes thus acquire ways of behaving that help them to survive. Now there are several modes of learning, which contrast with instinctive ways of behaviour [157]. Earlier in this chapter we considered the sign language that seems to unlock appropriate behaviour at particular times. The capacity to make these signals appears to be inborn. If, for instance, a male stickleback or a male guppy is reared in isolation from birth, each, when mature, will court a female in the ways of its kind. This kind of test seems convincing, but it is impossible to isolate a fish from every kind of experience. The fish may see its own reflection, make contact with its food in the aquarium, and sense the by-products of its metabolism. Moreover, one ethologist found that male guppies reared from birth in separate containers would attempt to mate as readily with a male as with a female. On the other hand, isolated guppies would not school with other species such as *Mollienisia sphenops*, but did so readily with their own kind [149]. It is evident that a good many more tests are needed before we can be sure that particular patterns of behaviour are instinctive. In the meantime, it might be best to suspect that some learning or conditioning may be involved until proved otherwise.

Like the grey snapper, fishes learn by trial and error. When a bass (*Morone americanus*) was first separated by a glass plate from a group of minnows, it collided with the partition on making an attack. Eventually, it ceased attacking minnows in that part of the aquarium, even after the glass plate was removed. Such tests show that fishes – and we have seen how they come to know their territories – are particularly adept at place-learning [157]. This capacity is well shown in detour experiments, when fish have to find a roundabout way to a feeding place. Here, to take a simple but revealing instance, is what Dr E. S. Russell found with sticklebacks.

> A year or two ago I spent some time training sticklebacks (*Gasterosteus aculeatus*) to take food out of a small glass jar placed at the bottom of their aquarium. This sounds an easy problem, but it proved difficult at first, for the reason that sticklebacks, being visual feeders, were strongly attracted by the sight of the food, and spent a long time during the first few trials in fruitless efforts to seize the food through the glass. This prevented them at first from making the simple detour over the rim of the jar which would lead them to the food. With one exception they all solved the problem by chance after one or two tests. Having desisted from the fruitless direct attacks, they swam about through the tank, and happening to pass over the mouth saw the food in the jar from above and darted straight down into the jar and ate it. After a very few chance successes of this kind their behaviour changed; they alternated with a direct attack through the glass, definite rises towards the rim of the jar, going in over the edge after a few of these rises; as time went on entry was effected more rapidly, until in the end the fish sometimes went directly to the mouth of the jar and straight in. They had learned to resist the direct attraction of the food and to take the roundabout way [158].

Moreover, this kind of learning was retained for about three months.

Some species are better at making detours than others, as was revealed during tests with cichlid fishes. The obstacle was a pane of glass, the upper edge being about 2 centimetres below the surface of the aquarium. The size of the fish was such that they had to swim on the side if they were to move over the pane without breaking surface.

Even after a fortnight, two *Tilapia zillii* had not managed to surmount the obstacle, whereas a male *Geophages braziliensis* immediately swam over the pane and was thereafter able to solve the problem. Females of the same species were unsuccessful, as were two species of *Cichlasoma*. Another successful species was *Acaropsis nassa* [157].

Experienced fishes may have a detailed as well as a general awareness of the fixed parts of their surroundings. A striking indication of this was considered earlier (p. 226): the association by minnows of a feeding place with fine marks and patterns on the side of the tank. This agrees well with conditioning tests on these species, which have shown that they can learn to distinguish between quite subtle differences of size and pattern. Sticklebacks can be readily trained to discriminate between a triangle and a square and they will continue to do so after the figures have been rotated through an angle of thirty degrees. At an angle of 45° their judgement is greatly reduced and at greater angles it vanishes. Evidently the fish were not simply responding to the figures, but to the combination of figures and background.

Fishes in a group learn quicker than isolated individuals, which can be seen when they have to find their way through a maze to a feeding place. Several pairs of eyes are better than one and a group is more ready to explore. Moreover, certain members of a group may be quicker to learn than others, who follow the brighter ones. At all events, the learning of mazes is quickened by adding a trained fish to a group under test. When green sun-fishes (*Lepomis cyanellus*) were solving a maze, there was some change of leader from one trial to the next, but the largest individual was often the one who took the lead.

The study of learning in fishes is not yet very advanced. To turn to brain structure and its connexion with learning is to find an even more open field. Tests have shown that visual learning is by way of the optic centres of the mid-brain, while olfactory learning is through the centres of the forebrain. But if the optic centres are removed, as was done in gold-fish (*Carassius auratus*), olfactory learning is abolished. The mid-brain then plays an important intermediary part. After removal of the forebrain, balance, vision, and movements are unimpaired, yet the fishes lack initiative, the capacity to respond by active body movement other than by reflex ones to changes in their surroundings. Feeding was not affected by the operation, for it is largely a reflex activity, but, unlike intact fishes, those lacking a forebrain were much less wary of new and unusual conditions. Both normal and operated individuals learned to associate a colour with feeding, though the deprived ones took a little longer to train and were less certain in making their choice. Some kind of pattern is thus beginning to emerge regarding the divisions of the brain and their rôle in learning [115]. Perhaps the findings concerning wariness and initiative will prove to be the most interesting ones.

To end these selected aspects of behaviour, it seems fair to say that we are beginning to gain certain new insights to the world of fishes. Yet in some ways we are well endowed to study animal behaviour, so long as our interest is not overwhelmed by solemn learning and nomenclature. The use of gesture by animals, whether inborn or acquired, so as to release apt ways of acting, is regarded by Professor Otto Koehler as a kind of 'thinking without words' [159]. We do this too. 'A man can think in images alone, without words, in diagrams and models, in gestures, as in using the deaf and dumb alphabet; and in muscular movements.' [160] Indeed, our repertoire of gestures is much enhanced by our faces. 'The human face', as Joseph Campbell wrote, '. . . has become an organ of exquisite mobility, capable of a range and refinement of social signalling infinitely more versatile than the social releasers (the bird cries, flourished antlers and tail flashes) of the animal kingdom.' [161] We are thus naturally endowed, provided we do not pursue the mirage of complete objectivity, to study sympathetically the ways of fishes.

Elsewhere in this book I have said that the senses abstract and the animal may act. The senses are the first filter between an animal and its changing surroundings. The second filter is passed after the animal reacts in ways that help the survival of its kind. We use words, and 'Words are the third filter between nature and mind. Words can never exhaust a matter entirely. The best always remains unsaid. Words are like a veil before the wordless images seen before our inner eyes, but they are never the image itself, and even the image is far from being reality.' [159] Professor Koehler's words are an apt ending to a chapter that insisted on growing.

Life Histories

Chapter 14

Aspects of Reproduction

LIFE is a continual coming to terms with existing conditions. By such adaptability, a fish may eventually further the history of its kind; and the word 'may' is essential, for not every mature fish succeeds in mating. Life from day to day, which has been our main concern until now, is a prelude to the initiation of new lives.

Life histories, the medium time-scales of existence, begin in fishes when eggs have been fertilized. In the two following chapters, we shall consider early life histories in marine and freshwater habitats. Here we turn to aspects of reproduction.

Germ cells and hormones

The future testes or ovaries appear as genital ridges in the embryo, one on each side of the mesentery that suspends the gut from the roof of the body cavity (in some fishes only a single gonad is formed). Each egg develops in a capsule of follicular cells, the ovary being made of egg-follicles bound together by connective tissue (figure 59). Before the breeding season, the egg cells are very small and translucent. After passing through a series of growth stages, when the nuclei mature and yolk is acquired, all or many of the eggs become mature. When ready to be laid, the eggs are released from the follicles. Maturing testes are formed of a great many cyst-like units containing cells that produce sperm (figure 59). In most fishes these sperm-cysts develop in elaborate systems of small tubes or lobes [162].

In lampreys and hag-fishes the eggs and sperm are shed into the body cavity and are then passed to the exterior through genital pores that open near the anus. The sexual products of most other fishes enter some kind of duct. Sharks, rays, chimaeras, lung-fishes, sturgeons and the bowfin shed their eggs into the body cavity, where they enter the funnel of the appropriate oviduct. The garpikes and teleost fishes have ovaries that open directly into oviducts (figure 24), though in some kinds of teleosts, notably the Salmonidae, the oviducts are reduced and not attached to the ovaries. Each testis of a teleost has its own duct, and this leads to the genital opening. In most other fishes, certain tubules derived from the kidney convey the sperm to the kidney duct or to a separate spermatic duct.

The eggs of cyclostomes are fertilized after they are laid, the lampreys producing many small eggs, the hag-fishes fewer and larger eggs which have horny membranes. Internal fertilization is the rule

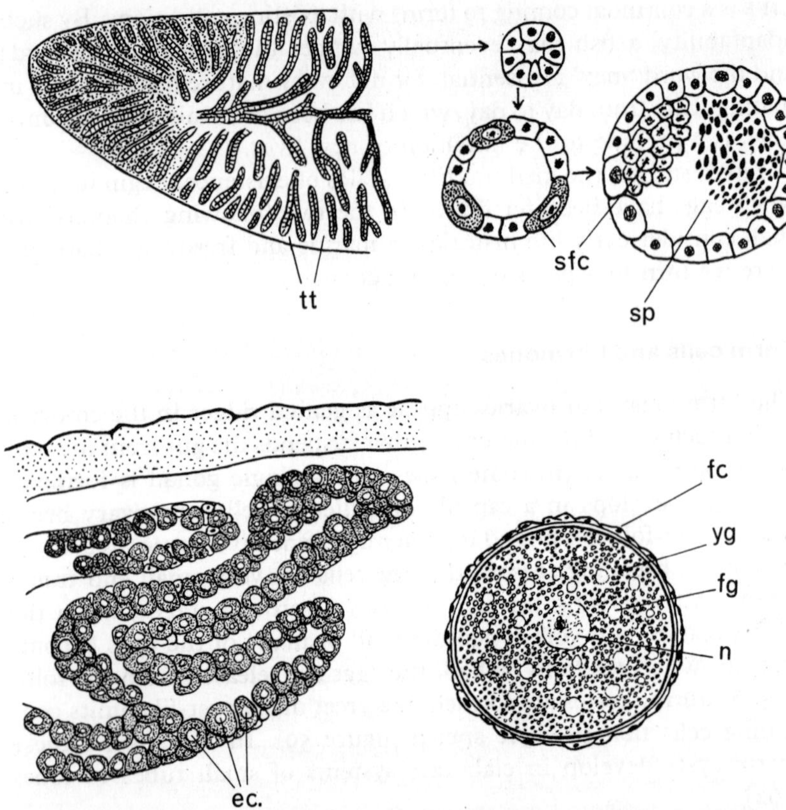

Figure 59. Gonads of a teleost fish. Above: part of an unripe testis showing the tubules (tt). Between these are connective and interstitial tissues. As the testes ripen these tubules become larger and form large cells (sfc), which divide again and again to form the sperm (sp). Below, left: Part of a ripening ovary showing the folds of ovarian tissue and the egg cells (ec). Right: a ripening egg showing the follicle cells (fc), yolk granules (yg), fat-globules (fg) and the nucleus (n).

among the cartilaginous fishes, some of which are viviparous. All the survivors of the archaic groups of bony fishes: lung-fishes, *Latimeria*, bichirs, sturgeons, garpikes and the bowfin, produce eggs that are fertilized externally. This is true of most teleosts, but we shall see that viviparity has evolved in a number of groups.

The reproductive systems of fishes are not simply the producers of germ cells. Fishes deprived of their testes or ovaries fail to acquire the secondary sexual characters and behaviour patterns that are proper for their kind. Male or female colour patterns are not developed; nor are other sexual characters such as the pearl organs in male cyprinid fishes (p. 235); the nest-building, thread-secreting part of the kidney of male sticklebacks; the seasonally enlarged liver in female rice-fishes and so forth [64].

These tests and others indicate that the gonads of fishes produce sex hormones. Indeed, extracts of the testes of Pacific salmon will produce comb growth in capons. If the testes of rice-fishes are transplanted into spayed females, the latter acquire the secondary sexual characters of males. Moreover, the cellular structure and seasonal activity of the testes suggest that the male hormones are secreted by interstitial cells comparable to those of the higher vertebrates. The female hormones may well come from transformed follicle cells.

Secretions stored in the pituitary gland, which is attached to the floor of the tween brain or diencephalon (figure 60), are also intimately involved in sexual affairs [64]. If the pituitary is excised, the testes or ovaries eventually regress and the development of nuptial colours is also halted. Moreover, products of the pituitary gland regulate the final stages of egg-ripening and stimulate the release of the eggs from the ovaries.

The environment and breeding season

While pituitary secretions have definite effects on the gonads, the reverse hormonal interplay, which obtains in the higher vertebrates, has still to be proved or disproved in fishes. The release of ripe eggs after apt pituitary activity reminds us that fishes spawn at certain times and seasons. Naturalists have long realized that breeding begins at certain levels of temperature, which vary from species to species. In the North Sea, for instance, most fishes spawn during the spring and summer when temperatures are rising. The dab, sole, lumpsucker and haddock spawn during the first half of the year, mainly in the spring. Red mullet, horse-mackerel, shanny, weevers and mackerel are summer spawners. Herring breed in the spring or autumn, while the plaice and armed bullhead do much of their spawning during the winter months.

Most kinds of European freshwater fishes also spawn in the spring or summer. Perch, bullhead, bitterling, pike, grayling and lampreys spawn in spring, while carp, gudgeon, tench, minnow and chub breed in the summer. Salmon and trout spawn from October to January.

For each species, there is a certain range of temperature, usually rather restricted, within which it can spawn. Here, for instance, are such temperature ranges of a few North Atlantic fishes: cod, 0·4° to

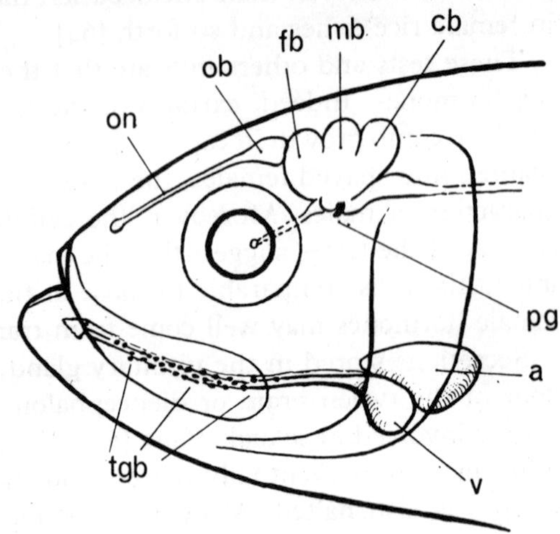

Figure 60. The pituitary and thyroid glands of a herring. The former (pg) is shown as a black body under the brain, which shows an olfactory nerve (on), an olfactory bulb (ob), the forebrain (fb), the midbrain (mb) and cerebellum (cb). The thyroid bodies (tgb) lie along the ventral aorta, which carries blood to the gills. (After Schnackenbeck.)

7°C; plaice, 4° to 7°C; pilchard, 9° to 16·5°C; mackerel, 10° to 15°C; herring, spring spawners, 3·7° to 9·3°C; autumn spawners, 9·1° to 13·3°C. All five species have a wide distribution and their spawning temperatures vary somewhat from place to place [163].

Fishes of moderately warm seas are also seasonal breeders. In the Gulf of Mexico, temperatures off the coast of Texas range from about 30°C in summer to about 13°C in winter. Here Dr Gordon Gunther found that, '. . . over six times as many fishes spawned in the spring and summer as spawned in the fall and winter. Most of the fishes spawning in the fall and winter were of temperate and subtropical distribution, while the warm month spawners included a majority of species that are tropical and subtropical in distribution.' [164] The same kind of relationship is true of cool temperate waters. Summer spawning fishes in the North Sea have more southerly living spaces than those that spawn during the colder seasons.

In tropical seas the annual range of temperature is much more restricted. Off the coasts of Kenya and Tanganyika, not far south of the equator, surface temperatures from May to October are from 24° to 25°C. From November to April the range is 27° to 29°C. Study of the perch-like snappers (Lutianidae) that live in these waters showed Frank Talbot that: 'No sharply marked breeding seasons were found in any of the species studied, although sometimes a single sample would contain many ripe fishes of both sexes. In general, all species seemed to breed over a large part of the year, but mostly in the warm north-east monsoon period.' [165] Spawning seasons are more closely linked to the monsoons in certain Indian waters. Off the coast of Madras many fishes breed during the cold season following the onset of the north-east monsoon in September and October. The Indian oil sardine enters coastal waters to spawn after the heavy rains of June and July, when the south-west monsoon is blowing.

While our knowledge of the breeding seasons of fishes is by no means adequate, it would seem that in any reach of the sea, the less the seasonal changes, especially in temperature, the less definite are the mating periods. Is this true of polar waters? Most antarctic fishes, for instance, live at temperatures between 0° and −2·0°C. Many are faced with temperatures of −1·9°C for the whole of their existence. Certain kinds undoubtedly have a main spawning period in spring and summer, but the young of others hatch during the other seasons. A main spawning in spring and summer is to be expected; these are the seasons when the plankton can flourish and young, small-mouthed larvae need such food if they are to survive.

How are breeding seasons governed in the depths of the ocean? Many mid-water species migrate upward to the food-rich surface waters each day, when they will experience changes in temperature that some fishes only undergo during the course of a year. Others, like the deep-sea angler-fishes, stay below at depths between 1,000 and 3,000 metres, where local temperatures are virtually constant at values between about 2° and 5°C. Yet larval angler-fishes appear in the surface waters of the North Atlantic at definite seasons. The majority of species are summer spawners, but the linophyrynid anglers breed in the spring [104]. The tentative rule expressed in the first sentence of the preceeding paragraph would thus appear to apply only to fishes of shallow waters.

Cave fishes and the deeper dwelling oceanic fishes reproduce in sunless waters. But the reproductive rhythms of those that live in lighted surroundings may be decidedly influenced by the amount of sunlight they receive. Aquarists who keep tropical freshwater fishes know well that in spring there is more sexual activity, even in species that reproduce all through the year. Controlled tests involving species belonging to seven different families have shown that changes in the quantity of light lead to changes concerned with reproduction [64]. Under normal conditions the quantity of light falling on North American brook trout (*Salvelinus fontinalis*) decreases after mid-summer and they begin to spawn in the early part of November. If these fishes are kept so that the light they receive increases by one hour each day over natural daylight from January to March, thereafter decreasing by one hour each day until the sexual products are mature, there will be ripe fishes by the end of July, three months before the onset of the natural spawning season.

Comparable experiments with other fishes have given somewhat similar results. During any such test the temperature of the water is kept at a constant level. Alternatively, the illumination may be held steady and the range of temperature varied from one batch of fish to the next. For instance, three groups of three-spined sticklebacks were kept at temperatures of 4° to 6°C, 9° to 13°C and 20° to 22°C under constant illumination. Nuptial colours began to appear in the warmest fishes in thirteen days, while spawning occurred in twenty days. There was no change in the ovaries of females kept at the two lower ranges of temperature. Fishes kept in total darkness and at high temperatures did not ripen, but they did so when exposed to natural daylight. When three batches of females, which were kept at temperatures of 16·5°C, 12° to 13°C and 10°C, were injected with pituitary hormones they were able to ripen. A proper level of temperature, as well as adequate illumination, is thus essential if the pituitary is to initiate the ripening and release of the sexual cells.

In nature, no doubt, the pituitary of a fish will not produce the necessary sexual triggers until its possessor has been exposed to certain changes of light and heat. These changes are registered through the eyes and temperature-sensitive nerve endings in the skin, receptors that presumably influence the pituitary by way of the brain. But light and heat are not the only stimuli to spawning. A number of attractively coloured top-minnows, which live in parts of tropical Africa and South America, are known as annual fishes. During the hot dry season their habitat is liable to dry up, but before the water entirely vanishes these fish mate and bury their eggs in the mud. The urge to reproduce seemingly arises when these fishes sense the shrinking cover of their environment. On the other hand, flooding is thought to provide a spawning stimulus to other tropical freshwater fishes. According to Dr Hugh M. Smith:

> The annual inundation of the vast central plains of Thailand and of the various minor plains is an event of great importance in the life of all the fishes. As the streams begin to rise and fill their beds, together with the connecting canals, and the tributary ponds, lakes, swamps and marshes that had become reduced during the protracted dry season, the fishes follow the flood waters, into the rice fields, into the lakes, and into the swamps being converted into lakes, and by the time the inundation has reached its height the vast majority of the free-swimming fishes has spawned. With the falling of the flood waters, the adults move back into the river channels leaving the young to follow in accordance with their rate of growth and respective needs. The flood waters give protection to the young, which in general have a rapid growth and many attain maturity in one year [166].

Similar sequences of events have been observed in India, Africa and South America. Even so, this is not to say that an expansion or shrinkage of water cover is the sole influence behind spawning. Some species that live in hot, dry regions must evidently seize their chance to multiply when the rains come. But such fishes have been exposed to certain regimes of light and heat, which play some part in the ripening of the sexual products.

This review of the influence of environmental factors on spawning seasons may be ended, but not completed, with the remarkable breeding habits of the grunion (*Leuresthes tenuis*). This fish, which is a silverside (family Atherinidae) and grows to a length of about 7 inches, lives along the shores of Southern and Lower California. The breeding season lasts from late February to early September, but the spawning runs are restricted to certain lunar times. For three or four nights following a new or full moon, when high tides occur, the grunion move in with the surf and leap on to the beach (plate 36). There they spawn over a period of one to three hours as the tide is beginning to ebb. The females – and each may have one or two males curled round her – burrow into the sand and shed their eggs, which are promptly inseminated. The grunion now skitter back into the sea. Their eggs, which get an extra covering of surf-washed sand, are usually ready to hatch in a week. When, after a fortnight, the next high tide exposes the eggs, the young emerge and are washed out to sea. In the laboratory, eggs that are about to hatch will not do so until they are shaken, which in nature is no doubt done by the roll of the surf.

This kind of grunion spawns only at night, but in the Gulf of California a related species, *Hubbsiella sardina*, is known to spawn during the day, when they fall an easy prey to flocks of sea birds. Again, these fishes spawn on the beach, just after the high spring tides.

How do grunion sense when it is time to move for the shore? Perhaps they have an intimate and accurate awareness of tidal rhythms; for it seems most unlikely that they can see just when the moon is new or full. Close observations at sea and well-designed tests might reveal much. And need we be surprised if we discover that some fishes know the tides as well as we know night and day?

Modes of reproduction

Reference to Chapter 13 will show that pre-spawning procedures vary a great deal among fishes. But whatever the simplicity or complexity of such behaviour, most fishes reproduce by eggs that are made fertile after they are laid. This is true despite the fact that fertilization is internal in sharks, rays and chimaeras. Even if one includes the teleosts with intromittant organs, in no more than one kind of fish out of twenty-five are the eggs made fertile within the female.

Nearly all bony fishes lay eggs that are less than ¼ inch in diameter (notable exceptions are *Latimeria* and certain cat-fishes that use the mouth as an incubating chamber). A great many marine species produce floating, planktonic eggs, whereas very few freshwater fishes have evolved such eggs. Many simply scatter their spawn over the bottom of the river or lake. Others bury their eggs, or attach them to weeds or lay them in some kind of nest. Numerous inshore fishes also produce non-buoyant kinds of eggs, which are also the rule in polar coastal waters. Taking fishes of comparable size, those that form floating spawn generally lay smaller and more numerous eggs than do the species with non-buoyant kinds of eggs. These diverse ways of disposing of particular forms of eggs are at the very centre of the life of fishes. But how each way enables a fish to survive in certain surroundings may be left to Chapters 15 and 16.

Fishes are not invariably male and female. Individuals of some kinds of sea-perch (Serranidae), maenids and sea-bream (Sparidae) are functional hermaphrodites, which is to say that one fish can produce both ripe sperm and eggs. One part of the gonad (usually the inner) forms the male cells: the other elaborates the eggs, though one part may ripen before the other. But in some sea-perches belonging to the grouper subfamily Epinephelinae the gonad is not always divided into male and female parts. Instead there is a complete change from one sex to the other. As the fishes grow larger they turn from males to females. In the transition period they have ovotestes [167].

In a small sea-perch (*Serranellus subligarius*) that lives among Floridan coral reefs, each mature individual, regardless of size, bears both ripe eggs and sperm. During the spawning season there may be cross fertilization between the hermaphrodites, each fish alternatively playing male and female rôles. But one fish may simultaneously release both eggs and sperm, and from such an individual, embryos and larvae are produced [168]. Self fertilization is evidently possible; it has been discovered in a small top-minnow (*Rivulus marmoratus*) that lives in brackish waters and tidal flood plains from Florida to the West Indies [169]. This species usually sheds the eggs, but some individuals retain them in the ovaries. It may also form functional ovotestes. In this connexion, some individuals produced fertile eggs after thirty-two months in captivity. An egg from one fish was incubated and the young one reared in isolation. After six months this fish laid eggs and again one was separated and reared; it produced fertile eggs after a period of four months. For two generations then, this top-minnow produced self-fertilized and perfectly fertile eggs.

Ovotestes have also been found in certain deep-sea fishes belonging to the order Iniomi. These are a lancet-fish (*Alepisaurus ferox*) and a barracudina (*Lestidium pseudosphyraenoides*), which are mid-water species; also six bottom-dwelling species belonging to the tripod-fish family Bathypteroidae and the Chlorophthalmidae. The male and female components of the ovotestes were in advanced stages of maturation. One tripod fish (*Benthosaurus grallator*) contained a mature ovotestis and a sperm-filled vas deferens: a chlorophthalmid (*Parasudis truculentus*) had mature eggs in both the gonad and oviduct and sperm in the ducts [170]. Cross fertilization, as in *Serranellus*, might thus be possible. After all, meetings between the sexes are not easily contrived in the deep sea. Encounters between hermaphrodites would be infinitely more fruitful than those between individuals of the same sex.

Live-bearing fishes

In the top-minnow (*Rivulus marmoratus*) self-fertilized eggs may develop for a while in the ovotestes before they are laid. The normally fertilized eggs of another top-minnow, the Japanese medaka (*Oryzias latipes*) may also be kept a little time within the female. But both species are essentially oviparous. Even so, it is interesting to find incipient live-bearing in two egg-laying top-minnows, for some

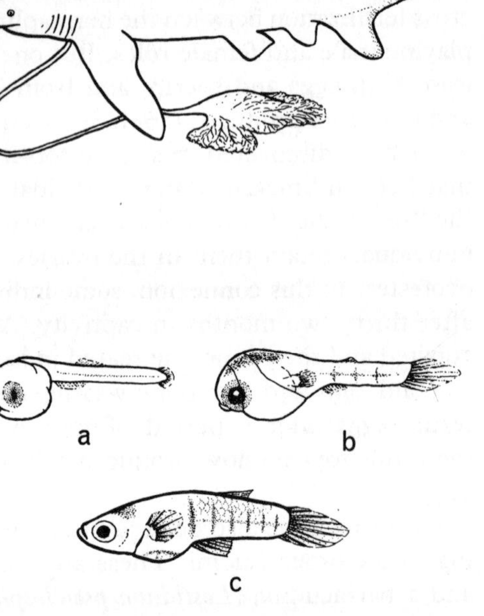

Figure 61. Above: A 2-foot embryo white-tip shark (*Carcharchinus longimanus*) taken from a 7-foot female. Note the long umbilical cord and the yolk-sac placenta, which makes intimate contact with folds in the oviduct. (After Bigelow and Schroeder.) Three stages in the development of a viviparous top-minnow, *Heterandria formosa*, which shows superfoetation (p. 256). (a) An early embryo (3·5 millimetres) with the head enveloped in a pericardial sac. (b) A later stage (5·0 millimetres) when the head has pushed through the pericardial sac. (c) A fish (6·5 millimetres) near the end of gestation. It is approaching the adult form. (After C. L. Turner.)

of their relatives are true live-bearers. Most of these species belong to one family, Poeciliidae but there are also the four-eyed fishes (Anablepidae), the Goodeidae and the Jenynsiidae.

Live-bearing teleosts have a single oviduct and usually a single ovary. In male live-bearing top-minnows the forward rays of the anal fin are set apart or modified to form a kind of penis or gonopodium (figure 62). Of the poeciliids, the guppy (*Lebistes reticulatus*) is the best known instance of an ovoviviparous species. The eggs are fertilized within their follicles, where they develop for a time. On rupture of the follicles, the embryos are released into the cavity of the ovary where they complete their development, sustained by the yolk and, seemingly, by secretions of special follicular cells.

In the viviparous species, the young are nourished through some kind of placental connexion with the mother. The dwarf top-minnow (*Heterandria formosa*) has been well studied (figure 61). The embryos, which have a small supply of yolk, are retained within the ovarian follicles. The walls of each gestating follicle acquire an elaborate bed of blood capillaries, which extend into small processes

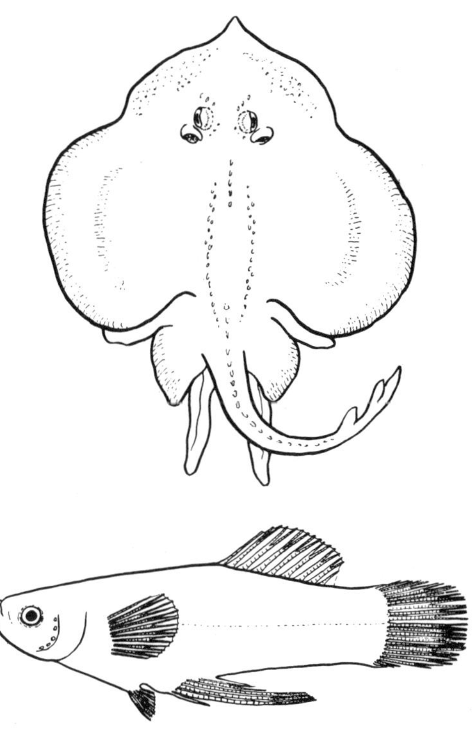

Figure 62. The sperm-transmitting organs in cartilaginous fishes represented by a skate (*Psammobatis mira*), and in viviparous top-minnows, such as *Brachyraphis hartwegi* (lower figure). The former have two claspers which are formed from the inner sides of the pelvic fins. The intromittent organ of the viviparous top-minnows is formed from the forward rays of the anal fin. (Lower figure after D. E. Rosen and Reeve M. Bailey.)

known as villi. These make intimate contact with the external surface of the embryo, where the maternal and embryonic blood system forms a placenta through which the embryo is nourished. In the four-eyed fishes (*Anableps*) the embryos are also gestated in follicles, but there is a different kind of 'placenta', formed by a great expansion of the pericardial wall. This bears vascular bulbs, which are associated with villi that grow from the walls of the follicles. The pouch-like belly is mostly filled by an enlarged part of the gut, and this is said to digest nutritive fluids that are taken in through the mouth.

In the top-minnows of the families Jenynsiidae and Goodeidae, the embryos, which carry a small store of yolk, are discharged from the follicles into the ovarian cavity, where they undergo their development. The embryos of *Jenynsia* are at first sustained by secretions of the ovary. Later, the lining of the ovary forms highly vascular folds, which make contact with the gills of the young, thereby forming a branchial placenta. The later nutrition of some goodeid fishes is by way of food-absorbing processes that grow out of the hind gut, a way of feeding that may well be used by the young of all viviparous top-minnows.

Live-bearing fishes have no definite period of gestation comparable to those of mammals. Most mammals keep the heat of the body at a constant level, but the temperature of a fish is that of its surroundings. If, for instance, the water is unusually warm, the period of gestation is somewhat reduced. The size of the brood depends on the size of the female. While young female guppies give birth to about six offspring, large females can produce sixty or more. Large sword-tails (*Xiphophorus helleri*), which can drop from a hundred to two hundred young, are also more prolific than are the younger females. After all, the larger the mother, the more young she can carry.

Like certain other live-bearing top-minnows, *Heterandria* is able to store sperm, which becomes embedded in the lining of the ovary. Ten months after mating, one female still carried live sperm. At given times over many months, sperm are thus available to fertilize succeeding batches of eggs. It is this facility that has presumably led to superfetation, the bearing of two or more broods in the ovary. In *Heterandria*, for instance, which give birth to young at intervals of about five days, as many as nine broods, each at a different stage of formation, have been found.

Thousands of young are produced by certain live-bearing species of the scorpion-fish family, but they are much larger than top-minnows. Moreover, the young are born in a larval stage, not fully formed, as are newly born top-minnows. The red-fish or Norway haddock produces from 18 to 150 thousand eggs, which may be in various stages of maturity. At the other extreme are surf perches, living along the Pacific Coast of North America. The number in the brood, which depends on the species and the size of the mother, varies from three to about eighty. In the species *Cymatogaster aggregatus*, the range is from three to twenty eggs and male fishes may remain in the ovary until they are sexually mature. The eggs are fertilized in the follicles but are soon released into the cavity of the ovary where development is completed, the embryos being nourished by ovarian secretions.

Most of the half-beaks (Hemirhamphidae) are marine, but certain genera (*Dermogenys*, *Nomorhamphus*, *Hemirhamphodon* and *Zenarchopterus*) live in fresh and brackish waters from India through Malayan and Indonesian regions to the Philippine Islands. Unlike their marine relatives, which are oviparous, these genera are ovoviviparous and the young are born in an advanced state of development. In *Nomorhamphus hageni*, each ovary may bear seven or eight embryos, while eighteen have been recorded from *Dermogenys pusillus*; in this species the young emerge head first.

The members of a diverse family of (mostly) deep-sea fishes, the Brotulidae, are also both oviparous and ovoviviparous. In the live-bearing species the embryos probably feed on nutrient fluids secreted into the ovary. Certain kinds, at least, have long food-absorbing processes called trophonemata that emerge from the anal region.

Lastly, the young of the European eel-pout (*Zoarces viviparus*) are nourished in the ovary by the secretions of follicles that become suitably modified after discharging the eggs. Young females give birth to twenty to forty young, whereas older mothers can hold three hundred or more. The newly-born are about 1½ inches in length and are miniatures of their parents.

Teleost fishes have thus evolved diverse means to produce living young. In some, such as the red-fishes *Sebastes* and *Sebastomus*, the ovary is simply an incubator: the larvae are born soon after hatching. The more advanced live-bearers nourish the young in the ovary. Such species produce more advanced young, which in many instances are miniatures of the adults. The last observation is also true of the live-bearing sharks and rays, which are more diverse than those that lay eggs. Live-bearing teleosts, on the other hand, are overwhelmingly outnumbered by the egg-laying species. But all the cartilaginous fishes have, so to speak, taken a fundamental step towards becoming live-bearers: the males bear a pair of grooved claspers that are formed from the inner parts of the pelvic fins (figure 62). It is these organs that transmit the sperm to the oviducts of the female.

There are over three hundred species of rays and nearly half are classified in the skate family (Rajidae). The skates lay very large yolky eggs (plate 15), which acquire a horny covering (the mermaid's purse) on their way down the oviducts. Of the European species, the cuckoo-skate (*Raja circularis*) produces the smallest eggs (60 × 35 millimetres), while the largest (180×135 millimetres) are those of the bottle-nosed skate (*Raja alba*). Some species lay their eggs in pairs at certain intervals, which in the laboratory, at least, take from four and a half months to over a year to hatch. The newly hatched young look much like their parents, but a number of changes are required before they acquire the adult form.

All other rays, so far as we know, are ovoviviparous. The embryos are nurtured in the oviducts of the mother, but there is no 'placental' association. As the yolk supply is limited, the embryos of sting-rays, electric-rays and eagle-rays are nourished by the milky secretions of vascular villi that emerge from the wall of the 'uterus'. The young draw milk into their mouth and spiracles, though some milk may be absorbed by elongated gill filaments that grow out of the gill clefts of the embryo. Each embryo of the butterfly-ray (*Gymnura*) – and there is generally one in each oviduct – is fed through long uterine villi that enter the spiracles and extend into the foregut.

Most sharks are ovoviviparous; some are oviparous, while a few are viviparous. The egg-laying kinds include the dog-fishes, Port Jackson sharks (Heterodontidae), certain carpet-sharks (Orectolobidae), such as *Chiloscyllium* and *Stegostoma*, and the whale-shark. Like the skates, the oviparous sharks produce large yolky eggs. Those of the whale-shark measure a foot in length.

In ovoviviparous sharks, the embryos are soon liberated from the egg capsules to finish their development in the oviducts. Apart from their own reserves of yolk, they are nourished on fluids secreted by filaments or villi of the mother's oviducts. These nutrients are absorbed by the yolk sac of the embryo. In some species outgrowths of the yolk-sac stalk – the 'umbilical cord' – provide extra surface for the absorption of nutrients. The young of mackerel-sharks and sand-sharks (*Odontaspis*) also feed on unfertilized eggs that lie near them in the oviduct.

Six-gilled and seven-gilled sharks, spiny-finned sharks (Squaloidea) and diverse galeoid sharks are ovoviviparous. The viviparous species include certain blue sharks (*Prionace*, *Scoliodon* spp.), hammerheads and smooth dog-fishes (*Mustelus*). The young are carried in pockets of the uterus and their connexion with the mother is by way of a yolk sac 'placenta' (figure 61). Outgrowths of the yolk sac interdigitate with vascular villi that emerge from the uterine walls. By such intimate association the young are provided with food and oxygen. The pups are fully formed at birth. The blue shark (*Prionace glauca*) produces from twenty-eight to fifty-four while the complement in two smooth dog-fishes (*Mustelus canis* and *M. mustelus*) is from ten to twenty.

Most live-bearing sharks and rays thus provide their embryos with yolk and later with a nutrient medium ('milk') that is secreted by villi in the uterine part of the oviducts. This pabulam is absorbed by the epithelium of the yolk sac. Only a few sharks have taken the further step of evolving a yolk-sac placenta. It will be evident that ovoviviparous reproduction has played a great part in the evolutionary history of the sharks and rays. Even so, the skate genus *Raja*, comprised entirely of about a hundred egg-laying species, is the most diverse genus of cartilaginous fishes. The chimaeras (Holocephali) also produce large yolky eggs, but they are the relics of a once diverse group. Most of the really successful fishes, the Teleostii, lay many small eggs. Through such reproductive means, they have become vertebrate masters of the hydrosphere.

Chapter 15

The Early Life of Marine Fishes

MOST KINDS of marine fishes start life as eggs that float in the sea. A good many, principally species that live near the shore, lay their eggs on the bottom, or in nests, or fasten them to rocks, shells, sea weeds, sponges and so forth. Some of these fishes guard their eggs, but parental care is more advanced in those other kinds that carry their eggs or use part of the body as a brood chamber. In still more advanced species, represented by many sharks and rays and some teleosts, the eggs develop within the female, who nourishes the embryos after they hatch. But whatever the beginnings of fish life, the end is the continued existence of species. During their evolutionary history, the early life of fishes has surely been shaped to this end by the sheer pressure of natural selection. To survive, a fish must come to fruitful terms with its physical and biological environments.

More immediately, the early way of life is directed towards the maintenance of regional populations. Close study of the plaice, which lives on sandy stretches of the European continental shelf from Iceland and the White Sea south to the Western Mediterranean, has shown this very well. In the North Sea there are two main spawning grounds, one in the Southern Bight, the other in the Heligoland Bight (figure 63). Knowing the mean fecundity of plaice – the eggs produced vary from 16,000 to 350,000, the number increasing with the size of the female – and having made a census of the eggs caught in plankton nets, naturalists of the Fisheries Laboratory, Lowestoft, estimated that some 50 million plaice take part in the spawning in the Southern Bight, which is about equal to the spawning stock off Heligoland. Most of the females are from five to seven years old, the males from four to six. The breeding season lasts from December to early March [173].

During the final stages of ripening, the eggs of plaice, and of other species that lay floating eggs (plate 32), are made buoyant and much larger by acquiring watery fluids of low density from the follicle cells.

The eggs float on being laid, when they become perfect spheres, as are the eggs of most bony fishes. In spanning 1·65 to 2·2 millimetres, plaice eggs are larger than those of most teleosts that produce buoy-

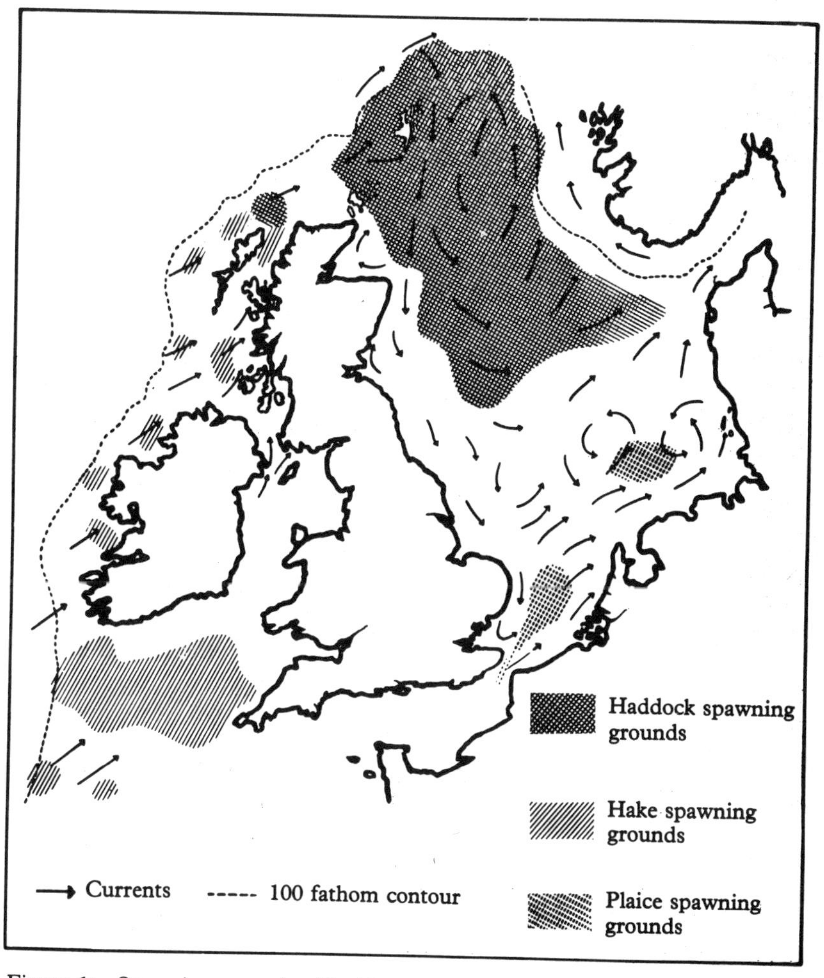

Figure 63. Spawning grounds of haddock, hake and plaice in the North Sea and waters west of the British Isles. The arrows indicate the direction of the main currents. (After Parrish [175].)

ant eggs, which range from 0·5 to 1·5 millimetres in diameter. The turbot (*Scophthalmus maximus*), for instance, lays several million eggs with a diameter range of 0·9 to 1·2 millimetres, while the corresponding figures for the flounder (*Platichthys flesus*) are from $\frac{1}{2}$ to $1\frac{1}{2}$

million and 0·8 to 1·15 millimetres. For every 10 grams of their gutted weight, the plaice, turbot and flounder produce 1,500, 16,000 to 22,000, and 12,700 eggs in this order.

Apart from differences of size, turbot eggs can be distinguished from those of the plaice and flounder in that they have an oil-globule in the yolk. The eggs of the related brill (*Scophthalmus laevis*) also bear an oil-globule, but they are distinctly larger (1·15 to 1·5 millimetres) than turbot eggs. More generally, the features used to assign buoyant kinds of eggs to particular species are their size-range; their shape (usually spherical, though anchovy eggs are ellipsoidal); the presence or absence of oil-globules; and the appearance of the yolk mass, which is most often entire, but may be segmented. In most of these eggs, the surface of the yolk lies close to the transparent egg membrane, though in some, such as the pilchards and *Sardinella*, there is a considerable gap or perivitelline space between the yolk and membrane. The latter is usually smooth, but exceptions include dragonets and rat-tailed fishes in which the surface of the membrane is stamped with a fine hexagonal pattern.

Drifting in the sea, where they are tossed and spun by turbulent motion, many billions of plaice eggs are gradually dispersed from the spawning grounds. At the winter temperatures of about 7°C to which they are exposed, the delicate, translucent larvae hatch from the eggs in about seventeen days. They have large black eyes, a bulbous head, a short trunk and a long tail (figure 64). Their length is from 6 to 7 millimetres. The future muscle segments or myotomes can be seen clearly along the trunk and tail. A high and filmy vertical fin extends from the back of the head and trunk to the tip of the tail. Between the head and anus is a globular mass of yolk, which sustains the larva until it has the muscles and jaws to fend for itself. In the region around the delicate pectoral fins the skin of the larva is separated from the underlying tissues by a large subdermal space. This is filled with a watery, somewhat gelatinous fluid that is less dense than sea water, the whole space forming a buoyancy chamber [174] (figure 64).

Most kinds of floating eggs hatch into larvae that look rather like those of the plaice (figure 64). They also carry buoyancy chambers, but the larvae of nearly all teleosts that spend their lives in aquatic space soon acquire a swimbladder. The plaice and its right-handed relatives, such as the dab, lemon-sole, witch and flounder never develop this organ, though it is formed by the larvae of some sinistral flat-fishes (Bothidae), well known instances being the turbot and brill. But the swimbladder is rudimentary or absent in all adult flatfishes, as it is in other fishes that spend most of their lives on the sea floor.

Upheld by their buoyancy chambers, the larval plaice drift in the

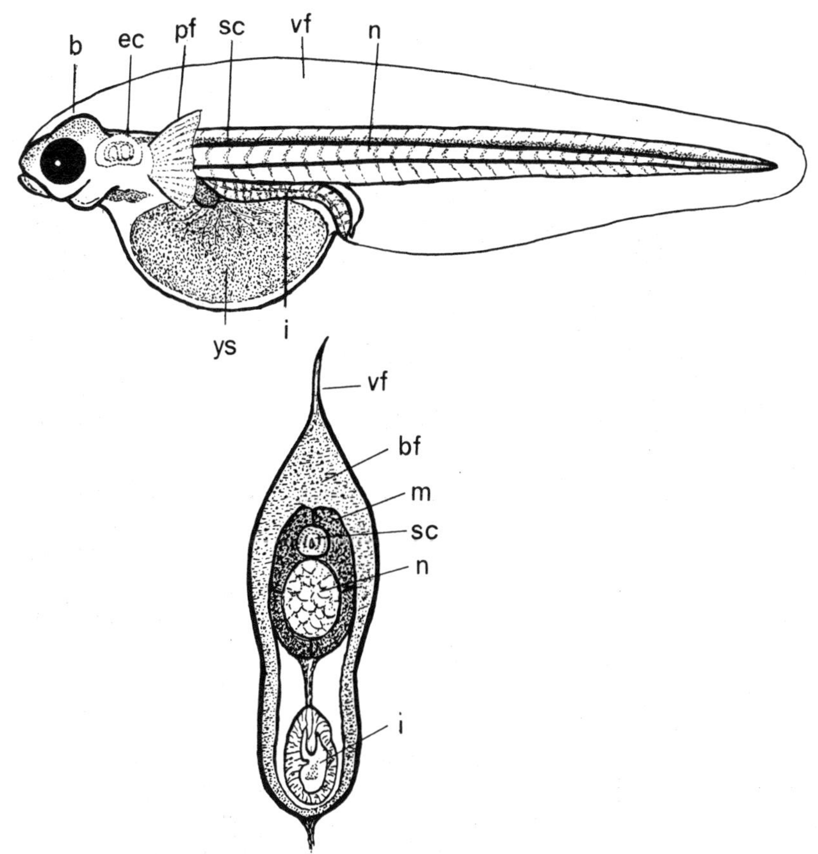

Figure 64. Above: A newly hatched plaice larva, showing the brain (b), ear capsule (ec), spinal cord (sc), notochord (n), muscle segments or myotomes (m), yolk-sac (ys), intestine (i), pectoral fin (pf) and vertical fin-fold (vf). The lower figure represents a cross section through the trunk of a somewhat older larva. Between the skin and the inner organs, labelled as above, are dilute body fluids (bf) which assist greatly in flotation. (After Shelbourne [174].)

plankton. After eight or nine days, when their yolk reserves are nearly exhausted, they begin to feed on small diatoms. Later, having grown a little more, they coil and spring on appendicularians (*Oikopleura*) and the young stages of copepods and molluscs. But even three weeks after hatching, they are little or no more than 10 millimetres in length.

After a month of this drifting life, the young plaice, which is perfectly symmetrical, begins its metamophosis into a flat-fish. The most striking changes, and they take about two and a half weeks, are the deepening of the body and the gradual migration of the left eye to the right side of the head. Towards the end of metamorphosis, the two eyes are close together and now the young plaice tends to swim with its right side uppermost. It is now denser than its surroundings, the time having come when it must descend to the bed of the sea. During the settling period, the right side of the body acquires a more definite colour pattern, while the blind side gradually loses its pigmentation. All flat-fishes undergo a similar kind of metamorphosis. In young tongue-soles (Cynoglossidae) and flounders (Bothidae) it is the right eye that moves to the other side of the head, while the true soles (Soleidae), like the plaice and its relatives (Pleuronectidae), eventually have the eyes on the right side of the body. The entire series of changes whereby a bilaterally symmetrical larva becomes a 'one-sided' flat-fish is more involved than that just described. But, as Michael Graham says, '. . . however drastic and disturbing the bodily changes may seem at this stage, the end of the process is a harmony between fish and the bed of the sea, when the plaice, its thickness hardly rising above the surface of the sand, rests almost invisible to predator or hunter; or leaves its bed to swim in a gliding motion as it searches for its food' [173] (plate 25).

The main nursery grounds for young North Sea plaice are sandy stretches off the coast of Holland, Germany and Denmark. Here the growing 'postage-stamp' stages find plenty of small worms, crustaceans and molluscs. But to get an idea of how they come to reach these grounds, we must remember that the planktonic existence of plaice, first as eggs and then as larvae, lasts for seven to eight weeks. During this time the distance of drift is likely to be well over one hundred miles, assuming the currents run about 3 miles in a day. Moreover, the set of the currents is usually from the spawning grounds to the nursery areas (figure 63). In other words, North Sea plaice seem to have chosen their spawning areas so that the young will be carried to favourable feeding grounds. As the southern North Sea and the English Channel were land during the late-glacial and early post-glacial periods, the choosing must have taken place during the last fifty thousand years or so. Clearly, the 'wiser' choosers were likely to have left the most offspring. The rest is speculation, but we do know that modern plaice stay on their continental nursery grounds until they are about two years old. After this, they move into off-shore waters. Three to five year olds, for instance, are common over the Dogger Bank. Marking tests have shown that some of these older plaice migrate to the Southern Bight spawning grounds from October to March. After spawning, the fish move back to the north. The spawning migration is thus against the current that will carry the young stages to the nursery grounds. The spent fish go with the current.

Fishes, it would seem, have evolved breeding habits and taken spawning grounds to give their young a good start in life. Fishery naturalists have done much to trace such natural provision for the stocks of cod, haddock and hake in the seas around Britain [175]. From Cape Hatteras on the western Atlantic sea board and the Bay of Biscay on the eastern side, the cod ranges northwards to subarctic waters. The distribution of the haddock (*Melanogrammus aeglefinus*) is quite similar, though the centres of concentration lie more to the south. The hake lives in rather deeper waters, which extend southward from the latitude of the Faroe Islands to the Moroccan coast of Africa. A smaller race of hake lives in the Mediterranean. All three species lay floating, spherical eggs which contain a small oil-globule, the size ranges being 1·15 to 1·6 millimetres for the cod; 1·2 to 1·65 millimetres for the haddock; and 0·95 to 1·05 millimetres for the hake. A fair sized hake lays about a million eggs, while the fecundity ranges of the cod and haddock are 2 to 9 million and 12 thousand to 3 million eggs respectively.

Cod breed in the North Sea from February to April, their chief spawning grounds lying over banks to the north of Flamborough Head. The main stocks of spawning haddock, which shed their eggs from February to May, are found in the deep basin in the northern North Sea (figure 63). Hake spawn from April to August in certain areas off the western coast of the British Isles (figure 63). The eggs of all three species tend to rise in the sea and they develop as they drift. Some ten to fourteen days after fertilization, the larvae, which measure about 4 millimetres, struggle out of the egg membranes, each, if healthy, bearing a good supply of yolk. This reserve sustains them for three or four weeks, after which they are active enough to tackle planktonic food. After growing a little more, they change into fry, when they acquire the look of their kind. Now they begin to seek deeper waters.

For the first six or more weeks of their life, cod, haddock and hake thus become part of the plankton. They move with the sea, and a glance at figure 63 will show that larval hake have a fair chance of being dispersed by north-easterly currents to inshore waters. These are their nursery grounds, where there is suitable food for larvae to become fry. At the end of a year, those that have made a good living have grown to a length of about 4 inches, which size is doubled during the next year. Now they behave like adult hake, moving over the sea floor during the day and swimming upwards at night.

As we have seen, the breeding grounds of the North Sea cod and haddock are mainly in the northern reaches. Here the spawning centres are largely encompassed by extensive eddy systems (figure 63), which will tend to keep the eggs and larvae moving within the shallow water (less than 100 fathom) confines of the northern North Sea. Those larvae that find suitable food, such as the young stages of molluscs and crustaceans, small copepods, and so forth, grow a little more and change into actively moving fry. These grow quickly, and within a year they are seeking the waters near the sea bed. Here they find a variety of small crustaceans, which form most of their food.

The convict surgeon-fish (*Acanthurus triostegus*) has a far wider range than plaice, cod, haddock or hake. It lives mainly among the coral reefs and atolls of the Indian and Pacific Oceans. In Hawaian waters, where it has been closely investigated [176], this species forms a separate race (*sandvicensis*), the members of which bear a diagnostic, sickle-shaped mark on the base of each pectoral fin. Like other surgeon-fishes, (*Acanthuridae*), this one is a vegetarian. Off Hawaii it lives in tide pools, bays, harbours and exposed reefs.

In the tropical parts of its range, the convict surgeon-fish may breed at any time of the year, but around the Hawaian Islands the spawning season is from December to July, during the colder season. The fertilized eggs, which tend to rise in the sea, measure from 0·66 to 0·70 millimetres in diameter and contain a single oil-globule. At a temperature of 26°C, the larvae hatch in twenty-six hours, a great contrast to the hatching time of plaice. The newly born larvae, measuring a little less than 2 millimetres in length, bear a large yolk sac (figure 65), but after sixteen hours of larval existence about half of this food is used. As the yolk reserve is consumed the larvae tend to sink, which they strive to counter by swimming upwards. After four days, when the swimbladder is well established, the young begin to dart after planktonic food.

The convict surgeon-fish is the most abundant acanthurid in Hawaian waters, where it is often seen in large schools. When spawning, a female may eventually release some forty thousand eggs and she may well do this more than once during the breeding season. We can imagine, then, that many millions of larvae will drift away from the Islands, particularly as the larval life lasts for about two and a

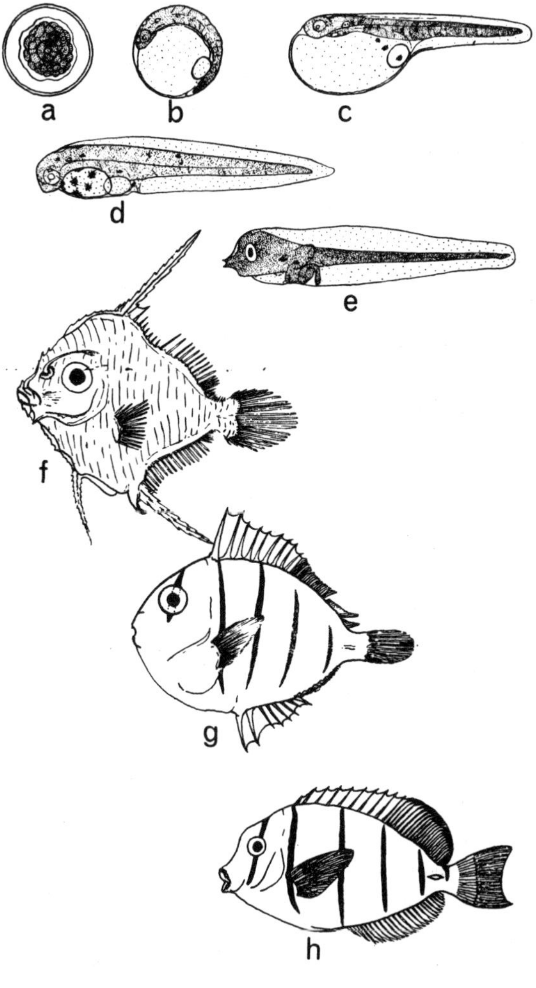

Figure 65. Stages in the life-history of the convict surgeon-fish (*Acanthurus triostegus*). After Randall [176] (a) egg, (b) egg with embryo, (c) larva soon after hatching, (d) larva just after losing yolk, (e) 5-day old larva, (f) larva changing to *acronurus* shape, (g) *acronurus* stage, (h) adult.

half months. But the local populations are maintained by advanced larvae that move inshore with the tides, an invasion that occurs at night and mainly when the moon is new. These *acronurus* larvae, as they are called, are disc-shaped, inch-long young with a translucent body that becomes silvery over the head and abdomen (figure 65). They are not entirely defenceless, for their second dorsal, second anal and pelvic spines are venomous. In four or five days the *acronuri*, which may be caught in tide-pools from February to early October, undergo a metamorphosis whereby they acquire the form and some of the features of their parents. One trenchant change concerns the digestive tract, which becomes three times as long, so being made ready for a vegetable diet. The *acronurus* stages feed on zooplankton organisms.

Tide-pools provide food and shelter for these transformed fishes. They grow quickly, adding half an inch to their length in a month (figure 65). Before long they acquire the size and power to move out to sea, where some will eventually reach sexual maturity [176].

These outlines of early life histories provide some inkling of how some local populations of particular species are maintained by buoyant eggs and larvae. Much is left to time and chance. Yet during their evolution, fishes have learned to make use of the moving ocean, which may hold the larvae and fry in a productive region or carry them to suitable nursery grounds.

To introduce the fishes that lay non-buoyant kinds of eggs, we turn to the winter flounder (*Pseudopleuronectes americanus*, plate 27), which lives along the Atlantic coast of North America from northern Labrador to North Carolina and Georgia. It is thickest on the grounds between Newfoundland and Chesapeake Bay and is seldom caught beyond a depth of 50 fathoms. This flat-fish gets its vernacular name from its habit of migrating into inshore waters during the winter. It spawns from January to May, particularly in shallow inlets and estuaries along the coast.

The females lay from ½ to 1½ million eggs (diameter 0·74 to 0·96 millimetres), which sink to the bottom and stick together in clusters. After fifteen to eighteen days, the larvae emerge; they are 3·0 to 3·5 millimetres in length and for a fortnight they live on their reserves of yolk. Some five to six weeks after hatching the vertical fin rays are beginning to appear, while the left eye is moving towards the right side of the head. Metamorphosis soon follows and young winter flounders, which measure no more than 10 millimetres in length, start their life as benthic fishes [177].

One population of winter flounders lives in Long Island Sound and one of their spawning grounds is in the Mystic River Estuary, Eastern Connecticut. In this estuary the early life history has been carefully surveyed. Most of the larvae, as the catches of plankton nets revealed, live near the bottom. Like the eggs, they are non-buoyant. Moreover, laboratory observations showed that the larvae have the habit of swimming upwards and then sinking. By staying on or close to the estuarine bed, the young are less likely to be carried away by tidal currents; and in the Mystic River Estuary about a third of the volume of water is removed seaward each day. By considering this tidal flushing and the near-bottom distribution of the larvae, Dr Pearcey [178] estimated that currents remove about three per cent of the larval population each day. If the larvae were evenly spread from top to bottom, the loss would be three times as great. Too many larvae would drift away from the nursery grounds, which are evidently much the same as the spawning beds. In fact, the food of larval winter flounders consists largely of nauplius stages of crustaceans, larval polychaet worms, invertebrate eggs and diatoms, and such organisms can be very abundant in estuarine waters.

Certain contrasts between this sort of early life history and that of species producing buoyant eggs will be evident. The plaice, for instance, breeds over off-shore spawning grounds and its eggs become part of the plankton. The larvae are also buoyant and will tend to drift towards inshore nursery areas, where, after metamorphosis, the fry settle on sandy stretches of the sea bed. By producing non-buoyant eggs and larvae, the young of the winter flounder tend to stay near the winter spawning beds, which are also the nursery grounds. Moreover, nearly all the larval fishes that are netted in the Mystic River Estuary belonged to species that lay non-buoyant eggs. Besides the winter flounder, these species include a sculpin (*Myoxocephalus aeneus*), the butter-fish (*Pholis gunnellus*), the lump-sucker (*Cyclopterus lumpus*) and a sand-eel (*Ammodytes americanus*). The larvae of these fishes were also most abundant near the bottom [178].

While thinking of this part of the world, it is relevant to consider a more northerly inlet, the Bay of Fundy, which opens into the northern Gulf of Maine. In this bay the tidal range is great and when the tide is rising and falling most rapidly, there are strong tidal currents into and out of the area. At such times, tide-rips and swirls mix and churn the water. As in the Mystic River Estuary, larvae that come from buoyant eggs are much scarcer than those hatched from non-buoyant kinds. The herring, for instance, which lays its eggs on seaweed and so on, breeds successfully in the Bay of Fundy. Fishes that shed floating eggs, such as cod, haddock, pollack, mackerel and cunner, do not spawn in the Bay, though they do so in the Gulf of St. Lawrence, where the tides are gentler and the water is stratified [179]. But in boisterous waters, buoyant larvae are more likely to be swept into regions unfavourable to their development. The capture of food organisms must be more difficult in turbulent waters, particularly when the larvae are just acquiring the 'sea legs' to fend for themselves. More active and advanced larvae try to keep station when faced with currents, which may well over-tax them in confused waters. Not enough of their hard won food would be available for growth.

Even the most cursory census of the great fecundity and limited numbers of adult fishes leads one to realize that myriads of eggs are needed to keep populations in being. There is a great mortality of eggs and larvae. Some die from bacterial and fungal diseases, while others are eaten, especially by medusae, jelly-fishes and arrow-worms. A great many eggs and larvae, particularly the buoyant kinds, may drift into waters that are physically or biologically unsuitable for their development. Such hazards were well shown during an extended survey of the early life history of the Atlantic mackerel (*Scomber scombrus*) [180]. In the western North Atlantic the spawning grounds are in coastal waters from Newfoundland to Chesapeake Bay, where the eggs are most numerous 10 to 30 miles from the shore. The larvae, which are about 3 millimetres long at birth, grow to 10 millimetres in about twenty-six days and to 50 millimetres in an additional forty days, when they assume adult form (figure 66).

South of Cape Cod, the catches of eggs in 1932 indicated that some sixty-four million million eggs were produced by a spawning population composed of one thousand million mackerel. During most of the larval life, the mortality was estimated to be 10 to 14 per cent each day, but was considerably more (30 to 45 per cent each day) in larvae measuring from eight to ten millimetres, at which stage they were rapidly acquiring their fins. By the time the young mackerel were about 50 millimetres long and about to end their planktonic existence, something of the order of one to ten fish had survived from each million of eggs that were laid.

Fishes recruited from the 1932 spawning season were scarce in the adult population of subsequent years. Why had so few survived? A female mackerel lays about five hundred thousand eggs in a year, and she can do this for four years, producing in all about two million eggs. To keep the adult population constant, one egg out of every million should produce a mature fish, but in 1932 only four fish per million eggs were left at the early age of three months, when the rate of mortality was about 10 per cent each day. Given this rate for no more than another thirty-five days, only one fish would survive from ten million eggs.

To know just what the sea and its life is doing from day to day is manifestly impossible, but south of Cape Cod there were two unusual features in 1932. The crops of plankton were poor and the winds were abnormal. Many of the larvae might have perished from starvation. At all events, the 1930- and 1931-year classes of mackerel were good ones, and during these years the plankton was more abundant than in 1932. In the same two years the winds were predominantly

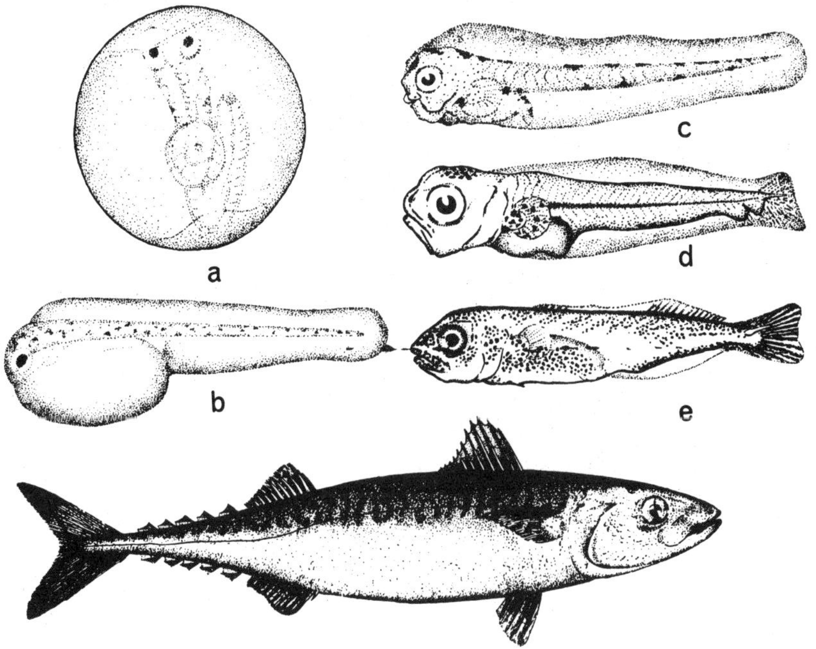

Figure 66. Stages in the life-history of the mackerel, *Scomber scombrus*. (After Bigelow and Schroeder [177].)
(a) egg and embryo, (b) newly hatched larva, (c) larva just after losing yolk sac, (d) a more advanced larva, (e) larva turning to fry.

from the south-west during the period of larval existence, but in 1932 they blew mostly from the north-east and were unusually strong. The sea-drift stirred by the north-easterly winds is likely to have carried the larval mackerel away from the juvenile nursery grounds, which are along the coast of southern New England to the eastern end of Long Island. But south-westerly winds are good for the welfare of young mackerel, blowing them in the sea to productive regions.

Reverting briefly to the winter flounder of the Mystic River Estuary, the total mortality during the larval and juvenile stages was estimated to be 99·98 to 99·9 per cent (one or two survivors from ten thousand newly hatched larvae) [178]. Here the death rate is largely due to tidal currents that remove the larvae from the spawning beds, which are also the nursery grounds. In the Vancouver area sea-going currents also remove many larval herring from benign inshore nursery grounds. But in the North Sea the spawning grounds of the herring are more off-shore, while the juvenile feeding grounds are more inshore. Each local population, it seems, must adapt its breeding habits to local conditions.

It is now time to look more generally at early life histories in the sea. Taking first the fishes that live in the seas over the continental shelves and produce floating eggs, there are many instances.* Fishes that range over the surface waters of the open ocean produce buoyant eggs, though certain flying-fishes attach their eggs to Sargassum weeds or floating debris and even to the feathers of birds. All of the deep-sea fishes that live at mid-water levels lay floating eggs, so far as we know. Considering that the mean depth of the ocean is about 4,000 metres and that the populations of most species are centred above a level of 1,000 metres, this is not altogether surprising. But even the deep-sea angler fishes which live below this level must produce buoyant eggs, for their early larval stages appear in the surface waters [104]. Perhaps all kinds of bathypelgic fishes spawn in the depths, from where their eggs float up to the surface, developing as they rise. At all events, the larval nursery grounds for these fishes is in the surface layers, which contain the greatest stores of planktonic food. Here the larvae drift and feed until they are about to change into fry, when there is a general descent towards the depths of the adult habitat.

Of the fishes that live on or near the deep-sea floor, the rat-tails, deep-sea eels and certain brotulids produce floating eggs. Some brotulids, however, are live-bearers. Virtually nothing is known of the early life history of tripod fishes (Bathypteroidae), spiny-eels (Notacanthidae), halosaurs, eel-pouts (Zoarcidae) and sea-snails

* Clupeidae (herrings, etc., most species), Apodes (eels), Anacanthini (cod-fishes, hake, etc., most species), Serranidae (sea-perches), Mullidae (red-mullets), Sciaenidae (drum-fishes, most species), Sparidae (sea-breams), Carangidae (horse-mackerel, jacks, yellow tails, etc.), Acanthuridae (surgeon-fishes), Scaridae (parrot-fishes), Sphyraenidae (barracudas), Scombridae (mackerel, tunny, etc.), Triglidae (gurnards), Heterosomata (flat-fishes, most species), Plectognathi (trigger-fishes, file-fishes, puffer-fishes, porcupine-fishes and sun-fishes, most species) and Lophiidae (angler-fishes).

(Liparidae). Perhaps the sea-snails, like their shallow water relatives, lay their eggs on fixed objects on the floor. More has been discovered concerning the larval life of the deep-sea eels. *Synaphobranchus kaupi*, the best-known species, is common in the North Atlantic, where its main spawning grounds are in the deep water to the north-east of the West Indies and Florida. The eggs, like those of other eels, hatch into glassy, leaf-like leptocephalus larvae, which are dispersed eastward across the Atlantic, feeding and growing as they drift. After a larval life of eighteen to twenty-two months, metamorphosis occurs and the young glass-eels make for the depths. The European freshwater eel, it will be recalled, has a larval life of about three years. Its Sargasso Sea spawning grounds are to the east of those of *Synaphobranchus kaupi*, and again the larvae are carried across the Atlantic [181].

Most of the fishes that lay non-buoyant, adhesive kinds of eggs tend to live in near-shore waters. Well-known instances are gobies, blennies, sand-eels, gunnels, bullheads, armed bullheads, lump-suckers, cling-fishes and toad-fishes (plates 23 and 26). A good many of these fishes spend all or part of their life between the tide marks. Sand-eels deposit their rather sticky eggs in the sand while empty shells may be used as a nest by gunnels, certain gobies and cling-fishes. Bullheads and lump-suckers leave clumps of eggs, which are commonly found in intertidal waters. The spawn of armed bull-heads may often be discovered adhering to the holdfasts of seaweeds and sea-snails also frequently use seaweeds and hydroids as anchor-ages for their eggs.

Compared to their size, these teleosts produce large eggs, which, not being diluted to buoyancy, carry a concentrated store of yolk (plate 37). The eggs of some species are non-spherical [182]. Gobies' eggs, for instance, range from a spherical to a glove-finger shape through elliptical and pear-drop forms. Some blennies and sand-eels produce slightly elliptical eggs, while cling-fishes eggs are broadly elliptical. As non-buoyant, attached kinds of eggs are relatively large, their producers – and many are small fishes – cannot be so fecund as the species that shed comparatively small, floating eggs. Fishes with non-buoyant eggs lay hundreds or thousands rather than tens or hundreds of thousands, or millions.

In coastal waters the tides are usually higher and the tidal currents stronger than those in the open ocean. Waves search and scour the shallows, keeping in motion long-shore currents and rip currents. Anchored, non-buoyant eggs are suited to such surroundings, but floating eggs, aside from other hazards, would often be cast ashore. Yet an attached egg mass makes excellent food. Perhaps it is primarily to thwart egg-eaters that some species have taken to guarding their eggs, which habit is common among blennies, gobies, lump-suckers, cling-fishes, bullheads and toad-fishes. The protector is nearly always the male, who threatens intruders by fearsome displays and advances, even by grunting if he has the means. Some gobies and cling-fishes also fan water over the eggs, which will help to keep them aerated and free of a coating of silt.

The larvae that hatch from non-buoyant eggs are generally more advanced in development than those that emerge from floating eggs. In a good many species the newly-born larvae bear a small supply of yolk and are nearly ready to find their own food. The yolk-sac stage, when mortality may be severe, is passed within the egg.

Larval winter flounders try to hug the spawning grounds, which may well be true of larval cling-fishes. The larvae of certain gobies also tend to keep near the sea floor, but this sort of behaviour is by no means universal among fishes laying non-buoyant eggs in near-shore waters. One Hawaian blenny, *Istiblennius zebra*, attaches its spherical eggs on the

> . . . underside of rocks or the walls of holes by an adhesive disc. The eggs hatch in about a week; the young *I. zebra* then become part of the plankton or drifting life of the open ocean. Here they grow to about an inch in length and ultimately return to shore as distinct postlarvae. These are glassily transparent fish, with teeth, digestive tracts and fins adapted to a much different way of life from the adult. Apparently most Hawaian blennies have these peculiar postlarval forms, many of which have been described as new genera and species. Upon reaching a suitable adult habitat the postlarval *I. zebra* metamorphose into small juveniles which resemble the adults except for size. Growth to breeding size requires about a year [183].

Some of the North Atlantic blennies also lead an off-shore larval life, as do certain greenlings (Hexagrammidae) in the north-western Pacific. Spurred by natural selection, each species has evolved its own way of maintaining its local populations. Each species has learned to use the sea as best it may.

Arctic and Antarctic coastal waters are near freezing and covered with ice for most of the year. All the resident fishes, so far as we know, lay large yolky eggs on the sea floor. Some species breed during the summer, when the melting of ice produces a surface layer of very low salinity; but this melted water, which might well damage the eggs, is not likely to reach them. If the eggs are not supercooled (p. 318), their being laid on the bottom will keep them from contact with ice crystals which float near the surface and might 'seed' the eggs, so making them freeze to a solid state. Thinking of egg size, perhaps a large and dense mass of yolk, as in certain frogs, enables the eggs to develop at a relatively high rate at very low temperatures. Moreover, the larvae that hatch from these eggs are comparatively large and advanced, which may well be biologically advantageous in near-shore polar waters, where the plankton is limited in quantity and season. For the larger the larva, the more capacious is its mouth and the quicker it can move: it can thus tackle a wider size-range of food organisms and range more widely in search of them.

There may be hundreds or even thousands of fish species in a complex of coral reefs, but very few kinds have been studied so well as the convict surgeon-fish. In northern temperate waters, where there are fewer species and great fisheries, more is known. Here fishery naturalists are beginning to discern how nicely the fecundity and early life histories of fishes enable them to survive in certain physical and biological surroundings [184]. The tunnies, sword-fish and sail-fish shed great quantities of floating eggs in the open Pacific. The incubation period is short and the larvae and fry grow quickly. Mackerel (*Pneumatophorus japonicus*), saury (*Cololabis saira*), sardine (*Sardinops sagax melanosticta*) and anchovy (*Engraulis japonicus*) are plentiful in the eastern Chinese and Yellow Seas, the sea of Japan, and off the shores of Japan. They produce floating eggs, but they spawn in coastal regions, where the currents are slow or form eddies. Their early life history is brief and the fry grow up in near-shore nursery grounds. Yet numerous species, including certain flat-fishes, lay non-buoyant or attached eggs, and the spawning grounds are often in regions with strong coastal currents. Failing more tranquil spawning grounds, such species have taken to grounded eggs to counter too great a dispersion and consequent mortality of their reproductive products. Moreover, fishes of the north-western Pacific tend to be more fecund than their counterparts in the North Atlantic. Evidently they face greater hazards during their early life and have made the necessary adjustments to fit their environment.

Despite the heavy mortality during their early life, marine teleosts have gained their great dominance in the ocean by means of floating or grounded eggs. While vital for the existence of numerous sharks and rays, nurture of the young by the female is confined to relatively few teleosts. Live-bearing has been covered in Chapter 14 (pp. 254–259), but this review of early life histories in the ocean would not be complete without some mention of the teleosts that retain their eggs after they are laid.

There are more kinds of 'mouth-breeding' teleosts in fresh waters than in the sea. The marine species are certain ariid cat-fishes, cardinal fishes (Apogonidae) and jaw-fishes (Opisthognathidae). Marine cat-fishes lay few eggs but they are large. In *Galeichthys felis*, which lives in the coastal waters of America from Cape Cod to Panama, the eggs are nearly 20 millimetres in diameter. After fertilization, the male retrieves the eggs and carries up to fifty-five in his mouth. In about a month the young emerge, but the male continues to shelter them in his mouth for the next two weeks. The eggs of mouth-breeding cardinal fishes vary a great deal in size. In a Mediterranean species, *Apogon imberbis*, the male incubates some twenty thousand eggs, each measuring about ½ millimetre in diameter. After swallowing the spawn, the male houses it in his pharynx, where the eggs are continually milled round so as to keep them properly aerated. Here the eggs stay until they hatch, but the larvae are not retained. Other cardinal fishes produce fewer and larger eggs, as in the Australian species *Apogon conspersus*, the male of which carries about a hundred and fifty eggs, each measuring about 4 millimetres in diameter. Oral incubation is evidently quite common among cardinal fishes and it is always the male, so it seems, who takes charge of the eggs. In those that mature at a small size, such as in species of *Siphamia*, the males have longer and deeper heads than the females.

Pipe-fishes, sea-horses and ghost pipe-fishes (Solenostomidae) have evolved special means of retaining the fertilized eggs. Unlike the members of the first two groups, the female ghost-fish takes charge of the spawn. She has an open brood pouch formed by the enlarged pelvic fins, the inner margins of which are fused to the wall of the abdomen. The eggs are fastened to short filaments that project from the inner surface of the pouch, but how this comes to be is not known.

During mating in pipe-fishes and sea-horses, the female passes the eggs to the male. The males of some pipe-fishes simply attach the eggs to the underside of the trunk or tail, but in other species and in sea-horses there is a definite brood pouch. In pipe-fishes of the genus *Syngnathus*, etc. the male has a pouch formed by two folds of skin that extend along the underside of the tail (figure 52). After he receives the eggs, these two folds come together to make a pouch with an opening at the front. The male of a common, north-eastern Atlantic, pipe-fish, *Siphostoma typhle*, may even secure the eggs of several

22. The Cornish sucker-fish (*Lepadogaster lepadogaster*) clinging to a rock on the sea shore.
23. Eggs of the Cornish sucker-fish attached to a rock on the shore.

24. Cod (*Gadus morhua*), the top two fishes, and pollack (*Pollachius*), the three bottom fishes.
25. Plaice (*Pleuronectes platessa*).

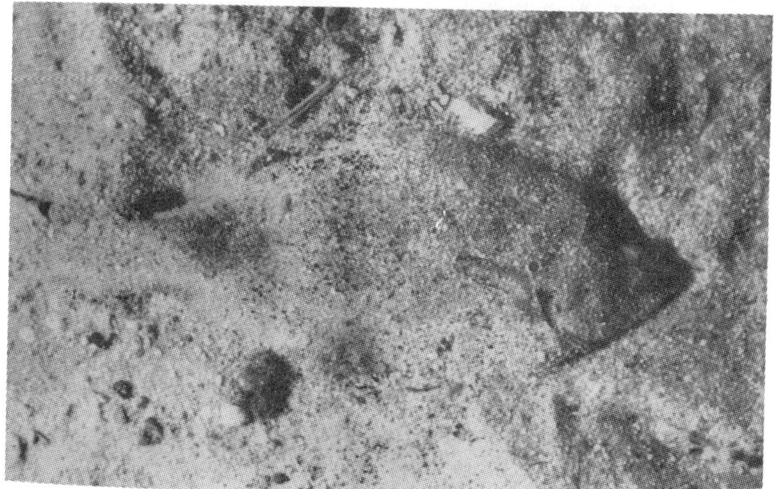

26. Nest of a toad fish (*Opsanus tau*) on a piece of drift-wood. The embryos are well developed (see also Plate 37).
27. Winter flounder (*Pseudopleuronectes americanus*) partly covered by sand.

28. Surgeon-fishes swimming over coral in an East African reef.
29. Bermudan coral fishes. Beside the angel-fish, sergeant-major, and the parrot-fish, seen in colour on plate 20, there are a squirrel-fish (lower right foreground), above which is swimming a yellow grunt (*Haemulon flavolineatus*). To the left of the angel-fish is a blue-head wrasse (*Thalassoma bifasciatum*).

30. A common European goby (*Gobius paganellus*).
31. A common European blenny (*Blennius gattorugine*) in an empty scallop shell.

32. Eggs of a moray-eel in the plankton.

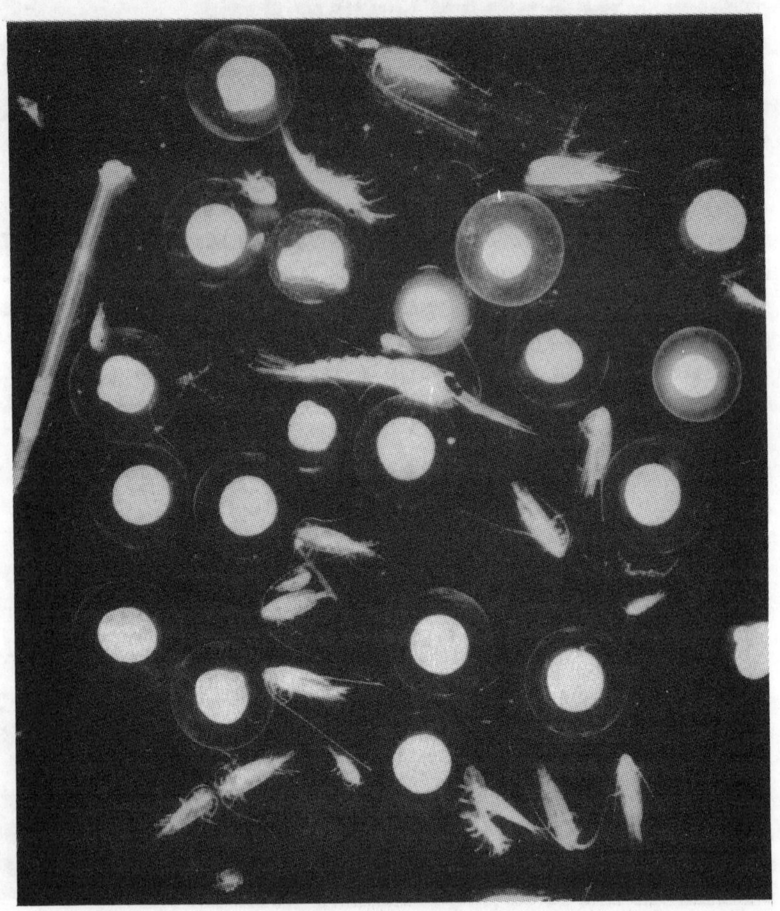

33. Atlantic moray-eels (*Muraena helena*). The head of a conger-eel is emerging from a hole in the rock.

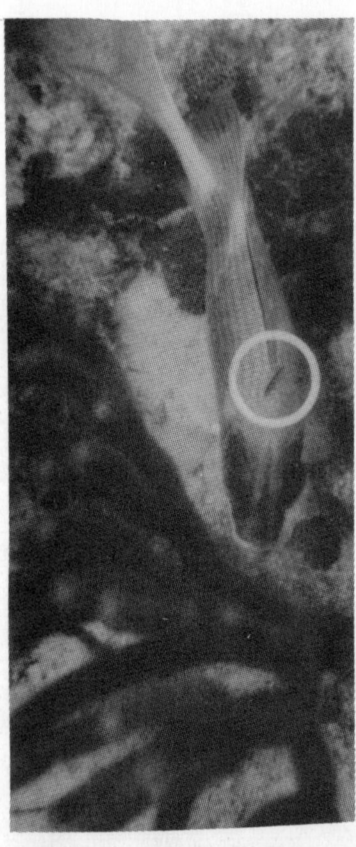

34. A neon-goby (*Elecatinus oceanops*), which is ringed in white, on the head of a squirrel-fish (*Holocentrus ascensionis*). The goby cleans other fishes.

35. Head of a shark-sucker (*Echeneis*) showing the suction-plate.

36. Grunion
(*Leuresthes tenuis*)
coming ashore to spawn.

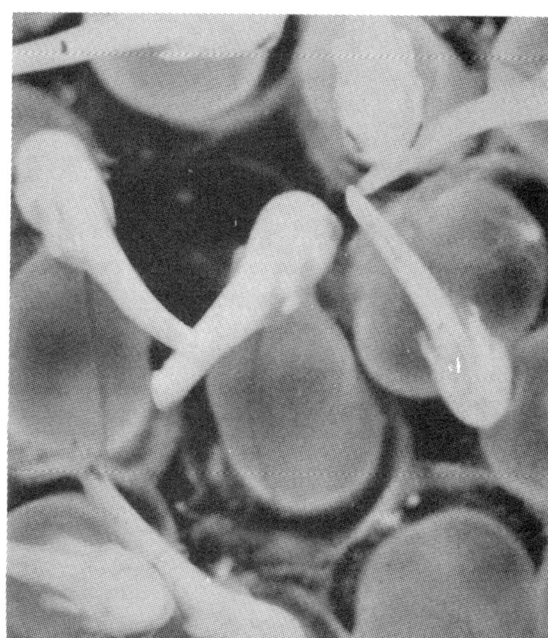

37. Embryo toad-fishes
(*Opsanus tau*) length
about 6 millimetres.
Note the large yolk-sac.

38. A trumpet-fish (*Aulostomus* sp.).

39. Worm pipe-fishes (*Nerophis lumbriciformis*) found exposed at low tide under a flat stone.

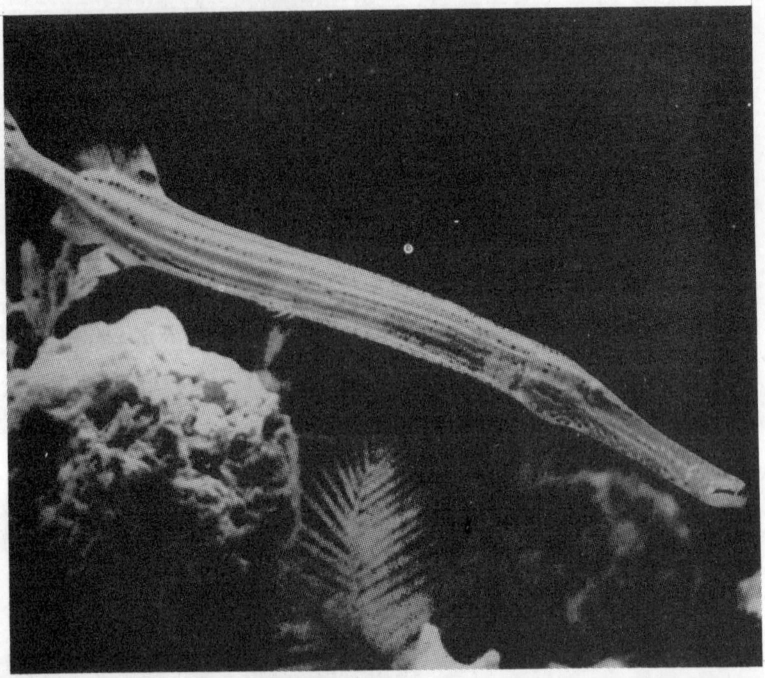

40. Sea-horse (*Hippocampus hudsonius*) among sea-weed.
41. Boar-fish (*Capros aper*).

42. A school of mackerel (*Scomber scombrus*).
43. A turning school of grey mullet (*Mugil labrosus*).

females if there is room in his pouch. As he takes charge, the male fertilizes the eggs, which start to develop. Meanwhile, the lining of the pouch thickens and becomes charged with blood vessels. Folds and processes of this lining, which carry a network of blood capillaries, grow between the eggs, so conveying oxygen to the developing embryos. These outgrowths also produce a nutrient secretion. When the young are fully formed, which may require three weeks, the pouch opens and the male gives birth to his brood. After this, he sheds the placenta-like lining of the pouch. In all essentials the young are like their parents and they begin to feed soon after emergence. For a while they stay near the male and, if need be, are said to take shelter in his pouch.

Male sea-horses have the brood pouch under the tail and it is entirely closed except for an opening opposite the dorsal fin. The smallest sea-horse, *Hippocampus zosterae*, which lives off eastern Florida and in the Gulf of Mexico, breeds during nine months of the year. The brood pouch holds fifty-five eggs at most. Two days after delivering the brood, the male is ready to receive another clutch of eggs. The young grow quickly to maturity in two or three months, the entire life span being no more than a year [118].

There is a placenta-like lining in the brood pouch of male seahorses, which must have the same functions as that of a pipe-fish. When the young are ready to emerge, the male sea-horse expels them by flexing his body. Newly born sea-horses are miniatures of their parents, and, like young pipe-fishes, are ready to make their own living. One of their first acts is to swim to the surface, where they swallow air to fill the developing swimbladder. When this organ is well established, they acquire the balance and precision of their kind.

There is one last aspect. Eggs that develop in a female or in a brood pouch are not exposed to the light. This may seem obvious, but there is growing evidence that ultra-violet light or visible blue light can harm fish eggs [185]. Hatchery experts know that direct sunlight may kill salmon and trout eggs in a few minutes. The eggs of plaice hatch most successfully in shaded conditions, but the best yields from cod eggs are obtained in lighted surroundings. Indeed, the eggs of some fishes, such as the gold-fish, will not hatch in total darkness. Clearly, one kind of egg responds differently from the next. But it may well be that too great an exposure to blue light is harmful, or even lethal, to the fertilized eggs and developing embryos of fishes.

If this is so, we can begin to appreciate the very different optical properties of floating and grounded eggs. Floating eggs, which have a crystalline transparency, may often be found near the sea surface, particularly in calm conditions. If it is also sunny, such eggs will be exposed to bright light, which could kill those that are closest to the surface. Those that are not so near may well escape injury because of their transparency. Most of the harmful rays will pass right through them, which would also apply to the larvae. Moreover, transparent eggs and larvae are presumably not easily detected by predators with eyes.

Grounded eggs are much more opaque, frequently containing yellow or orange pigments, and they tend to be laid in shaded surroundings, under stones or shells, or in seaweeds and so forth. In shallow intertidal reaches, particularly when the tide is out, such eggs may need all the protection they can get from strong sunlight. After all, they are in one place, not spread out in depth, as are the more numerous, buoyant kinds of eggs. Tough, opaque membranes and yellowish colour filters will help to shelter the developing fish from injury by the rays at the blue end of the spectrum [185].

Besides the hazards considered earlier in this chapter (p. 270), fish eggs and larvae may thus face injury from the blue and ultraviolet rays of sunlight. Much is still uncertain, but if present indications are sound, it will be fascinating to look more closely at the means by which eggs and young fishes are protected from the sun. Our understanding of the adaptability of fishes, which is so severely tested during their early life, will thereby be enlarged.

Chapter 16

The Early Life of Freshwater Fishes

IN THE open ocean most kinds of fishes lay floating eggs. In coastal seas, especially in near-shore waters, many species have taken to non-buoyant eggs. Such eggs are produced by nearly all of the fishes that live in estuaries, where sea and land waters meet. On moving inland to true fresh waters, we find that non-buoyant eggs are almost the invariable rule.

In the preceding chapter we saw how the fecundity and early life of marine fishes are fitted to their physical and biological surroundings. Before trying to appreciate such adaptability in freshwater fishes, it will be better to look rather closely at the early life of certain selected species. These are the brown trout (*Salmo trutta*) [186], the bullhead or miller's thumb (*Cottus gobio*) [187] and the perch (*Perca fluviatilis*) [187].

The brown trout ranges from Iceland and the northern coasts of Europe to Mediterranean regions. The silvery sea-going form extends from Iceland to the Bay of Biscay, and there are other races of migratory trout in the Black Sea, the Caspian Sea and Aral Sea. In the British Isles the brown trout breeds from October to February. Not long after reaching the spawning grounds, the females, which arrive before the males, select a site and hold their ground. When cutting the redd where the eggs will be laid, the female excavates the gravel by using her tail as a besom. She is now attended by the male, who drives away small male rivals or threatens those that are bulkier than himself. After the eggs are fertilized, they are covered with gravel. For every pound of her weight, the female lays from five hundred to twelve hundred orange-yellow eggs, which have tough membranes and measure from four to five and a half millimetres in diameter. They rest under briskly flowing waters, where currents run clear through the gravel. During the long incubation period, usually

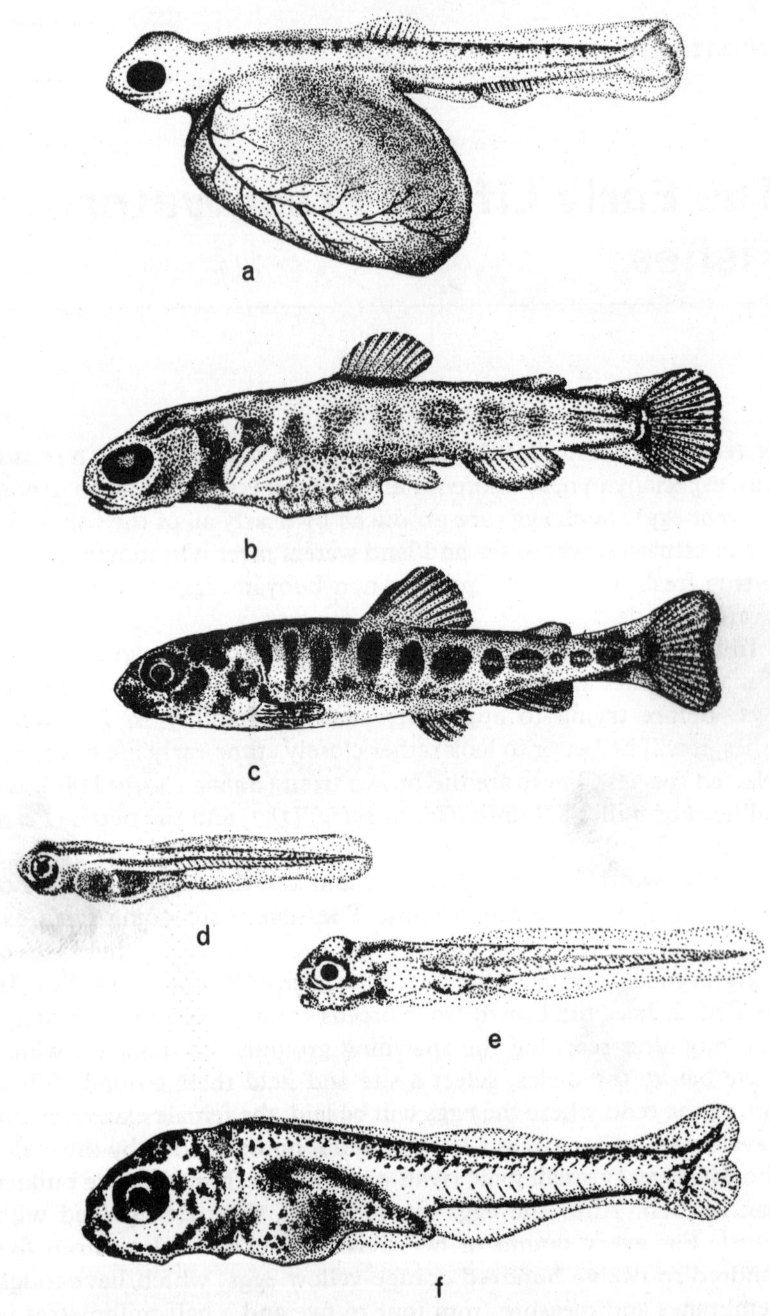

Figure 67. Stages in the life history of brown trout (a–c) and perch (d–f). The newly hatched trout (a) has a large yolk-sac which has almost disappeared in (b). The newly-hatched perch (d) has a small yolk-sac. (After Otto Schindler.)

two to three months, the eggs must be properly aerated. Moreover, as trout return to spawn in their natal waters, the more perfect the redds the more should be the returns. The best redds are in reaches where their stony covering is not likely to be scoured and scattered by winter spates.

The new-born larvae carry a large, globular yolk sac and are slightly more than a centimetre in length (figure 67). Three to four days after hatching, during which time the yolk sac acquires an elongated and compressed form, the young rest quietly in the gravel. Having stream-lined the yolk sac, the larvae now wriggle into the gravel away from the redds. If they become exposed to light, they immediately burrow into the bottom, an action that persists while the yolk is being absorbed. After three weeks, or even three months if the weather is very cold, the fry begin to emerge from the redds. Now they make for the light and they also display a marked tendency to head into currents. The fry have the elegant adult form and are fully finned (figure 67). For a few days after emergence, they are unable to swim in mid-water, but they acquire this facility when the swimbladder is functional. Their food consists largely of small insect larvae, small worms and crustaceans. Those that feed well get their parr markings in a few months, and in three years, perhaps, some are ready to spawn.

While developing the shape, the power and the jaws to make their own living, larval trout are thus well sheltered from predators. Water trickling through the stones brings them oxygen, but it may also bring silt. Now eggs that become smothered by fine particles of silt are not likely to hatch, yet newly hatched larvae throw off silt by fanning their pectoral fins and flexing the tail. Moreover, in keeping clear water around them, the mass activity of the larvae will also keep silt from falling on some of the unhatched eggs.

Bullheads live in cool, hurrying streams, where they keep close to the gravel or lurk behind stones. Except for Spain and Greece, they are found over most parts of Europe. Their average length is from 3 to 4 inches and they spawn during the spring. The eggs (diameter 2·0 to 2·5 millimetres) adhere in pinkish clusters. They are generally laid in a nest that is dug by the male, often under stones. He also guards the spawn, which consists of a few hundred eggs. During the three or four weeks before hatching, the male fans the spawn with his pectoral fins, so keeping the eggs free from silt, which is caught by the eddies that curl behind stones and where it tends to accumulate.

The newly emerged larvae bear a globular yolk sac and are about

6 millimetres in length. While absorbing their yolk reserves, which takes ten days or more, they spend most of their time resting between stones. At the end of this time they have large pectoral fins and are nearly 10 millimetres in length. Now they begin to feed along the bottom on small invertebrate animals. Unlike young trout, they need not swim to the surface to pass air into the developing swimbladder, which is absent in bullheads. They stay, if they can, close to the gravel, where the currents are slowest.

The perch ranges over Europe and much of Russia. According to A. F. Magri MacMahon; 'In rivers it favours spots where the water flows slowly and where banks of weeds, snags, bridge piles, tangles of roots and the like give it suitable hiding places from which to dart on shoals of fry or small fish.' [55] Perch become sexually mature at the age of two or three years, when they spawn in the shallows during the spring. Eggs (2·0 to 2·5 millimetres in diameter) are produced in the proportion of a hundred thousand per pound of weight. They are laid in strings, which adhere to plants, stones and water-logged drift wood. At an ambient temperature of 10°C the eggs hatch in eighteen days, a time which is halved at 20°C. The newly emerged larvae (length about 6 millimetres) bear rather small yolk reserves (figure 67). Unlike larval trout, they are not averse to the light and wriggle up to the surface, where they move up and down, alternately swimming upwards and sinking. After swallowing air to fill the developing swimbladder, they begin to swim in the horizontal plane. When the yolk is gone they feed on protozoans and other small organisms. In a fortnight or so the spines of the dorsal and anal fins appear and the body becomes deeper and broader. Not long after, the fry are found in schools in the shallows, where they hunt for small insect larvae and crustaceans. They grow quickly to a length of 3 to 4 inches in the course of a year.

Trout bury their spawn; bullheads guard and tend it; perch leave it exposed to predators, but they are much more fecund than the other two species. During the yolk-sac stage, young trout stay in the gravel, while bullheads shelter among stones. Larval perch, however, climb to the surface and are dispersed away from their birthplace, but they are born in gently flowing waters. There might well be mass starvation if thousands of larvae were clustered in one small part of the stream. To end this introductory section, we may simply note how well the early life of these fishes is adapted to their surroundings. Breeding grounds, spawning habits, egg disposal and larval behaviour are intimately interrelated, and are directed, as it were, to one end: the survival of enough individuals to maintain the local populations.

Looking more generally at the early life of freshwater fishes, the largest group, which comprises species like the perch, take no care of their spawn, but this consists of great many eggs. Most of the cyprinids and characins, some cat-fishes and pike are included here. Many of these fishes simply scatter their eggs over the bottom or among weeds. Most European cyprinid fishes, for instance, lay their eggs in vegetation. But a good many cyprinids and characins produce non-adhesive eggs that fall to the bottom. While the yolk-sac is being absorbed the larvae rest on the river bed or attach themselves to plants. Larval carp and pike, for instance, have adhesive organs on the head. When the swimbladder has begun to function, the larvae swim with ease in the horizontal plane. Before this time they are heavier than water and swim in an awkward way. Unlike the larvae of marine fishes that hatch from floating eggs, the larvae of freshwater fishes are not equipped with buoyancy chambers.

Egg buriers

We have seen that salmon and trout bury their eggs in gravel. So do their relatives, such as grayling (*Thymallus*) and char (*Salvelinus*). The European barbel (*Barbus barbus*) and certain North American darters such as *Percina caprodes*, *Etheostoma caeruleum* and *E. spectabile* also spawn in gravel or sand. All of these fishes lay relatively few, large eggs in briskly flowing streams. The fitness of such disposal of the spawn in these habitats has already been discussed.

The annual fishes, which are small top-minnows belonging to genera such as *Aphyosemion* and *Cynolebias*, bury their eggs to conform to a much harsher environment. They live in parts of South America and Africa, where rainy seasons alternate with hot, dry ones, when nothing is left of some streams but caked tracks of mud. Before this happens these fishes mate and bury their spawn in the mud. When the rains come again, the tough-coated eggs are not long in hatching, a process which is said to require the presence of certain bacteria. At all events, aquarists have found that the eggs of *Aphyosemion arnoldi* can be made to hatch by adding a pinch of powdered milk to the water.

Nest makers

Diverse freshwater fishes lay their eggs in some kind of nest. Sticklebacks, which elaborate a weedy nest (figure 57), have already been discussed. An electric fish, the mormyroid *Gymnarchus niloticus*, makes a floating, sac-like nest of grasses. This nest holds about a thousand eggs (diameter 10 millimetres), which hatch in about five days. J. A. Budgett describes the sequel as follows:

> Immediately after hatching, the larvae commence their characteristic movements, throwing the head and the fore part of the body from side to side incessantly. The larvae are at first so small in proportion to the yolk sac that they are quite unable to move it. By this constant movement the larvae tend towards the surface, and the weight of the yolk tending downwards, the yolk sac becomes gradually drawn out. . . . About three days after hatching, the larvae are strong enough by their movements to raise the yolk sac off the bottom of the nest for a moment, but it is quickly drawn back by its weight.
>
> By the tenth day after hatching, the larvae are able to drag their yolk sac to the surface of the water, where they take in a gulp of air into their lung-like swimbladder and fall again to the bottom, on reaching which they start again for the surface with unceasing regularity, so that when looked at from above the nest of *Gymnarchus* with its swarm of scarlet-bearded yolk-hampered larvae (figure 68) presents a most amazing spectacle [188].

The 'scarlet-beards' are the two long bunches of gill filaments on each side of the head, which are lost by the time the young leave the nest and the yolk reserves have been exhausted. They now measure 3 inches in length and are adult in form.

The larvae of African lung-fishes (*Protopterus*) also bear external gills (figure 68). The eggs are laid in depressions among the reeds and are guarded by the male fish. *Lepidosiren paradoxus*, the South American lung-fish, spawns in deep holes. Once more, the larvae develop feathery external gills (figure 68) and the nest is guarded by the male, whose pelvic fins (and sometimes the pectorals) bear temporary gills. By growing tufts of respiratory filaments on these fins the male need not leave its charge and swim to the surface to gulp air, which is its normal means of respiration. Like the larvae with temporary external gills, the male *Lepidosiren* is endowed with a large gill surface for capturing the limited amounts of oxygen in swampy waters. Indeed, young African and South American lung-fish do not begin to breathe air until they have almost lost their external gills, which is after they have left the nest.

The American bowfin and the African osteoglossid (*Clupisudis niloticus*) clear a nesting site among weeds by flailing the tail. Larval *Clupisudis* have external gills, as do the young of African bichirs (figure 68). These fishes also converge in bearing cement glands on the snout, the secretions of which enable the larvae to adhere to weeds etc. Cement organs are also found on the larvae of the bowfin, the South American and African lung-fishes, sturgeons and diverse teleosts. The primary use of these glands can hardly be to anchor the

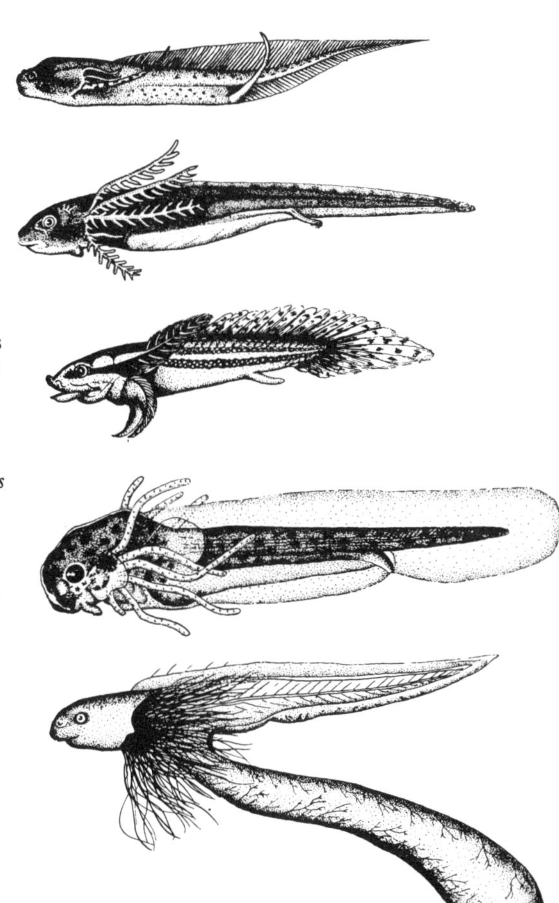

Figure 68. Young fishes with external gills. Top to bottom: African lung-fish; South American lung-fish; bichir; loach (*Misgurnus fossilis*); and the mormyroid fish *Gymnarchus niloticus*. (After G. V. Nikolsky.)

larvae in the face of currents, for they occur in fishes that spawn in still waters. But anchored larvae can rest while consuming their yolk, which is a relatively heavy mass to support. Moreover, by keeping still, such larvae are less likely to be noticed by their enemies. More often than not, it is by moving that fishes give their lives to predators.

Some fishes make a nest on the bottom. The sun-fishes and basses of North American fresh waters, which form the percoid family Centrarchidae, generally construct their nests on gravel or sand [189]. By fanning his fins and using the tail member as a besom, the male clears a circular depression. Stones and other large objects are removed by his jaws. When he is ready, the male finds a mate, who lays her slightly adhesive eggs in the nest and then departs. Now the male guards the eggs, chasing away intruders. He also fans the eggs, thereby keeping them from being smothered in silt. Care of the young is most marked in the basses. To take one instance, the recently hatched larvae of the small-mouthed black bass (*Micropterus dolomieu*) lie for a time among the stones of the nest. After they leave the nest, the male starts to herd the larvae, which are described as weaving about in a dense school.

Though not making nests for their eggs, which are stuck on hard surfaces, some cichlids, such as the jewel-fish (*Hemichromis bimaculatus*), *Aequidens latifrons* and certain species of *Cichlasoma*, dig nursery pits for the young [190]. As the larvae hatch they are inhaled by the parent and blown into the pit. For a few days after hatching, the larvae, which have cement glands, cling to the substrate or plants and by massed wriggling movements they cause water to flow through and aerate the pit. Next, they leave their anchorages and move on over the floor of the pit. Before long, they are strong enough to swim freely, when they leave the pit and develop the habit of dogging their parents. The young of each species acquire an individual colour pattern, which is evidently recognized by the parents. After a week or more, the parents are no longer shadowed by their offspring, who go their own way in a school, feeding and growing together until they become mature.

Certain freshwater fishes which live in still and stagnant waters of tropical regions suspend their spawn at the surface in a nest of bubbles. Most of these belong to the anabantoid group of spiny-finned fishes, which breathe air through a labyrinthine organ in each gill chamber (p. 96). The Siamese fighting-fish (*Betta splendens*), the croaking gourami (*Trichopsis vittatus*), the thick-lipped gourami (*Colisa labiosa*), the dwarf gourami (*Colisa lalia*) and the paradise-fish (*Macropodus opercularis*) belong to this group of bubble-nest blowers. The first and last species lay non-buoyant eggs, but the eggs of gouramis are buoyant. Certain air-breathing, armed cat-fishes (*Callichthys* and *Hoplosternum*) have also taken to bubble nests. So has an African characin (*Hydrocyanoides odoe*). Suspended among bubbles of air, the eggs obtain adequate oxygen for respiration.

Moreover, in swampy tropical waters the surface film contains the only sure supply of oxygen. In this connexion it is significant that two anabantoid fishes, the climbing perch (*Anabas*) and the kissing gourami (*Helostoma temmincki*) do not make a bubble nest. They produce large, lighter-than-water eggs that float to the surface as soon as they are laid.

Other air-breathing fishes, the snake heads (*Ophicephalus*), also lay such eggs. A widely distributed species, *O. striatus*, which ranges from China to India and Ceylon, sheds eggs (diameter 1·25 to 1·5 millimetres) containing a large oil globule. The eggs float flush with the surface and are guarded by the male fish.

The bubble nests of the anabantoid fishes and the armoured cat-fish (*Callichthys callichthys*) are the work of the males. The bubbles are coated with a mucous secretion from the mouth. Having made the nest, which consists of thousands of cohering bubbles, the male seeks a mate and entices her close to the raft. The eggs fall or float from the female as they are fertilized. The male retrieves them, spits them into the nest, and then stands guard. If a non-buoyant egg should fall from the nest, it is caught by the male and replaced. He may also repair the nest by blowing more bubbles. After a few days the larvae emerge and these too may be blown back into the nest should they become dislodged. But three or four days after hatching, they are strong enough to go their own way.

Oral incubators

Like certain marine species, some freshwater fishes use their mouth and pharynx as a brood chamber. An African cichlid (*Tilapia mossambica*), may be taken as a model [190] (other mouth-breeding cichlids include other species of *Tilapia* and numerous species of *Haplochromis*). The eggs, which are laid in batches containing twenty to fifty, are fertilized by the male and promptly mouthed by the female. Having acquired her brood, the female seeks shelter for the next ten to twelve days. She rests quietly, alternately respiring and 'gargling' the eggs, which soon hatch. If the eggs are taken from her they cannot be reared: they become mouldy and die. At the end of her time of retreat, the larvae begin to leave her mouth, having used up their reserves of yolk. On emergence, they swarm round the female's head, particularly seeking her mouth, eyes and the dark recesses of the paired fins. When they become stronger, the young form a school and follow their mother, who calls them to her mouth by backing movements after any undue disturbance. After four to five days the female is no longer prepared to shelter the young in her mouth, but they are now active enough to fend for themselves.

While the Siamese fighting-fish (*Betta splendens*) blows a bubble nest, an Indian relative (*B. brederi*) is an oral incubator. As the eggs fall from the female they are caught by the male in his anal fin and fertilized. The female now retrieves the eggs and spits them at her mate, who catches them in his mouth. Having received the entire clutch, he incubates it for four or five days.

Lastly, some of the Tachysuridae, a family of Indo-Pacific cat-fishes which live in salt and freshwater habitats, practise oral incubation. Again, the eggs are incubated by the male, who has a longer head and more capacious oral cavity than his mate. According to H. M. Smith,

> During the protracted period of hatching, which may cover six to eight weeks, the male takes no food, and his fasting is further prolonged by the retention of the young in his mouth until the complete or partial absorption of the yolk sac. By the time the young have left the shelter of the parental mouth, the male undergoes considerable emaciation, and it may be assumed that in fishes so ravenous and gluttonous as the cat-fishes the spawning season is a time of great stress for the males [166].

Egg carriers

Certain cat-fishes of tropical America have evolved other means of looking after their eggs. The males of some species of mailed cat-fishes (Loricariidae) carry the eggs in folds of skin near the lips. The bunocephalid cat-fishes (*Aspredo* and *Bunocephalus*) have gone to more remarkable lengths. After the eggs have been shed and fertilized, the female lies on them, when they sink into the soft and spongy skin of her underparts. In this way the lower part of the head and abdomen become covered with eggs. Each egg is eventually carried on a stalked cup that grows from the skin. Lastly, in one genus of perch-like fishes (*Kurtus*), found in freshwater and brackish habitats in the Indo-Australian region, the males bear a bony hook on the forehead. The hook is used to carry the eggs, which are usually in two bunches and connected by a fibrous thread.

Live-bearing species having been considered in Chapter 14, this review may be ended with the early life of the splashing tetra (*Copeina arnoldi*), a characin that ranges from Venezuela to the Amazon Basin. Male and female leap out of the water together and deposit the spawn on an overhanging leaf or stone. Until they hatch, which takes two or three days, the male splashes water on the eggs. When the larvae emerge, they slip into the water and swim away.

We may now see the contrasts between the early life histories of marine and freshwater fishes. The outstanding difference, which has been touched on before, is that nearly all freshwater fishes lay non-buoyant eggs. Except in near-shore waters, floating eggs are common to most marine fishes.

Before they mature, the eggs of marine teleosts are made buoyant through the addition of dilute fluids that are lighter than sea water. Sea water is, of course, denser than fresh water, which means that extra means, such as a high content of oil, must be found to make eggs float in a river or lake. This is biologically feasible, for we have seen that gourami eggs, which contain a large oil globule, float to the surface as soon as they are shed. Certain cyprinids, such as the grass-carp (*Ctenopharyngodon idellus*) and the ziege (*Pelecus cultratus*) also lay buoyant eggs. It would seem, though, that non-buoyant eggs are suited to most freshwater habitats.

The ocean is permanent and vast, forming about 99 per cent of the hydrosphere. For the most part, sea surface currents are slow, drifting a few miles in a day. The Gulf Stream and Kuroshio Currents, which may flow at a few miles an hour, are the exceptions. By far the greater part of the earth's fresh waters are frozen, which means that inland waters form much less than 1 per cent of the hydrosphere. Most of the world's rivers eventually reach the ocean, while all but a few lakes may be regarded as temporary places for water on its way to the sea. Mountain streams in spate run at a velocity of 5 miles per hour or more, but most reaches of rivers flow a mile or less an hour. It is understandable that fishes spawning in quickly flowing streams must either bury their eggs or secure them firmly in some way. Buoyant eggs are out of place too in more gently flowing waters. Let us suppose that a long stretch of river moves at a mean velocity of half a mile in an hour and that the time from spawning until the time when the young are able to face currents is ten days. During this period buoyant eggs and larvae would have drifted a hundred and twenty miles from the spawning grounds. On the other hand, to maintain local populations there must be some dispersal of the young, otherwise swarms of larvae would be seeking limited supplies of food in restricted spaces. Too great a dispersal would not only leave few recruits for local stocks, but is likely to increase the mortality rate of the young. Many would drift into waters that are deficient in food of the right kind.

By producing non-buoyant eggs – and many kinds are adhesive – the larvae of freshwater fishes are born on or not far from the spawning grounds. Moreover, the newly hatched larvae, unlike those of many marine fishes, are also heavier than water. While absorbing their heavy reserves of yolk, they stay in the gravel, rest on the bottom, cling to weeds, and so forth. And the spawning beds are often in the shallows, where the water runs slowly. During rainy tropical seasons, these shallows are extended, as it were, over the surrounding countryside. Fishes range over the flooded plains and spawn. The young grow quickly and make their way back to the rivers as the floods recede.

We saw that most freshwater fishes lay many small eggs that are left to take their chance. These species are like the marine teleosts that produce myriads of floating eggs. By and large, the freshwater species tend to lay somewhat larger eggs than their marine counterparts, which means that they are not so fecund as marine species of the same size and shape. But local populations of marine species occupy a greater living space than do those of most freshwater fishes. There is, so to say, more ground to be covered.

Like their marine relatives, freshwater fishes that exercise some form of parental care lay relatively few, large eggs. Perhaps the greatest danger to large, adhesive eggs is not the lack of oxygen, at any rate in moving waters, but rather the ever-present hazard of their becoming covered in silt. By fanning and rubbing the eggs, they can be kept clean. Considering the entire fish fauna, parental care is commoner in fresh waters than in the sea. If the comparison is made between the faunas of inshore marine waters and fresh waters, the former may well contain a greater proportion of fishes that tend their spawn.

To move from the open ocean to inland waters is thus to pass from a world where most fishes start life as floating eggs to one where the young, with very few exceptions, emerge from grounded eggs. Inshore waters may be seen as a transitional region. There is much to be done if we are to appreciate the biological implications of these two seemingly simple statements. To comprehend these aspects will be to come close to the very life of fishes.

Living Spaces

Chapter 17

The Deep Ocean

THE DEEP ocean begins to unfold at the edge of the continental shelf. Here, at depths between 100 and 150 metres, the gentle gradients of the shelf give way to the steeper declivities of the continental slope. These reaches of the deep-sea floor form the largest unbroken escarpments on the face of the earth; they may descend for a hundred miles before the abyssal floor of the ocean is attained, where the sediments lie at depths of 4,000 metres or more.

Sea water is some two thousand seven hundred times more abundant on the earth than impounded fresh water. This may be appreciated when we realize that the deep ocean covers nearly two-thirds of the earth and has a mean depth of about 4,000 metres. But no more than one in seven of the fifteen thousand or so kinds of marine fishes live in the immense water masses beyond the land-fringing seas. The more diverse fishes of shelf waters will be considered in the next chapter. Here we turn first to the surface-dwelling fishes of the open ocean, then to those that live in the deeps.

Fishes of the turbulent upper waters

In the tropical and subtropical oceans, the surface, sun-warmed waters move over underlying cooler waters. A thin transition layer of rapidly diminishing temperatures marks the passage from surface to lower waters. This thermocline, which lies at depths between 20 and 150 metres, tends to persist throughout the year. Its persistence profoundly influences the productivity of the warm ocean and thus the abundance of fishes; for it damps out mixing between waters above and below its level. Nutrient salts, particularly phosphates and nitrates, and growth factors such as vitamin B 12, which are required for the vigorous flowering of the surface-dwelling plants, tend to accumulate below the thermocline. Dead animals and plants sink, and most of the processes of decay, when phosphates and nitrates

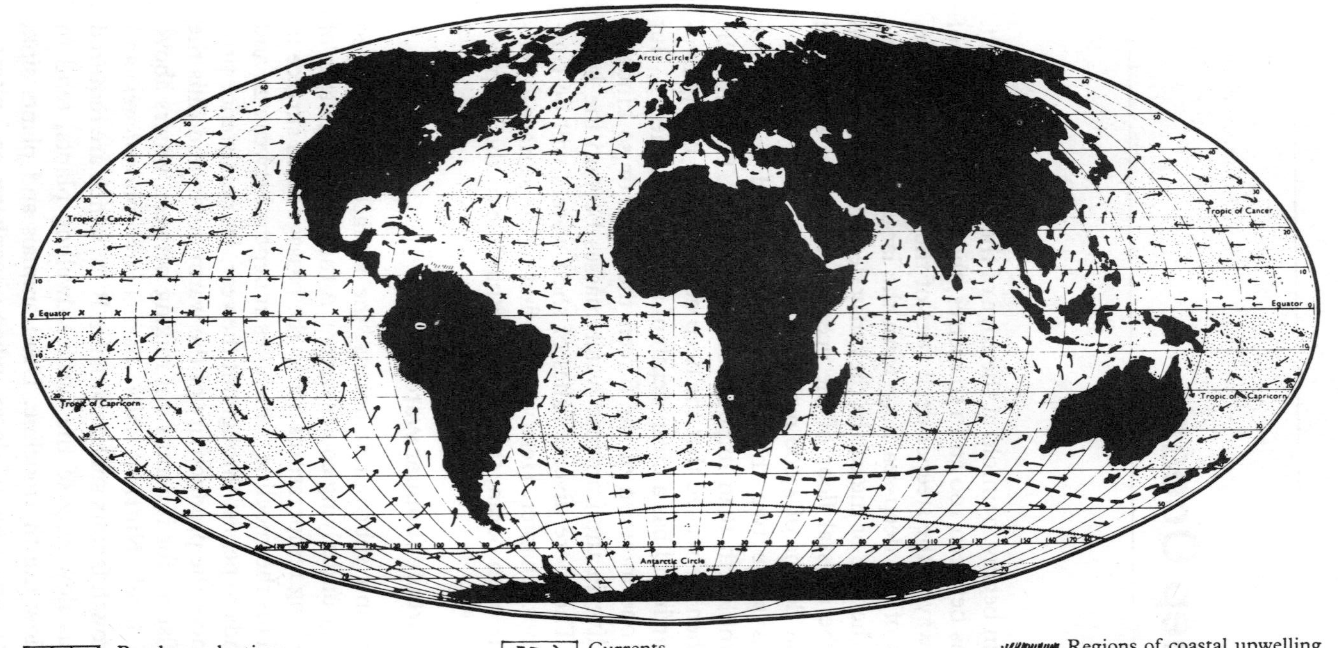

Figure 69. The circulation and productivity of the ocean. The most barren areas, shown dotted, are in the central water masses to the north and south of the equator. In equatorial regions divergences between currents lead to upwelling and an increase in plant production. Upwelling and great productivity also occur off America and Africa and Arabia, shown cross-hatched. There is also an upward circulation of nutrient-rich water South of the Antarctic Convergence. To the north of this is shown the Subtropical Convergence.

are released by bacterial activity, evidently take place below the thermocline. During the growing season, which is more or less continuous in the tropics – though there may be times when flowering is more vigorous – the productivity of the oceanic phytoplankton depends on the supplies of available nutrients over a given period of time. If the thermocline is breached and water wells up from below, the surface, plant-producing waters will be refreshed with nutrients. Indeed, the productivity of any part of the ocean is directly linked with such upward movements of water.

In the equatorial regions of all three oceans there are latitudinal divergences between the currents. The Atlantic and Pacific divergences are near the equator and along a line about latitude 10° north. During the north-east monsoon from October to April, the lines of current divergence across the Indian Ocean are along the equator and latitude 8° south. After the onset of the south-west monsoon, which blows from April to October, they are more to the north. These divergences are between the westerly equatorial currents and the countercurrent systems, the salient point being that there is an upwelling of nutrient-rich water below the cleavage. Except for any oceanic effects of nutrient renewal in certain coastal waters (p. 328), such upwelling makes these equatorial regions the most productive in the tropical and subtropical ocean (figure 69). The phytoplankton flourishes, providing food for swarms of herbivorous planktonic animals such as copepods, which directly or indirectly support the life of fishes. Copepods, for instance, are taken by flying-fishes, which are the prey of dolphin-fish (*Coryphaena*), snake-mackerel (*Gempylus*) and other predatory fishes.

North and south of the equatorial current systems, extending to latitudes of about 40°, are the great central water masses which circulate in immense, slowly-moving eddies. Mixing between the layers above and below the thermocline is so limited that these central waters are the least productive of life in the warm ocean (figure 69).

Let us now look at the linkage between this great pattern of oceanic productivity and the abundance of fishes in the tropics. The Pacific equatorial belt and the central waters of the north Pacific have been well explored. Centred about the equatorial divergence, particularly in the easterly half of the ocean, there are large populations of yellow-fin tunny (*Neothunnus*) and skip-jacks, together with abounding wahoo, dolphin-fish, marlin and flying-fishes. Oceanic sharks, such as the white-tip (*Carcharhinus longimanus*) and the silky-shark (*Eulamia floridanus*) are also commoner in the easterly equatorial belt. The same is true of the blue-shark, which is not, however, confined to tropical waters. Catches of white-tips and silky-sharks are low in the food-poor central water masses between latitudes 10° and 30° North, as are those of blue sharks (figure 69).

The catches of Japanese tunny boats give ample proof that there are corresponding aggregations of tunny and other pelagic fishes in the equatorial regions of the Atlantic and Indian Oceans. During the winter months in the Indian Ocean, for instance, yellow-fin tunny are concentrated about latitude 8° South, which marks the divergence between the Equatorial Counter Current and the South Equatorial Current. In the Atlantic Ocean the numbers in schools of flying-fish are greatest near the equator, whereas in the less productive central water masses to the north and south the schools are smaller.

Oceanic fishes are good indicators of the temperature pattern of the surface waters. In the Pacific Ocean, silky and white-tip sharks, as we saw, are confined to the tropical belt: to waters warmer than 25°C. Yellow-fin tunny and skip-jacks range over the surface waters of both tropical and subtropical regions. They are most abundant where the sea is 20°C or warmer, though skip-jack may sometimes appear in somewhat cooler waters. Yellow-fins and albacore (*Thunnus alalunga*) also swim near the thermocline, together with big-eye tunny (*Parathunnus sibi*). Albacore range also into the surface waters of the transition zone, between the subtropical waters and the temperate or subarctic waters of the North Pacific Ocean, where great blue sharks also abound.

Oceanic sharks subsist mainly on fishes and squids. When following ships they will swallow all manner of rubbish, such as tin cans, paper cartons, and pieces of rag, as well as food scraps from the galley. Dr Stewart Springer remarks that species of the blue-shark family (Carcharhinidae) are

> . . . opportunists in feeding and necessarily so. Some . . . are relatively ineffective at catching uninjured fish in open water or even at finding such slow moving objects as crabs. The shark's ability to exist on a regimen of feast and famine, imposed by its ineptness in catching food at will, is probably made possible through its unique digestion and fat-storage organs. . . . It is pertinent to point out that their digestion is rapid and thorough, and that the shark's liver with its high percentage of oil is a good index of its metabolic well-being. The larger, fatter livers are found in sharks in good condition while smaller livers with little oil are frequently found in sharks having severe injuries, sharks in obviously poor condition, or in males at the end of the mating season [191].

The liver of a shark in fine condition may scale a quarter of its total weight. The quantity of oil in such a liver is not only a good reserve of food, tiding the shark over lean periods, but it will enhance its owner's buoyancy. A marine fish without a swimbladder must have a fat content equal to nearly a third of its weight if it is to be exactly buoyant.

Except for such kinds as makos, mackerel-sharks and the great white shark (*Carcharodon*), sharks are rather slow-moving beasts. If, during a burst of speed, they miss their prey, they are unable to suddenly stop, turn and renew their chase. Tunny-fishes, marlins and sword-fish use their pectoral fins as brakes and their other fins are also more mobile than those of a shark. Tunny and sword-fish also bear a well-formed keel (marlins have two) on each side of the caudal peduncle, not far ahead of their powerful means of propulsion, the large, lunate, quickly-sculling tail fin. Mr Edmund Watts, who has put several outstanding features of fish design to good account in the design of his ships, decided to fit a 'caudal keel', shaped like that of a tunny, on each side and before the propellor. Trials at sea revealed that the turning-circle of a ship could be halved after the fitting of fish keels. Perhaps the caudal keels of a tunny or sword-fish also enable them to make tight turns. After all, they feed on small fishes and squid, which can soon change direction. Makos, mackerel-sharks and the great white shark also have caudal keels. Being fast movers, they may well need extra manoeuvrability.

In the more productive parts of the open ocean, the smallest fishes that live and swim swiftly in the surface waters are skippers, up to eighteen inches, sauries and flying-fishes, from seven to sixteen inches. Even Melvilles' 'sleek little pilot-fish, azure and slim' that were so 'alert in attendance' about the Maldive shark, grow to a length of 2 feet. Besides accompanying sharks and giant devil-rays (*Manta*), pilot-fishes (*Naucrates ductor*) swim with schools of large tunny. The Leviathan of fishes, the whale-shark, which reaches a length of 60 feet or more, is circumtropical in distribution in the surface waters, where it must find enough plankton and small fishes to sustain its great bulk. Little tunny (*Euthynnus*) and bonito (*Sarda*) are generally less than 7 pounds in weight and each a length of 30 inches. The skip-jack comes close to a length of 40 inches and a weight of 50 pounds. Yellow-fins reach nearly 300 pounds. Blue-fin tunnies may grow to over 800 pounds. Certain marlins exceed a length of 11 feet and a weight of 1,200 pounds, while the sword-fish almost reaches the 1,000-pound mark and a total length of 15 feet.

The common ocean-sun-fish attains a length of 10 feet and a weight of more than a ton.

There are smaller, surface-dwelling fishes in the tropical ocean, but they live in special habitats. The Portuguese-man-of-war fish (*Nomeus gronovii*) has a maximum span of 6 inches. A number of small fishes live in the floating tangles of Sargassum weed: frog-fishes (*Histrio*), pipe-fishes and file-fishes. But to swim at large and to make a living in the surface waters of the open ocean, most adult fishes evidently have to be 6 or more inches long.*

The production of life in temperate waters is much more tied to seasons than it is in the tropics. As the chilled surface waters sink during the winter, there are compensating up-drafts or convection currents that renew the supply of nutrient salts in the upper layer. In spring, when the light is strong enough and the surface waters are sufficiently stable, there is a great flowering of the phytoplankton, which leads to the production of large populations of zooplankton species. Fish life is thereby renewed. It is in the summer season that blue-fin tunny, albacore, sword-fish, mako-sharks, the blue-shark, ocean-sun-fish and skippers move into the temperate waters of the open ocean, where they find good feeding. Basking-sharks also take advantage of these seasonal outbursts of plankton, but they lose their gill rakers and stop feeding during the winter. The opah or moon-fish (*Lampris*), ribbon-fishes (Trachypteridae) (plate 6), the oar-fish or king of the herrings (*Regalecus*) and bramid fishes (Ray's bream and so on) also migrate to temperate waters during the summer. Except for mackerel-sharks, there are no resident populations of surface-dwelling, oceanic fishes in temperate waters. Not every species has the capacity of true temperate fishes to live at a low metabolic level during the unfavourable season.

The vast circumpolar belt of the Antarctic Ocean is rich in planktonic life throughout the short summer season from January to March. The northern limit of Antarctic water is an encircling convergence between latitudes of about 50° to 60° South (figure 69). Here the cold surface waters that drift away from Antarctica meet warmer, subantarctic waters and sink below them, eventually to a depth of about 1,000 metres. Below this fall of water a warm deep current rises towards Antarctica, bringing nutrient-rich water to the surface. Hence the great outburst of planktonic plants and animals when conditions became favourable. But no large pelagic fishes cross

* There is, however, one or more dwarf species of saury, which live in the central water masses.

the convergence to feast on the fruits of the Antarctic summer. Indeed, few of the resident fishes take advantage of a prime product of the antarctic plankton, the shrimp-like euphausiid crustaceans known to whalers as krill. Krill (*Euphausia superba*) is the exclusive food of the great whale-bone whales and it forms most or all of the food of crab-eater seals, which is by far the most abundant seal in antarctic waters. Sea birds, particularly penguins, depend a great deal on krill. In fact, krill has been largely appropriated by the warm-blooded animals. Here, then, is a great part of the ocean where the numbers of fish are but a poor reflection of planktonic productivity.

Deep-sea fishes [100]

Living below the fine, blue-backed species of the surface waters are very different assemblages of fishes. Deep-sea fishes are more than ten times as diverse as the surface dwellers. The two thousand-odd species are roughly partitioned between the mid-waters (bathypelagic fishes) and the deep-sea floor (benthic fishes) (figure 70). They live at depths between 250 and 7,000 metres, even, perhaps, down to the very deepest part of the ocean.

Looking first at the mid-water dwellers (plates 8 and 9), about a third of them – stomiatoids, deep-sea salmonoids and alepocephaloids – belong to the herring and salmon order (Isospondyli). Another third, most of which are lantern-fishes (plates 16 and 17 and figure 85) are classified in the order Iniomi. Besides these, there are about a hundred species of deep-sea angler-fishes (plate 7), deep-sea eels and berycomorph fishes, mainly belonging to the family Melamphaidae (plate 6). Groups such as the whale-fishes (Cetunculi), giganturoids, gulper-eels, Miripinnati (figure 85) and great-swallowers (Chiasmodontidae), consist of fewer species.

Most of the mid-water fishes are less than 6 inches long when adult. Indeed, Dr Johan Hjort justly called them a 'Lilliputian' fauna [192]. The 'Gullivers', such as gulper-eels, lancet-fishes (figure 85) and the javelin-fish (*Anotopterus*), reach a length of three to six feet, but they are long and lean looking fishes. Certain of the barracudinas (*Notolepis coatsi*, *Paralepis barysoma*), which are more shapely and muscular, grow to at least 2 feet. One deep-sea angler-fish (*Ceratias holboelli*) sometimes strays into far northern or southern waters, where the females reach a length of about 4 feet. But in sub-polar waters, where it may find enough food to grow to an unusually large size, it is well outside its reproductive living space.

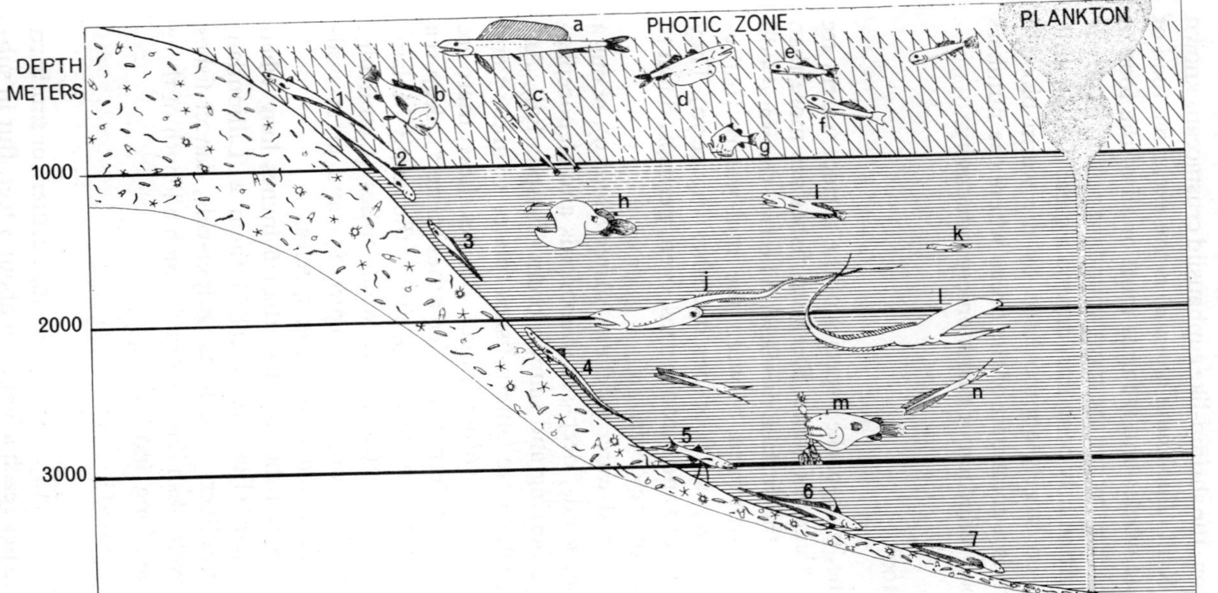

Figure 70. Vertical disposition of deep-sea fishes in relation to the amount of plankton, shown to the right, the penetration of light (the twilight zone is represented by lattice-work and the sunless region by horizontal lines) and the amount of food on the bottom (shown immediately below the profile of the deep-sea floor). (a–m) midwater fishes: (a) lancet-fish (*Alepisaurus*); (b) *Anoplogaster*; (c) barracudina (*Paralepis*); (d) giant swallower (*Chiasmodon*); (e) lantern-fish; (f) star-eater (*Astronesthes*); (g) hatchet-fish (*Argyropelecus*); (h) deep-sea angler-fish (*Melanocetus*); (i) whale-fish (*Cetomimus*); (j) gulper-eel (*Saccopharynx*); (k) bristle-mouth (*Cyclothone*); (l) gulper-eel (*Eurypharynx*); (m) deep-sea angler-fish (*Linophryne*). 1–7 Bottom-dwelling fishes which are most abundant between depths of 200 and 2,000 metres. (1) rat-tail; (2) notacanth; (3) brotulid; (4) halosaur; (5) tripod-fish; (6) rat-tail (*Gadomus*); (7) sea-snail (*Careproctus*).

For most bathypelagic fishes this fertile living space is in tropical and subtropical regions. By day, at least, they live at various levels below the thermocline, well out of reach of the waves. Though a number range into deep temperate waters, few manage to reproduce in such surroundings. The antarctic fauna of bathypelagic fishes is even smaller (about thirty species), but certain species, like the lantern-fish (*Electrona antarctica*) and the barracudina (*Notolepis coatsi*), pass their entire life history south of the Antarctic Convergence. High arctic waters, so far as we know, are without a resident fauna of bathypelagic fishes.

Each kind of mid-water fish is spread through a certain volume of deep-sea space. Even though much of the deep ocean has still to be explored, it is fair to say that no species of deep-sea fish is entirely cosmopolitan. A delicate stomiatoid fish, the black bristle-mouth, *Cyclothone microdon*, comes closest to having such a distribution. It swims in the deep waters of the tropical and subtropical ocean and ranges into temperate and antarctic waters. In the North Atlantic it goes as far as the subarctic reach of the warm, northerly fanning of the Gulf Stream. Quite a number (about forty species) have a circumglobal distribution, but mainly or entirely within the tropical or subtropical ocean. Dr J. C. Briggs [193] lists eighteen species of lantern-fishes, ten species of deep-sea angler-fishes, four bristle-mouths (including *Cyclothone microdon*), a hatchet-fish (*Sternoptyx diaphana*), a pearl-eye (*Scopelarchus*), and so on. But most species, so far as we know, have a distribution limited to two oceans, one ocean, or part of an ocean. All the same, even a quarter of an ocean is a large living space.

Thinking now of disposition in depth, the populations of most bathypelagic fishes, which include the majority of stomiatoids, deep-sea salmonoids, lantern-fishes, alepisauroids (lancet-fish, javelin-fish, barracudinas etc.), great-swallowers and melamphaids, are centred between depths of 200 and 1,000 metres. These levels encompass the twilight zone of the ocean, the region where the sun's rays become very dim (figure 70). At the threshold of submarine sunlight, which in clear tropical waters is probably between depths of 750 and 1,000 metres, the twilight merges with an underlying blackness, lit only by luminescent flashes from the light organs of siphonophores, copepods, euphausiid shrimps, deep-sea prawns, squids, fishes and so on. In these sunless, deeper reaches of the ocean, at depths of 1,000 metres or well below, live the deep-sea angler-fishes, gulper-eels, giganturoids, and a few stomiatoids, notably the black species of bristle-mouths (plate 7) (figure 70). These deeper waters are cold, having temperatures of 5°C or less. The fishes of the upper mid-waters live in the transition layer; for just below the near-surface thermocline to a level of 1,000 metres, temperatures in the warm ocean drop rapidly from values near 20°C to about 5°C.

But all deep-sea fishes spend part or all of their adult life beyond the reach of the seasons, so far as changes of temperature are concerned. (The yearly fluctuations of temperature do not penetrate much beyond the depth of 100 metres.) The part-time deep-sea fishes comprise the many kinds from the twilight zone that move into the surface layers at night, diving below before dawn. These species bear large sensitive eyes capable, we imagine, of signalling the approach of sunset and sunrise, the times of their upward and downward migrations.

Perhaps as many as three-quarters of the thousand-odd kinds of bathypelagic fishes have light-organ systems. In a census of individuals this fraction may be nine-tenths or more, judging by the catches of large mid-water nets. Most of these luminous species and individuals are lantern-fishes and stomiatoids. Any kind of lantern-fish, and there are more than two hundred, can be recognized by the pattern of round, pearly light-organs studding its sides (plate 16). Each such photophore receives a nerve, which must serve, perhaps in concert with hormones, to switch on the small mass of glandular, light-producing cells. The light is reflected forward by a silvery backing and focussed through a lens-like thickening of the scale overlying the light organ (figure 44). Lantern-fishes seem to form small schools and it may well be that the members of a species recognize one another by displaying their individual constellation of lights. But besides the pearl-button lights, certain lantern-fishes, belonging to *Diaphus* and related genera, bear large light organs on the head (plate 17), which could be used, like a miner's head-lamp, to shine light in dim surroundings. In diverse species there are also large plate-like light organs on the caudal peduncle, which are differently developed and placed in the two sexes. The male usually has the larger lights, which are housed along the upper surface of the peduncle; the female's lesser lights – and in certain species there are none – are generally set along the underside. The light organs emit brilliant flashes, which presumably play a leading part in courtship and mating activities. Perhaps the larger males are able to intimidate smaller rivals by displaying their more powerful lights, which may also be more impressive to the female.

Most kinds of stomiatoid fishes have many more light organs than do lantern fishes. There is usually a double row of large photophores along each side of the underparts and in nearly all species one or more lights shine directly into each eye. In the viper-fishes (*Chauliodus*), scaly dragon-fishes (*Stomias* spp.), star-eaters (Astronesthidae) and black dragon-fishes (*Melanostomiatidae*) the sides of the trunk and tail also bear many smaller kinds of light organs. The light organs of most stomiatoids not only outnumber those of lantern fishes, but the light-emitting gland, at least in the ventral lights, is larger than that in the pearl-button organs of a lantern-fish (figure 44). When all the photophores flash, there should be a brilliant display. Perhaps the small 'pilot lights' that shine into the eyes adapt the highly sensitive retinae to a concerted display of light. It would not do for a stomiatoid fish to be blinded by its own luminescence, for the deep ocean contains many predatory fishes and squids.

At all events, the pattern of light organs of most stomiatoid fishes distinguishes the species and could thus be used as recognition signs. But what are we to make of the luminous glands on the chin-barbels of star-eaters and scaly and black dragon-fishes? According to William Beebe and Jocelyn Crane, a slight disturbance of the water near the barbel of one black dragon-fish (*Echiostoma tanneri*) goaded this species '. . . to the utmost so that it thrashed about and snapped, striving to reach and bite the source of irritation. Again and again we proved the astonishing sensitiveness of this organ.' [194] The twitching of a lighted barbel may be a bait to attract prey with well formed eyes, such as other fishes, squids and the larger crustaceans. If the barbels of all stomiatoids are as sensitive as that of *Echiostoma*, the approach of sizeable prey should be readily detected.

Instead of a barbel, the viper-fishes have a very long second dorsal ray which bears a luminous tip. One of these fishes was seen from a bathyscaphe by Dr J. M. Pérès. The *Chauliodus* was hovering head upwards, the long axis of its body making an angle of about 45° to the horizontal plane. The whip-like dorsal ray was inclined forwards so that the tip dangled in front of the mouth (plate 9). Here, surely, is good circumstantial evidence for deep-sea angling.

Predatory kinds of stomiatoids* which can seize and swallow sizeable prey, have a large light organ below or behind each eye (figure 45). The light-emitting mass of cells is backed by an inner tunic, which may contain silvery reflecting pigments, and an outer

* Scaly dragon-fishes, black dragon-fishes, viper-fishes, star-eaters, idiacanthids and malacosteids.

layer of black colour-cells. The tendon of a long strip of muscle is inserted on the outer curved face of the light organ and the muscle runs below and behind this photophore to a setting on a suspensory bone, the hyomandibular, of the upper jaw. When this muscle contracts, the outer face of the light organ slips downwards, so that the light-emitting surface is hidden behind a sheath of black pigment cells (in the genus *Idiacanthus* the muscle pulls a screen of black pigment cells over the light organ). By the rhythmic contraction and relaxation of the muscle the light could be dowsed and flashed like that of a lighthouse. If need be, it could be left to shine a steady beam.

One fine observation at sea by E. R. Günther revealed the power of these lights in a silvery, eel-shaped stomiatoid fish. The fish was feeding on krill near pack-ice east of the South Sandwich Islands. Its two ocular photophores

> . . . emitted a beam, of varying intensity, of strong blue light, which shone directly forwards for a distance of about 2 feet. The fish had the habit of lurking at a depth of about 2 to 6 feet below the surface, poised at an angle of about 35° to 40° from the horizontal – this gave the beam an upward tilt; occasionally the fish swam round and with a quick action snapped at the cloud of krill above it [195].

Now to human eyes, peering through several feet of water, the range of the beam was at least twice the length of the fish, which may well have more sensitive eyes than ours. This stomiatoid would seem, then, to have a powerful visual aid; yet there is a further possibility. Predatory stomiatoids, and there is some supporting evidence, probably hunt alone. A flashing beam may not only light up the visual field for the fish, but warn off other predatory fishes. Dr J. A. C. Nicol, who has made a special study of luminescence in deep sea animals, wrote thus of stomiatoid fishes:

> The large subocular organs may be used as torches, for illuminating the surrounding water. The serially arranged photophores may permit the animals to recognize each other. The result may be mutual repulsion, thus keeping the fish spread out in hunting territories delimited by the intensity of the light and the distance at which it can be seen; or mutual attraction when the animals differ in sex [196].

Concerning the last suggestion, there are sexual differences in the ocular or cheek lights of some stomiatoid fishes. The males of *Idiacanthus*, though much smaller than the females, have much larger ocular lights. Perhaps the females, who must swim much faster than their mates, are guided and courted by male headlights. In some black dragon-fishes the post-ocular lights are small or absent in the females and large in the males. Again, in a malacosteid fish (*Photostomias guerni*) the corresponding cheek photophores are larger in the males.

To recapitulate, lantern-fishes and stomiatoids are the most diverse groups of luminous fishes in the upper mid-waters of the ocean. In the lower reaches, well beyond a depth of 1,000 metres, their place is taken by the ceratioid angler-fishes (plate 8, figure 70). All but two of the hundred-odd species bear a bulbous light-lure, which is restricted to the females. The lure is attached to a rod known as the illicium which is hinged to a basal bone. The illicium projects from the top or forepart of the snout; it is a modified fin-ray and immediately behind it in young and adolescent fishes can be seen a small second ray. In the males the illicium is vestigial at all stages.

The lure holds a light gland, the secretions of which contain minute bodies that are thought to be luminous bacteria. The gland receives blood vessels and as oxygen is needed for luminescence, Dr E. Bertelsen [104] concluded that the activity of the bacteria could be controlled by the flow of blood to the lure. If oxygenated arterial blood is fed to the gland the bacteria will luminesce, but they will stop soon after such blood is withheld. Whatever the precise nature and control of the light, observers have noticed that it ranges in colour from orange to yellow, to yellowish-green and blue-green. It seems to be emitted in a series of flashes.

Deep-sea angler fishes have yet to be seen taking their prey, but it seems certain that they took to angling millions of years before man. Their relatives from shallow seas, such as the frog-fishes and goose-fishes, angle with a non-luminous bait. In numerous kinds, like *Ceratias* and *Centrophryne*, and diverse oneirodid anglers, the rod is attached to a long and moveable basal bone, which slides in a groove along the roof of the skull and the back. The basal bone is moved by the two main pairs of muscles, one pair pulling it forwards, the other backwards. The rod and lure can thus be extended well in front of the enormous head and jaws. If the winking lure attracts a gullible fish, squid or prawn, the lure and prey could be drawn towards the large and powerful jaws, which bear long pointed teeth. The rest may be imagined.

The lure of each kind of deep-sea angler-fish generally bears an individual dressing of filaments, tags or flaps of skin. The movements of these attachments no doubt help to lure the prey, but why should each species have its own arrangement? Perhaps the smaller males, who have well-formed eyes, recognize their proper mates by the pattern of bunting that flies near their lights.

Concerning eyes, we may recall that deep-sea fishes of the twilight zone have large, highly sensitive eyes. The mosaics of rods in the retinae bear visual gold, pigments that are most sensitive to the rays at the blue end of the spectrum, which penetrate deepest into the ocean. Light organs also emit bluish rays and the range of wave lengths produced by a lantern-fish (*Myctophum punctatum*) measured about 410 to 600 mμ, the light being most intense at about 470 mμ. Now this range of the spectrum is very close to that absorbed by the visual gold of the lantern-fish, a correspondence that is most probably found in many other species [197]. Like the other luminous animals that they see, fishes have produced living-light to match deep-sea sunlight. Their eyes make the best of both worlds, which is a fine instance of nature's parsimonious ways.

The sparks and flashes of light organs can be detected to a depth of at least 4,000 metres. In the twilight zone the frequency of flashing can be such as to fuse into a virtually continuous background of luminescent light. With this in mind, Dr J. A. C. Nicol has made a new approach towards understanding the colours of deep-sea animals. Some of the fishes have silvery, iridescent flanks, especially the hatchet-fishes and many kinds of lantern-fishes. A black or dark brown skin, looking like velvet, invests black dragon-fishes, viper-fishes, star-eaters, gulper-eels, deep-sea angler-fishes and others. Scarlet or red colouring, which is found in so many mid-water invertebrates such as squids, octopods, prawns, copepods, jelly-fishes and so on, is rare in deep-sea fishes. The most striking exceptions are the whale-fishes. Dr Nicol found that the surfaces of red prawns and black fishes reflect very little blue light, such as that found in luminescent displays. 'There is little doubt that these are concealing colours that make the fish or other animals difficult to see. Light in the ocean depths comes from animal luminescence, and black surfaces serve to reduce tell-tale gleams to the minimum. The carotenoid pigment responsible for the red colouration of pelagic crustacea reflects very little blue light, which forms such a preponderant part of animal luminescence, and is as effective as a black covering [198]. The last sentence must also be true of the whale-fishes. Prey-seeking stomiatoids and angling female ceratioids thus have excellent velvet-black cloaks in more than one sense.

But, as reflection tests of a hatchet-fish (*Argyropelecus olfersii*) showed, a silvery fish is a good reflector of luminescent light. The biological meaning of this is less obvious, yet the oscillating mirrored sides of a swimming fish could catch the light and thus dazzle and deceive. At all events, silvery-sided fishes and those with light coloured sides have dark backs. Evidently the downward component of sunlight is sufficient, at least in the upper reaches of the twilight zone, for countershading to be effective.

Benthic deep-sea fishes

The most diverse of the fishes that live close to the deep-sea floor are rat-tails and brotulids. The deep-sea cods, eel-pouts, sea-snails, halosaurs and notacanths, alepocephalids and tripod-fishes are less prominent but fairly diverse (figure 70 and plate 13).* Besides these groups of teleosts, there are a number of families containing deep-sea sharks. The related chimaeroid fishes are most common over the upper reaches of the continental slopes.

Benthic deep-sea fishes are quite unlike or not very closely related to those of the mid-waters, most of which are relatively primitive, soft-rayed fishes (Isospondyli and Iniomi). Most of the benthic fishes belong to orders that are specialized in certain structural respects.

While nearly all kinds of mid-water fishes, when adult, span 6 inches or less, most of the benthic species measure more than this length. Indeed, a good many reach more than a foot. But much of the length of a rat-tail, halosaur, or a notacanth consists of a long tapering tail, the caudal fin being rudimentary or absent. Under the tail runs a long, many-rayed anal fin, which is usually more prominent than the dorsal members. When such a long under-fin is thrown into waves, the fish will not only be driven forwards or backwards, but the tail will tend to be raised above the head and trunk. As most of these fishes have the mouth under the snout, an inclination of the body axis should place the jaws in a good position for taking food from the deep-sea floor. Moreover, halosaurs, notacanths and most rat-tails have specially strengthened snouts. Moved and canted head-downwards by the anal fin, the fish can use its snout to root in the oozes, where it may uncover burrowing invertebrate animals. Yet

* The alepocephalids, sometimes called slick-heads, are related to the herring-like fishes of the order of Isospondyli. Tripod-fishes are classified with the lantern-fishes and other families in a suborder of the order Iniomi. Rat-tails, deep-sea cods and shelf-water cod-fishes form most of the order Anacanthini. Halosaurs and notacanths make two suborders of the order Heteromi, while eel-pouts and sea-snails are mail-cheeked teleosts (Scleroparei). The brotulids are placed in a suborder of the perch-like fishes (Percomophi).

rat-tailed fishes, at any rate those that live at a depth of a few hundred metres, also feed on freely swimming prey such as prawns, euphausiids and lantern-fishes. Like their relatives the cod-fishes, they take food on and above the ocean floor. When swimming away from the bottom, their long attenuated bodies may give them unusual powers to detect moving prey. The longer the tail the longer will be the lateral-line canal, which should increase the effective range of the sense organs in detecting the disturbances made by swimming animals. This, as Dr P. A. Orkin has argued, may partly account for the very elongated forms of so many oceanic fishes [154].

Brotulid fishes, eel-pouts and sea-snails have long dorsal and anal fins, which are continuous with a small caudal member. The tripod-fishes have well-muscled, spindle-shaped bodies. The name alludes to the very long stiff rays of the two pelvic fins and the caudal fin, which form a tripod undercarriage, used, as men in bathyscaphes have seen, when these fishes rest on the deep-sea floor (figure 85). Tripod-fishes have small eyes but well developed lateral line organs on the head. If such a fish sees a luminous flash or senses the vibrations of a swimming animal, one can imagine it swiftly taking off from the tripod to snap up an item of food.

Compared to the bathypelagic fish fauna, light organs are much less prevalent among the benthic fishes of the deep-sea floor. In fact, the rat-tails are the only group to have acquired such organs. In species of *Coelorhynchus* and *Hymenocephalus* the luminous gland lines a long duct that opens just before the anus (figure 46). The glands of *Hymenocephalus* and *Malacocephalus* shine through lenses which are placed between the pelvic fins (see also pp. 177–178).

Dr Y. Haneda [199] found that the glands hold luminous bacteria, which are responsible for the light. In *Coelorhynchus parallelus* he saw the luminosity as 'a beautiful filiform line on the central surface'. Concerning *Malacocephalus laevis*, Portuguese deep-sea fishermen discovered a good use for this species in dressing their bait, which they rub along the belly of the fish. This forces out a yellow viscid secretion from the opening of the light organ, which, when smeared on the bait, makes it glow with a sky-blue light for several hours. The bait thus becomes more attractive to the fish quarry of some really knowing deep-sea anglers.

The light organs of rat-tailed fishes may well play a leading part during courtship and mating. It will be recalled that the males of some kinds have a drumming swimbladder and their sound signals could be a long-distance means of attracting mates in dim or dark surroundings. But the light produced by a rat-tail is not likely to be seen much beyond a range of ten metres. Once a male has met a mate, he may, like a male cod-fish, lead her away from the ocean floor to a level suitable for spawning. If so, the glowing of his ventral light could stimulate her to follow. This is conjecture, of course, but it may not be long before such speculations can be tested. Deep-sea television is now in use, while deep-sea ships or bathyscaphes are becoming more versatile.

We saw that most of the mid-water, deep-sea fishes live below the warm surface waters of tropical and subtropical regions. The same is true of the bottom-dwelling species. But no species is known to live throughout the deep sea, not even in the circumtropical belt, in striking contrast to the bathypelagic species. Judging from the better explored areas, particularly the North Atlantic, many species are restricted to the continental slopes along one side of an ocean. Perhaps they are tied to certain good feeding grounds.

Again, like bathypelagic fishes, most of the bottom-dwellers live in the lesser depths of the ocean. They range over the upper parts of the continental slopes, largely between depths of 250 and 1,500 metres. While about forty species actually live in abyssal waters deeper than 2,000 metres, certain individuals of another two hundred-odd species have been trawled beyond this depth. The deepest trawlings, of a brotulid and a sea-snail, are from depths of a little over 7,000 metres. But during the dive of the bathyscaphe *Trieste* to the Challenger Deep in the Mariana Trench, which is the deepest known part of the ocean, Dr Jacques Piccard reported seeing a sole-like fish slithering over the oozes at the depth of about 10,300 metres. It would seem, then, that fishes, like numerous invertebrates are able to live at pressures of more than 1,000 atmospheres (over 6 tons to the square inch).

Living conditions

Deep-sea fishes, as will now be evident, live at levels from about 200 to 10,000 metres below the plant-bearing, surface waters of the ocean. Directly or indirectly, animal life depends on the synthesis of new living substances by plants, which are literally the drifting pastures of the seas. To appreciate the life of deep-sea fishes, an outline is needed of the quantities of potential food at various levels. By towing suitably designed nets through submarine space and along the deep-sea floor, a sample can be obtained of the standing stock of life. The quantity of mid-water life is often expressed as the weight or biomass in a cubic metre of the sea. The biomass of bottom-dwelling organisms is referred to a square metre of the deep-sea floor.

Since the plants grow and reproduce in the upper 100 metres of the ocean, this is the richest zone of planktonic life, which varies in quantity, in time and place. Hordes of herbivorous animals, particularly copepod crustaceans, feed on the plants and in turn are the food of animals, from medusae to fishes. Below the plant-producing waters to a depth of 1,000 metres, there is a rapid fall in the quantity of plankton, which between 500 and 1,000 metres is round about a tenth of that near the surface (figure 70). From 1,000 metres to the bottom layers, the standing stock of plankton continues to diminish but more slowly and regularly. As well as containing a more diverse fauna, the twilight zone thus holds more potential food than the sunless lower depths.

On moving down the continental slope to the abyssal ranges of the deep-sea floor, there are parallel decreases in the diversity and quantity of benthic animals. Figure 70 shows how the amount of bottom-dwelling life, the benthos, falls with increasing depth. But the figure does not show, and this is relevant to later discussion, that at any given depth the quantity of benthos (weight per square metre of the deep-sea floor) is much greater than the weight per cubic metre of the plankton.

Many fishes of the mid-water twilight zone move up and down each day. Soon after sunset, diverse lantern-fishes and smaller stomiatoids such as *Vinciguerria* (figure 84) may be netted in the surface waters. Before sunrise they dive into the depths. This we can infer from series of net-hauls and more recently from records taken by echo-sounders. Many of the most prominent scatterers of supersonic sound in the ocean are small fishes of the twilight zone. Most of these fishes, particularly lantern-fishes, have capacious swimbladders, which reflect a great deal of incident sound. By day, echograms show traces – deep scattering layers – between depths of about 250 and 700 metres. An hour or so before sunset, these sound scattering layers begin to move upwards, and soon after dark they enter the surface waters. Just before sunrise a particular layer will start to descend, reaching its day-time level in about an hour.

Lantern-fishes and small stomiatoids feed on planktonic animals and we have seen that there are many herbivorous animals, particularly copepods and larvae, in the productive, plant-bearing surface waters. There can be little doubt that the planktonic feeding fishes swim upwards each day to feed on the small animals that depend, directly or indirectly, on the plants. Indeed, Dr Donald Griffin has likened the vertical migrations of deep-sea animals to the twilight sorties of bats:

> Those bats which spend the daytime in caves fly out at night to prey upon insects that have grown by feeding on green plants. In much the same way many fishes and invertebrate animals retreat by day to the dimly lighted depths of the ocean, just below the photic zone, and then swim upward every evening to feed near the surface on the fruits of photosynthesis [84].

A good many predatory fishes also pay nightly visits to the surface waters. These kinds, such as dragon-fishes, viper-fishes and star-eaters, feed on other fishes, which may be as large as themselves, and on planktonic animals. By night they enter a layer that not only holds relatively abundant populations of resident animals, but also the diurnally migrating plankton-feeding fishes and invertebrates. Star-eaters, for instance, seem to have a particular liking for lantern-fishes. Perhaps these predatory fishes swim upward in company with the plankton-eating kinds, preying on them in transit.

The fishes of the sunless depths, those that live at 1,000 metres and below, do not migrate to the surface waters. If an angler-fish or a gulper-eel is found at the surface it is dead or dying and has met an accidental death. Angler-fishes do not swim in search of their prey, but lure it and certain species, at least, can swallow fishes as large as, or larger than, themselves. They also take copepods, deep-sea prawns and cephalopods. Gulper-eels have a somewhat similar diet, while the small black bristle-mouths feed on planktonic animals, particularly on copepods. Yet these fishes can also accommodate euphausiid shrimps of about half their own length and they also eat fishes.

The species that live near or on the deep-sea floor feed largely on invertebrate animals. The nourishment of such fishes can be traced to the plants and animals that live in submarine space. When these organisms die, they sink, decay and disintegrate. All remnants of this dissolution that are not intercepted by detritus-collecting members of the zooplankton, particularly by crustaceans, eventually come to rest on the deep-sea floor. Here these remains are attacked by bacteria that live near the surface of the oozes. The bacteria of the sediments, which may be as abundant as those in garden soil, are able to convert intractable materials, such as chitin and cellulose, to their own substance. Benthic invertebrates with the means to catch minute particles may well use the bacteria as food, while others that eat mud may carry an intestinal flora of bacteria, so aiding their digestion. At all events, bacteria seem to be the key organisms of the sediments, making life possible, at least in part, for many deep-sea invertebrates, which in turn form the food of fishes.

Using this biological background, we may now look at the organization of deep-sea fishes in relation to oceanic supplies of food. Besides their much greater diversity, fishes of the mid-water twilight zone are far more abundant than those that swim in the sunless deeper spaces of the ocean. This greater abundance is to be expected, simply because the twilight fishes live closer to the productive surface waters, many within migrating range. Bottom-dwelling fishes are also more diverse and abundant over the slopes than over the abyssal plains. Not only is there more chance of a greater fall of detritus to the shallower slope regions, but these declivities are closer to the outflows of organic material derived from inshore waters and the land, substances which may be washed down by turbidity currents. Hence the greater productivity of the slopes in invertebrate and fish life.

When studying the swimbladders of deep-sea fishes, I was led to consider the organization of species from the three main regions: the twilight mid-water zone, the sunless, mid-water zone and the ocean floor. About half the species from the twilight zone (about 200 to 1,000 metres), such as lantern-fishes, hatchet-fishes, melamphaids and so on, which are nearly all plankton eaters, have a well developed swimbladder. In the predatory kinds – scaly and black dragon-fishes, viper-fishes, alepisauroids and so on – the swimbladder is either reduced or absent. This organ is absent or rudimentary in *all* the fishes that live in the sunless mid-waters below a depth of about 1,000 metres, where the dominant groups are ceratioid angler-fishes, black bristle-mouths and gulper-eels. At least half of the fishes that live below them on the deeper ranges of the deep-sea floor (2,000 to 5,000 metres), fishes such as rat-tails, brotulids and halosaurs, have a well formed swimbladder. In the other bottom-dwelling groups, such as the tripod-fishes, eel-pouts and sea-snails, the swimbladder is absent.

Now the fishes of the sunless zone can hardly have lost the swimbladder because of the high pressures of their environment, for we have just seen that many of the benthic fishes that live at equal or greater pressures have a capacious swimbladder provided with

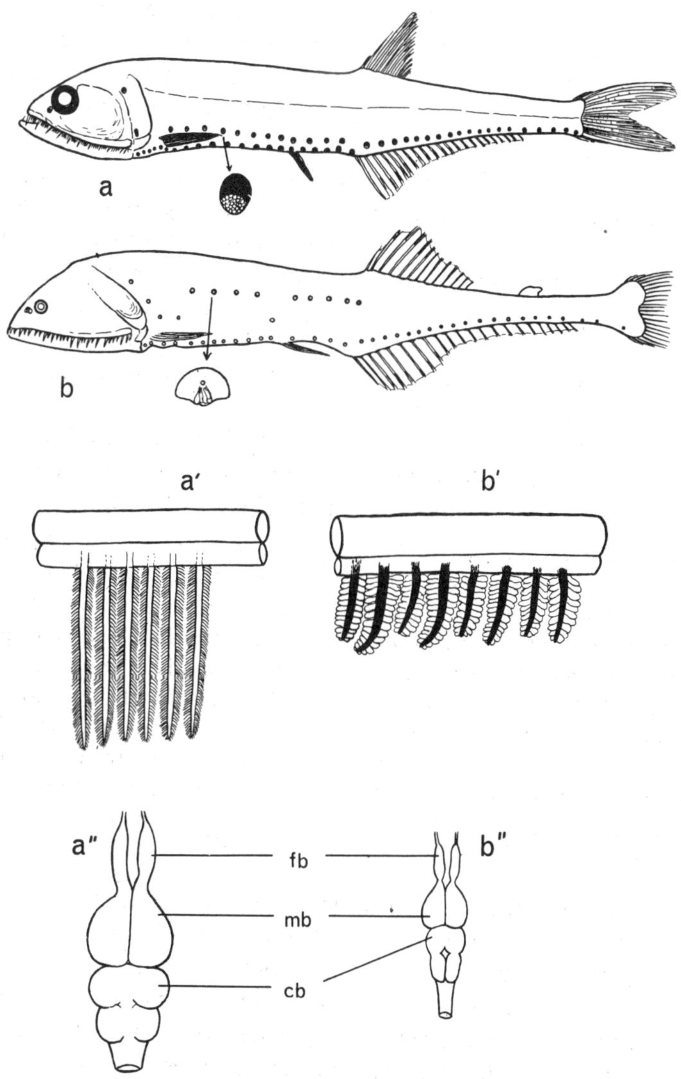

Figure 71. A good instance of the great differences between fishes of the twilight zone and the sunless depths of the ocean. *Gonostoma atlanticum* (a) lives in the twilight zone, while its relative, *Gonostoma bathyphilum* (b) is a 'sunless' fish. Besides having much larger eyes, larger light organs, a stronger skeleton and better developed muscles, *Gonostoma atlanticum* has larger gills (see a and b′) and a larger brain (see a″ and b″) than *G. bathyphilum*. *G. atlanticum* also has a swimbladder, which is absent in *G. bathyphilum*. (From Marshall [38].)

highly developed gas-producing tissues. But closer study of the 'sunless' fishes revealed that they have economized in all manner of tissues besides the swimbladder. Bristle-mouths, angler-fishes and gulper-eels have flimsy skeletons and weakly developed muscles. They have small eyes and brains, small kidneys, very small gills, and so forth (figure 71). The most reasonable explanation is that these economies have been forced on them because they live in the poorest food-containing part of the deep ocean. Having comparatively reduced tissue systems, particularly skeletal and muscular, these fishes must be neutrally buoyant, so being able to hover in the sea without undue effort. Poised angler-fishes wait for prey to approach their lures; they can swallow big meals and so can bristle-mouths and gulper-eels. They are not misfits, but are cunningly adapted to the exigencies of their living space.

Having a buoyant swimbladder, the plankton-feeding fishes of the twilight zone can carry more muscle and bone than those from the sunless depths. Elsewhere I described them as follows:

> They are muscular, active little fishes with large eyes and brains and highly developed light organs. Because of the extra propulsive power that a swimbladder allows, they are able to make daily visits to the food-rich surface waters, which they must do to maintain their highly organized bodies. Living with them are almost as many kinds of predatory fishes, nearly all of which have lost an internal float. In spite of this, those fishes come close to neutral buoyancy by reducing their muscular and skeletal tissues. All have well formed eyes and brain and many have complex batteries of light organs. Some of these fishes also migrate to the surface layers, compensating for their reduced myotomes by increased size of body [38].

If we now set side by side a fish from the sunless midwaters – a bristle-mouth, an angler-fish or a gulper-eel – with one from the abyssal plain – a rat-tail, a brotulid, a halosaur or a tripod-fish – the same kinds of contrasts are perfectly evident. The abyssal fishes are more highly organized, possessing better developed tissues of all kinds: muscular, skeletal, nervous, respiratory, excretory and so forth. But, once more, the abyssal fishes have more food at their disposal, enough to maintain their relatively well-built bodies. Yet the species from the sunless mid-waters show just how adaptable are teleost fishes. These fishes come closest to being vertebrates without backbones.

Chapter 18

Shallow Sea Fishes

MOST KINDS of marine fishes live in the coastal and coral seas of the tropics. At the other extreme, the faunas of coastal fishes are poorest in polar regions. On moving from ice-cold to more temperate seas, there is an increase in the number of species, which become still more numerous in warm temperate areas. Even so, temperate fish faunas are much less rich than are those of the tropics. Like other living organisms, fishes are most diverse where conditions are, and have been, both warm and stable. But coral formations, which provide many niches for fishes and invertebrates, only grow well in clear and shallow, tropical waters. If, as seems likely, there is some connexion between the diversity of living spaces and the adaptive multiplication of species, we can begin to understand why most kinds of shallow sea fishes live in coral reefs and atolls. We should also recall that deep-sea fishes, which live in cool and constant surroundings, are also most diverse in tropical regions. Stable environments do not necessarily have to be warm to permit diversity.

Polar fishes

There are over a hundred kinds of fishes in the waters fringing the Antarctic Continent and the off-lying islands. Nearly three-quarters of these belong to a division of perch-like fishes called the Nototheniiformes. Unlike most percoids, the fishes of this division have flexible fin-spines, pelvic fins on the throat, and a single nostril rather than two, one either side of the snout. For the most part they are sluggish, bottom-dwelling beasts with a large head and relatively short body that ends in a tail fin with a straight or convex margin (figure 72). Most of the remaining antarctic fishes are skates, eel-pouts and sea-snails.

Other kinds of skates, eel-pouts (figure 72) and sea-snails, but not nototheniiform fishes, are found in arctic coastal waters. In fact, the eel-pouts and sea-snails, together with species of cod-fishes, bullheads (figure 72), armed bullheads and flat-fishes, are the dominant fishes off arctic shores.

Like the antarctic fishes, most of these species are bottom-dwellers.

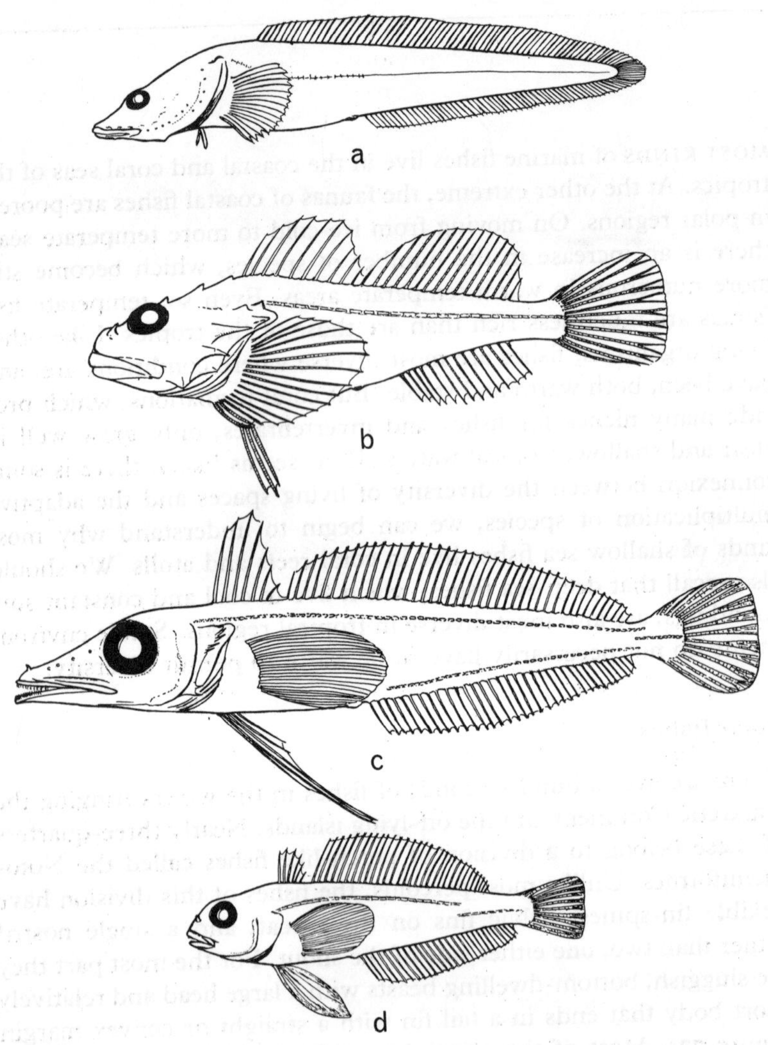

Figure 72. Arctic and antarctic fishes. (a) eel-pout, *Lycodes seminudus* (Barent Sea and other arctic waters); (b) bullhead, *Myoxocephalus scorpius* (arctic to temperate North Atlantic); (c) ice-fish, *Chaenodraco wilsoni* (off Antarctica); (d) *Trematomus pennellii* (off Antarctica). (a and b after A. P. Andriashev. c and d after C. T. Regan.)

And it is striking that the bullheads and armed-bullheads resemble many of the nototheniiforms in having these features: a large, frog-like head with capacious gill chambers; poorly developed lateral muscles; large, broad-based pectoral fins; and a square-cut or rounded tail fin with a reduced number of rays. But swimming in all three groups is largely by way of the pectoral fins, which are used as paddles.* There is thus no need for large lateral muscles and a well-developed tail fin. Moreover, these fishes spend much of their time lurking on the sea-floor; hence the capacious gill chambers to ensure adequate ventilation of the gills. Arctic and antarctic fishes are also alike in laying relatively large eggs in clusters on the bottom. Convergent evolution, whereby distantly related organisms in response to similar environmental pressures come to have similar features, is widespread among fishes.

Though living in near-freezing seas for most or all of the year, polar fishes are perfectly active, even at temperatures close to −2·0°C. Unlike temperate fishes, they feed and grow during the winter months. If one makes due allowance for the different temperatures of their living spaces, the pace of life, measured by the rate of oxygen consumption, is biologically higher in polar fishes than in those from warmer waters [200]. Polar fishes presumably have enzyme systems, such as those governing digestion and muscular activity, that are fitted to give a lively rate of metabolism in ice-cold waters.

Antarctic fishes of one nototheniiform family, the Chaenichthyidae (figure 72), are remarkable in another way. These are pallid, gelatinous-looking animals with colourless gills. One species (*Chaenocephalus aceratus*) which is abundant around South Georgia, is so translucent that it is known to British whalers as the 'ice-fish'. Norwegian whalers call it a 'bloodless fish'. In fact, blood taken from a chaenichthyid is nearly transparent. Red blood corpuscles, which contain haemoglobin, the iron-pigment that conveys oxygen to the cells of the body, are absent or very sparse. As might be expected, the oxygen-holding capacity of an ice-fish's blood is very low compared to that of its relatives with red blood. The plasma of a 'bloodless fish' simply holds oxygen in physical solution. But antarctic seas are well aerated, and most ice-fishes lead lurking, unharried lives [201].

A toad-fish (*Opsanus tau*), which is also a sluggish fish, still lives

* When these fishes rest, the pelvic fins and lower rays of the pectorals are used as an undercarriage.

after the haemoglobin of its blood has been put out of action by carbon monoxide. Yet three 'bloodless' fishes, *Champsocephalus gunnari*, *Pseudochaenichthys georgianus* and *Neopagetopis ionah*, must be quite active at times, for during the summer they feed near the surface on krill. Are they entirely without red blood corpuscles? And, to end, we should add that no one has yet reported seeing a 'bloodless' fish in arctic waters.

The blood and body fluids of fishes freeze at temperatures from −0·38° to −1·9°C. Holding solutes that are osmotically equivalent to those in the sea, the blood of hag-fishes and cartilaginous fishes freezes at −1·9°C, which figure is the same as that for average sea water. But the blood of marine teleosts, being less salty than the sea, freezes at higher temperatures (−0·5° to −1·0°C). How, then, do polar teleosts keep their blood from being frozen? Studies of antarctic species have yet to be made, but certain arctic fishes have been well investigated. Dr P. F. Scholander and his colleagues [202] used fishes taken from Hebron Fjord in northern Labrador. During their summer tests, the temperature of the shallow, inshore waters of the fjord varied from 4° to 7°C. The blood of fishes caught in these waters; fjord cod (*Gadus ogac*), sculpin (*Myoxocephalus scorpius*) and arctic char (*Salvelinus alpinus*), froze at temperatures between −0·75° and −0·85°C, which is about the normal figure for teleost blood. But the water at the bottom of Hebron Fjord stays at a temperature close to −1·75°C for the entire year. Species fished from these depths, such as polar cod (*Boreogadus saida*), eel-pout (*Lycodes turneri*), sea-snail (*Liparis koefoedi*) and bullheads (*Gymnacanthus tricuspis* and *Icelus spatula*) were found to have body fluids that froze at about −1·0°C. These fishes thus exist at temperatures below the freezing point of their blood. In other words, their blood is supercooled by about 0·75°C.

A supercooled fluid is not in a stable state. When fishes from the bottom of Hebron Fjord were put into sea water chilled to −1·7°C they froze and died. Yet tests with temperate fishes showed that they could survive in sea water supercooled to −3·0°C, though they froze solid as soon as the water was seeded with ice. The contact between fish and ice was enough to make them freeze and die. It thus looks as though the fishes at the bottom of Hebron Fjord are safe from freezing because they live below the reach of ice-crystals, which float near the surface. But what of the species that live in shallow, inshore waters during the winter months? Tests on two species, the fjord cod and sculpin, showed that the winter freezing points of their blood had fallen from the summer level of −0·8°C to about −1·5°C. Their body fluids are then slightly supercooled, by 0·2°C. Moreover they can survive after touching ice. It appears that the blood contains organic, anti-freeze compounds, which give them extra protection [203]. Have antarctic fishes evolved similar means of survival in ice-cold waters?

Tropical fishes

Fishes, as we saw, are most diverse in the tropical ocean. The greater part of this diversity is centred in seas where conditions allow reef-corals to thrive (figure 73). To grow properly, these corals need adequate heat and light. Flourishing coral communities are restricted to waters where the mean temperature does not fall below 18°C during the coldest part of the year. In fact, corals grow best in seas where the mean annual temperatures are from 23° to 25°C. Given clean, warm waters, reef corals thrive at depths down to about 50 metres. If the water is warm and shallow enough, but turbid, coral formation is severely restricted.

Corals reefs are made largely of the interlocked and encrusted skeletons of calcareous red algae and reef-building zoantharian corals. Such reefs are spread over an oceanic area of some 68 million square miles. They exist, to quote John W. Wells, '. . . wherever a suitable substratum lies within the lighted waters of the tropics beyond the influence of continental sediments, and away from the cool upwellings of the sea in the eastern part of the ocean basins.' [204]

Coral reefs and atolls are widespread over the Indian Ocean and western Pacific Ocean between latitudes 30° North and 30° South. In the tropical West Atlantic there are flourishing reefs in the Caribbean and around the West Indies, but they are not so strongly built as are those of the Indo-West Pacific region. One striking difference is that calcareous algae play a much greater part in the formation of Indo-Pacific coral reefs.

Having surveyed this immense oceanic scene, we now turn to the fishes that live in coral reefs and atolls (plates 20, 21, 28 and 29 and figures 74, 75 and 76). Considering first the families that are most closely associated with corals, these include moray-eels, squirrel-fishes, cardinal-fishes, butterfly-fishes and angel-fishes, damsel-fishes, surgeon-fishes, wrasses, parrot-fishes (Scaridae), gobioids (Gobiidae and Eleotridae), trigger-fishes, file-fishes and puffer-fishes. Of the fishes that find much of their food over bottoms covered with coral sand and so on, we may mention red-mullet (Mullidae),

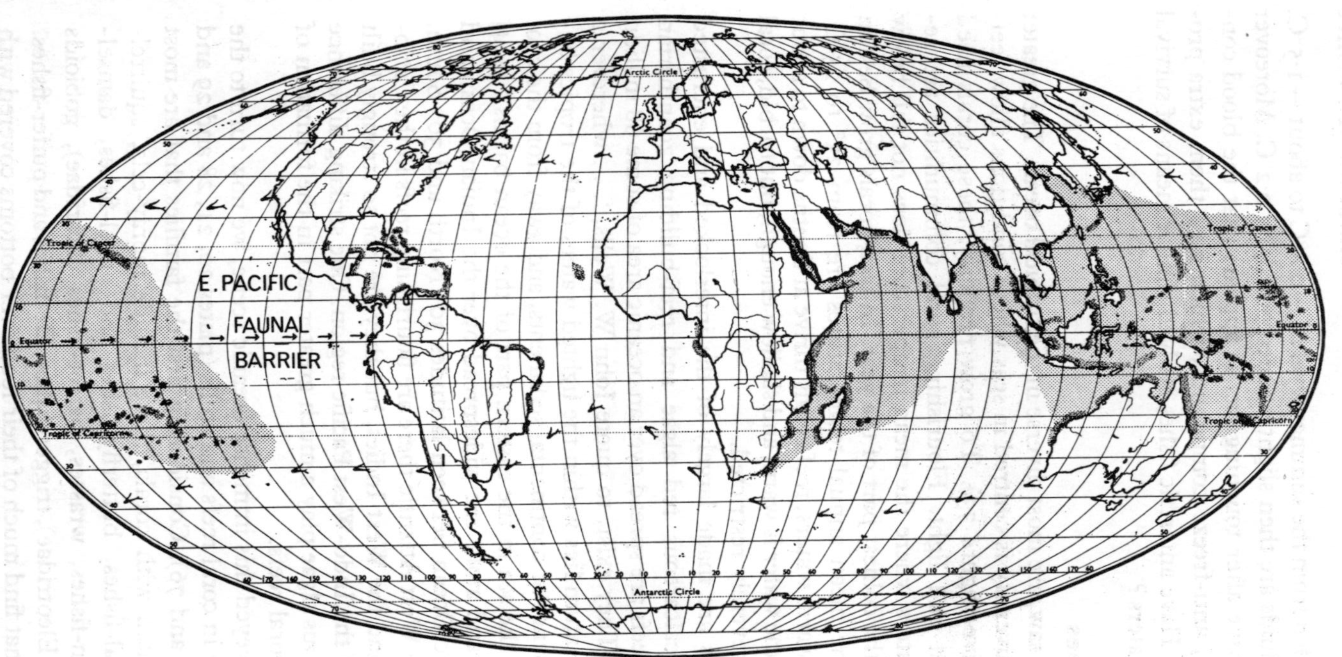

Figure 73. Coral forming parts of the ocean (dotted tint). Most of these are in the Indo-West Pacific faunal region (closer dots). Also shown are the northern and southern limits of distribution of flying-fishes and the East Pacific faunal barrier, which is bridged by the equatorial counter-current system (shown by arrows).

flounders, silver-perches (Leiognathidae) and flatheads. The small pelagic fishes that swim off coral reefs include round-herrings such as *Spratelloides* and *Jenkynsia*, silversides and half-beaks. Coral reefs are also full of predatory fishes, notably carcharhinid sharks, lizard-fishes,

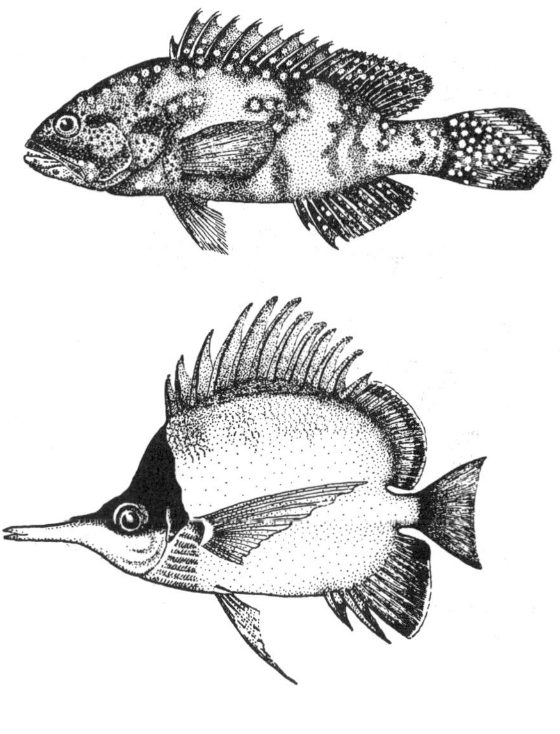

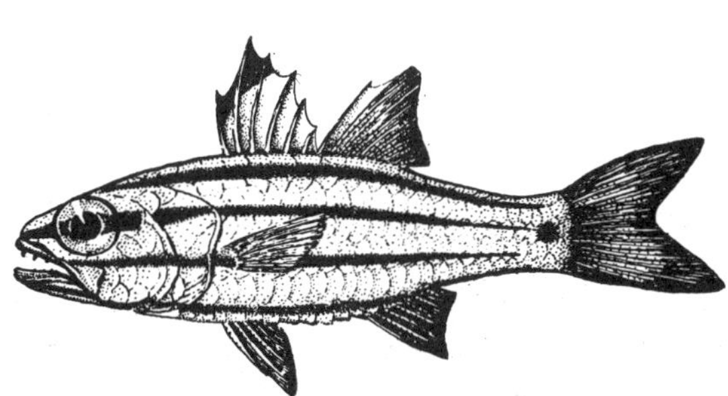

Figure 74. Three kinds of spiny-finned fishes found among Indo-Pacific coral reefs. Above: grouper, *Cephalopholis urodelus* Middle: butterfly-fish, *Forcipiger longirostris*. Bottom: cardinal fish, *Paramia quinquelineata*. (After L. P. Schultz.)

garfishes, barracuda, king-fishes (*Scomberomorus*), little tunny, amberjacks, groupers, snappers and scorpion-fishes. The entire list gives some impression of the diversity of fish life in coral reefs.

The shapes and colours of coral fishes reflect their diversity. The more intricate and profuse is a reef, the greater is this diversity and the total standing stock of fishes. There is, in fact, a great variation in the amount of fish life from one place to the next. In Hawaian waters, for instance, Dr Vernon Brock's surveys showed that the quantities of fishes ranged from over 2,000 pounds to as little as 7 pounds per acre.

Flourishing coral reefs seem to be among the most productive of all natural communities. Besides the unicellular plants (*Zoozanthellae*)

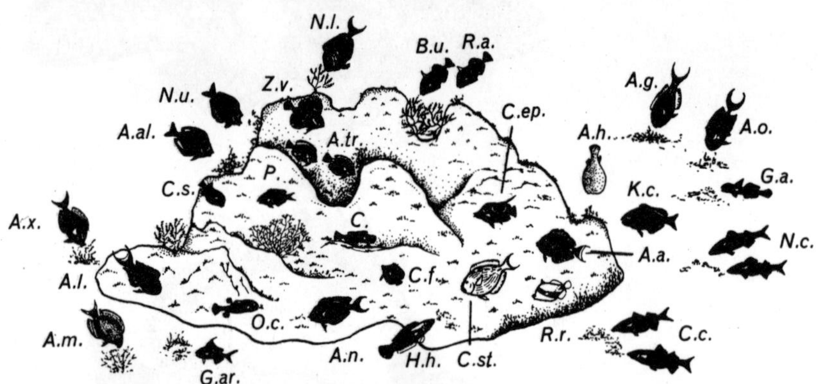

Figure 75. Plant-eating fishes of Indo-Pacific coral reefs (as seen in the Marshall Islands, by Hiatt and Strasburg [206]).
Surgeon-fishes: *Am, Acanthurus mata; Al, A. lineatus; Ax, A. xanthopterus; Aal, A. aliala; Ag, A. gahm; Ao, A. olivaceus; Atr, A. triostegus; Nu, Naso unicornis; Nl, N. lituratus; Zv, Zebrasoma veliferum; Cst, Ctenochaetus striatus.*
Puffer-fishes: *Cs, Canthigaster solandri; Ah, Arothron hispidus.*
Trigger-fishes: *Bu, Balistapus undulatus; Ra, Rhinecanthus aculeatus; Rr, R. rectangulus.*
Butterfly-fishes: *Cep, Chaetodon ephippium; Cf, C. flavissimus.*
Grey mullet: *Nc, Neomyxus chaptali; Cc, Crenimugil crenilabis.*
Other fishes: *Ga, Gnatholepis anjerensis; Kc, Kyphosus cinerascens* (rudder-fish); *Hh, Halichoeres loeveni* (wrasse)*; C, Cirripectus* (blenny)*; P,* Pomacentrid*; Oc, Ostracion cubicus* (box-fish); *Gar, Gerres argyreus.*

that are symbiotic with the coral polyps, filamentous algae grow on and within the calcareous skeleton. The coral community, according to Dr Eugene Odum [205], may consist of

three times as much plant as animal tissue. At night the polyps extend their tentacles and capture such plankton as may pass over the reef from oceanic waters; during the day the continuous sheet of algae manufactures food at a rapid rate, using nutrients released by consumers and decomposers. Food produced and dislodged in the shallow, swift water sections may be consumed by fish and other organisms in the quieter waters so that at least no more material is lost from the reef than is gained from adjacent oceanic communities. One reason there are so many fish on a reef (in addition to high primary productivity) is that so many of them are omnivorous or herbivorous and resort to 'grazing' on the algae and coral.

Surgeon-fishes (plate 28), which are circumtropical in range, are the leading group of herbivorous teleosts in coral reefs. They browse

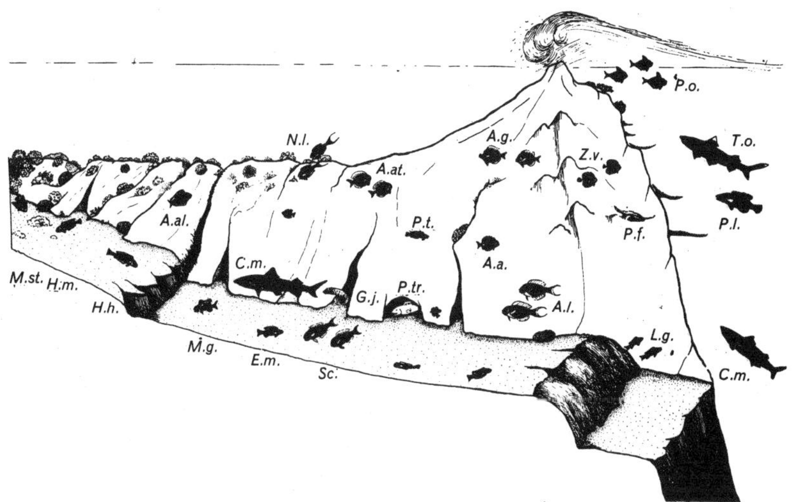

Figure 76. Fish fauna of the surf zone and surge channels of Indo-Pacific coral reefs (as seen in the Marshall Islands by Hiatt and Strasburg [206]).
Surgeon-fishes: *Ag*, *Acanthurus guttatus; Aa*, *A. achilles; Al*, *A. lineatus; Aal*, *A. aliala; Aat*, *A. atrimentatus; Zv*, *Zebrasoma veliferum; Nl*, *Naso lituratus.*
Sharks: *To*, *Triaenodon obesus* (white-tip shark)*; Cm*, *Carcharhinus melanopterus* (black-tip shark).
Wrasses: *Hh*, *Halichoeres hortulanus; Hm*, *H. margaritaceus.*
Other fishes: *Po*, *Pempheris oualensis* (sweepers)*; Pl*, *Plectropomus leopardus; Ptr*, *P. truncatus* (groupers)*; Pt*, *Paracirrhites forsteri* (hawk-fish)*; Lg*, *Lutianus gibber* (snapper)*; Sc*, *Scarus* sp. (parrot-fish)*; Em*, *Epinephelus merra* (grouper)*; Mg*, *Monotaxis grandoculis, Mst*, *Megaprotodon strigangulus* (butterfly-fish); *Gj*, *Gymnothorax javanicus* (moray-eel).

and graze on algae that grow over rocks, corals and sandy stretches of the sea floor. Their closely set teeth bear a cusped or serrated cutting edge, the entire dentition being thus well suited to the cropping of plants. All kinds of surgeon-fishes contain a long, coiled intestine, while some have a muscular gizzard. Rabbit-fishes (Siganidae) are also important plant-eaters in Indo-Pacific coral reefs. Like surgeon-fishes they have a small mouth armed with incisor-like teeth. Certain damsel-fishes, butterfly-fishes, blennies and gobies are also largely dependent on algal food, some of which is scraped from coral or rock surfaces (figure 75).

Besides cropping the seaweeds, a good many coral fishes depend also on invertebrate food, and may thus be classed as omnivores. Various damsel-fishes, gobies, trigger-fishes, puffers and trunk-fishes have such feeding habits. Trigger-fishes and trunk-fishes, for instance, take algae, crustaceans, echinoderms and molluscs. More often than not, the digestive tracts of omnivorous fishes contain both plant and animal food, but sometimes there are the remains of either plants or animals. The kind of food eaten may well be that most readily available.

Coral polyps bear tentacles that are armed with batteries of stinging cells. The planktonic food of corals, such as copepod crustaceans, is paralysed and caught by these cells. Now, one might suppose that fishes would be deterred from touching the polyps, which may well be true for many species, but certain kinds have taken to feeding on them. In fact, according to Robert Hiatt and Donald Strasburg:

> Three distinct evolutionary trends in dentition among a few families of fish have adapted them for accepting a live coral diet. For example, some butterfly-fish (Chaetodontidae) and the file-fish *Oxymonacanthus*, have developed pointed, produced snouts with very small terminal mouths containing fine, incisiform, protruding teeth for biting off individual polyps above the level of the corallite. Other butterfly-fishes crop the polyps so closely that they scrape off the tips of the individual corallites, and parrot-fish (Scaridae) have developed a strong protruding beak of fused teeth, taking not only the polyps but biting deeply into the calcareous corallus as well. Other groups, particularly the trigger-fish (Balistidae) and the puffers (Tetraodontidae) have developed very strong, heavy, protruding teeth (trigger-fish) or have them fused into a beak as in the puffers with which the ends of ramose and cespitose coral heads (*Acropora*, *Pocillopora*, *Stylophora*) are broken off and ingested. The comparatively small amount of animal tissue in relation to the mass of skeletal material requires that such fish keep the gut well crammed with these broken tips nearly all the time [206].

The pelagic fishes which feed on plankton are very numerous around reefs. The predatory species take other fishes and a variety of invertebrates. No more need be said about these kinds and we may pass on to consider some of the niches of coral fishes.

The tops of the windward ramparts of atolls and reefs are exposed to the pounding of breaking waves – waves that are set in motion by strong Trade Winds. The rush of the surf is broken by surge channels in the reef, which are themselves the creation of countless thrusting breakers. 'A view into one of these surge channels and out over the steep seaward slope is truly one of the most spectacular to be experienced by man. The reef edge and slope are almost solidly covered over by over one hundred species of corals of all shapes and hues, and the density and diversity of the fish fauna is exceedingly great.' In the Marshall Islands, continue Hiatt and Strasburg,

> Two species are so characteristic of the breaker zone, the sweeper, *Pempheris oualensis*, and the surgeon-fish, *Acanthurus guttatus*, that they are given special notation. *A. guttatus* particularly can be seen in large, fast-swimming schools in the white water of the breaking waves. The sweepers usually occupy positions near the head of shallow channels or just below the white water [206].

Other kinds of surgeon-fishes such as *Acanthurus achilles*, *Naso unicornis* and *Zebrasoma veliferum* live in the surge channels, together with sharks (*Carcharhinus melanopterus*, *C. menisorrah*), smooth dog-fishes (*Triakis*), moray-eels, groupers, parrot-fishes, blennies, trigger-fishes and so forth (figure 76).

Landward of the breakers, numerous fishes live in the reef flats: sharks, moray-eels, red-mullet, butterfly-fishes, surgeon-fishes, damsel-fishes, wrasses, parrot-fishes, gobies and blennies. The last two are characteristic of tide-pools. They are hardy fishes, able to withstand the heating of these pools by the tropical sun.

A coral community breaks the sea into a nexus of watery mazes. Such labyrinths form a good shelter for many fishes. Some of the smaller sleepers (eleotrids) and gobies, particularly the compressed, deep-bodied *Gobiodons*, rarely venture from their homes between the twigs of coral. The larger coral heads are attended by flocks of small damsel-fishes. They never go far from the coral, and take shelter in a flash if menaced by predators. If need be, the smaller wrasses, cardinal-fishes, trigger-fishes and file-fishes will also dart for nearby bolt-holes. During the daytime, squirrel-fishes hide in holes, caverns and ledges below nigger-heads of coral. They emerge at night – and they have large, wide open eyes – to feed on a variety of crustaceans, polychaet worms, molluscs and small fishes. Moray-eels squeeze their way into suitable coralline lairs. Certain species hunt at night, when they may well come across a sleeping wrasse or parrot fish.

Some of these fishes, at least, entirely invest themselves in a mucous sleeping-bag that is secreted by the skin. Now moray-eels have a very keen sense of smell, and it may well be that the sleeping-bag keeps the odours of its occupant from being wafted into the nostrils of these formidable predators.

Coral fishes are prominent among those species that swim by means that are not primarily centred on oscillations of the tail and fin. Moving their pectoral fins in a kind of breast-stroke, wrasses and parrot-fishes 'wing' themselves through the water. Trumpet-fishes and trigger-fishes move by flourishing their opposed dorsal and anal fins from side to side. Cruising file-fishes undulate these fins. Puffer-fishes and porcupine-fishes are impelled by side to side fannings of the dorsal and anal fins, but they also paddle with their pectoral fins. Using these precise and quickly adjustable ways of moving, all these fishes can neatly thread their way through coralline passages. During such intricate manoeuvres – and many of them are slim-bodied or deep, compressed fishes – the tail fin is more useful as a rudder. Is it not also significant that these species have rather small or tiny mouths ? This means that movements must be accurate and flexible if jaws are to be used effectively in feeding.

The colours of coral fishes have been considered elsewhere (pp. 185–186). We may now turn to the distribution of tropical fishes as a whole. There are four main faunas: the Indo-West Pacific, the Pacific American (Panamanian), the West Indian and the West African. Each contains many endemic species. A handful of species, such as the red-tailed trigger-fish (*Xanthichthys ringens*), the unicorn file-fish (*Alutera monoceros*) and the porcupine-fish (*Diodon hystrix*), are found in all four areas, which are much more alike at the generic level. Common genera include *Chaetodon* (butterfly-fishes), *Pomacentrus* and *Abudefduf* (damsel-fishes), *Gymnothorax* (moray-eels), *Fistularia* (tube-mouths), *Holocentrus* (squirrel-fishes), *Apogon* (cardinal-fishes), *Epinephelus* (groupers), *Lutianus* (snappers), *Upeneus* (red-mullets) and *Alutera* (file-fishes). There are many more.

The Indo-West Pacific shore fish fauna is much the richest of the four. More than a third of all known sea fishes live in this immense faunal region, which ranges from the Red Sea and the east coast of Africa across the Indian Ocean to the Western Pacific Ocean (figure 73). The richness of this 'great mother fish fauna of the tropics', as Dr George S. Myers calls it, may be seen in this contrast. 'On cold North Atlantic coasts there may be no more than fifty to two hundred species at any one locality, while at Amboina, in Indonesia, it is probable that fifteen hundred to two thousand will eventually be recorded from one small area.' [67] The West Indian fauna, which is the next most diverse, contains two-thirds or more of the truly tropical shore families that are spread over the Indo-West Pacific region.

The most easterly extent of the Indo-West Pacific fauna is along a line from the Hawaiian Islands south to the Tuamotu Archipelago (figure 73). Now a good many shore fishes – small kinds of damsel-fishes, cardinal-fishes, butterfly-fishes, gobies, blennies, and so forth, as well as larger forms – groupers, snappers, amberjacks, etc. – are found right from the African to the Polynesian limits of the Indo-West Pacific region. Some species are confined to a particular area, such as the Red Sea, the East African coast, Indian coasts, Indonesia, Hawaii, and so on. But few of the more widely spread species have managed to cross the Eastern Pacific and become established as part of the Panamanian fauna, which extends from the Gulf of California to Ecuador. The Eastern Pacific, stretching from the line between Hawaii and the Tuamotu Archipelago to the American Continent, and covering an east-west distance of about 3,000 miles (figure 73), is also a barrier to most of the marine invertebrates.

Besides the three circumtropical species already mentioned (p. 326) – and there are ten more – forty other tropical shore fishes have crossed the barrier. Some, like the tiger-shark (*Galeocerdo cuvieri*) the smooth hammerhead-shark (*Sphyrna zygaena*) and the spotted eagle-ray (*Aetobatus narinari*), are large and powerful fishes. But the smaller and more numerous species can hardly be suspected of having made their own way. It is more likely that most of them crossed the Eastern Pacific as planktonic eggs and larvae.

The planktonic life of most tropical fishes is probably less than three weeks. Even granted a drifting existence of thirty days, in order to beat the barrier, the currents would have to carry the young about a hundred miles per day, which is much more than the velocity of the Pacific Equatorial Countercurrent (figure 73) (the North and South Equatorial Currents flow away from America). But some of the barrier-crossers may well lead a long larval life. The bone-fish (*Albula vulpes*), and the six species of moray-eels (e.g. *Echidna zebra* and *Gymnothorax pictus*), have leptocephalus larvae, which can be adapted to very lengthy drifting existences (p. 273). Perhaps the two squirrel-fish crossers, like certain of their relatives, have spiny young that can live in the open ocean. We have already seen that the convict surgeon-fish has an advanced kind of larva, the *acronurus*, and a larval life of two and a half months. This is probably true of the three other surgeon-fishes that have breached the barrier. Butterfly-fishes also have an advanced and special kind of larva, the *tholichthys*, which perhaps applies to the two crossers, *Zanclus cornutus*, the Moorish Idol and *Forcipiger longirostris*, the long-beak butterfly-fish. On the other hand the three species of trigger-fishes and two species of file-fishes, some of whose relations can take to the open ocean, may have crossed as adults, perhaps swimming with drifting weeds and wreckage [207].

Even if it survives the crossing, an invading species still has to establish itself in strange surroundings. The adult must find a suitable living space and form a breeding stock. In this respect it is interesting that few of the invaders have settled along the mainland of tropical America. Most of them have become established around the offshore islands, such as the Galapagos, Cocos, Clipperton and Revillagigedos groups. Now the fish fauna of these islands is less diverse than that along the mainland, which may well mean that a newcomer will face less competition; an adaptable invader will stand a better chance of making an island living. On the land, too, island faunas are particularly liable to be overrun by adaptable intruders.

A good instance of this ecological principle is seen in the one-way passage of fishes through the Suez Canal, which is not due to a one-way flow of water. While over twenty species of Red Sea fishes have become established in the eastern Mediterranean, not one Mediterranean species is known to have gained a footing in the Red Sea. But there is a greater diversity of fishes in the Red Sea, where, *inter alia*, conditions are suitable for the growth of reef corals. Red Sea fishes would appear to have occupied most or all of the available living spaces.

At the beginning of this review of tropical shore fishes, mention was made of the cool upwellings of the sea along parts of the western coasts of the continents (figure 69). These upwellings occur in the Trade Wind belts, where the strong breezes blow surface waters away from the coasts. To replace these waters, cooler, underlying waters rise to the surface, bearing refreshing and rich supplies of nutrient salts. When the upwellings are most active, there is a great flowering of the plant plankton, which supports large standing stocks of animal plankton and fishes. Such conditions prevail off the coasts of Peru, California, Morocco, south-western Africa and north-western Australia. The monsoon system of the western Indian Ocean also stir upwellings of cool, nutrient-rich water off the coasts of

South Arabia and Somaliland. Though these upwellings are in tropical regions, local fish life is little or no more diverse than that in warm temperate waters. Apart from the paucity of coral, which provides few niches, conditions are changeable and much less stable than those in warm tropical waters. In the surface waters, conditions are particularly ideal for schools of plankton-feeding, clupeid fishes. In the Peru Current, for instance, there are immense stocks of anchovies (mostly of *Engraulis ringens*), which are preyed on by countless sea birds. Hence the rich guano deposits on land. In one year birds and man take at least four million tons of anchovies, which is about equal to one-seventh of the world's annual catch of fishes. Even more striking, these fishes live in a belt of coastal water less than 800 miles long and 30 miles wide – in an area that is no more than 0·02 per cent of the entire surface of the ocean [208].

Pilchards are the most prominent clupeid fishes off the coasts of California, south-western Africa and Morocco, the species being, respectively, *Sardinops caeruleus*, *S. sagax* and *Sardina pilchardus*. The second and third species, particularly the second, support important fisheries. There was a good fishery for the Californian species until 1952, when, for causes that are still not clear, the catches suddenly declined. Up to the present, there are no real signs of a recovery in the stocks.

The upwelling regions also support good stocks of demersal fishes. Just as herring-like fishes abound in the surface waters, so different species of hake are prominent in the waters near the bottom. The South African species *Merluccius capensis* supports a productive fishery; while *M. gayi* is an important resource in Chilean waters. *M. productus*, which lives off the Pacific coast of North America, might well repay exploitation.

Conditions in upwelling regions are cool and changeable enough to give a warm temperate character to parts of the tropical ocean. We are led, as it were, to look at conditions in the coastal waters of higher latitudes.

Fishes of temperate seas

In polar and tropical seas there is little change in temperature from one season to the next. Such change is greatest in temperate waters, which range from cold (subpolar) to warm temperate in character. North of Iceland, for instance, sea temperatures may be down to 0°C in February and up to 10°C in August. In the Mediterranean, which, except for the near-tropical, south-eastern part, may be regarded as a warm temperate sea, the annual range of temperature is from 12° to 13°C in February to 20° to 25°C in August.

The temperate (sub-antarctic) waters of the Southern Ocean lie between the Antarctic and Subtropical Convergences (figure 69). Most of the coastal waters in this region lie over the Chilean and Patagonian Shelves. But the entire shelf area, which includes that of the sub-antarctic islands and the south island of New Zealand, is very much less than that covered by the temperate shelf waters of the Northern Hemisphere. The southern fish fauna is also smaller.

The most diverse fishes on the Patagonian Shelf are nototheniids and eel-pouts (Zoarcidae); but herrings, cod-fishes and flat-fishes, which are so conspicuous a part of the fauna of north temperate seas, are poorly represented. Different species of nototheniids, eel-pouts and flat-fishes live in the Antarctic. In fact, only one species, a small, sculpin-like nototheniid called *Harpagifer bispinis*, lives in the coastal waters of both antarctic and sub-antarctic regions. The Antarctic Convergence, which virtually encircles the Southern Ocean, has evidently been a barrier for a long time to the free exchange of species between the two regions.

The basking shark and, so it seems, the Greenland shark, swim in north and south temperate waters. The hag-fish genus (*Myxine*), the skate genus (*Raja*) and the herring genus (*Clupea*) are also represented in both regions, but by quite distinct species. In fact, there is little resemblance, other than at the generic, or even the family level between the two temperate fish faunas.

Hydrographic similarities are also broad. One outstanding feature of the north is the meeting of warm north-going and cold south-going waters along the western sides of the continents. Just as the Gulf Stream impinges on the Labrador Current in the north-western Atlantic, so does the Kuroshio Current meet the Oyashio in the north-western Pacific (figure 69). Where these warm and cold currents meet, there are mixings and local divergences, which lead to upwellings that enrich the surface waters. Planktonic and fish life is thereby made abundant. The cold currents reach farthest in winter, but both warm currents swing eastward at much the same latitude: the Gulf Stream at Cape Hatteras, the Kuroshio at about 36° North along the coast of Japan.

In the north-western Atlantic and Pacific Oceans, the southernmost limit of the warm temperate fish fauna is close to the confluence of warm and cold currents. The northern limits extend to north Iceland and Scandinavia in the one ocean and to the Bering Sea and the Gulf of Alaska in the other. In the north-eastern parts of these oceans the warm temperate faunas reach to the Mediterranean and California.

The coastal fish fauna of the temperate North Pacific, which consists of more than a thousand species, is several times richer than that of the temperate North Atlantic. Rock-fishes (Scorpaenidae), sculpins (Cottidae), sea-poachers (Agonidae), sea-snails (Liparidae), kelp-fishes (Clinidae), eel-blennies (Lumpenidae) and flat-fishes are represented by different species in both areas, but they are more diverse in the North Pacific, which also contains certain endemic families, such as the greenlings and the surf-perches (Embiotocidae). However, there are more cod-fishes (Gadidae) in the North Atlantic. The two faunas are not closely related, though closely allied species of cod, herring and halibut live in both regions. Different populations of cod (*Gadus morhua*), herring (*Clupea harengus*) and halibut (*Hippoglossus hippoglossus*) occur on the American and European sides of the North Atlantic. Other species common to both sides include the spiny dog-fish, pollack, haddock, mackerel, lumpsucker and short-horn sculpin.

Temperate climates, whether in air or water, are neither hot nor cold. But in one way they are 'intemperate' compared to polar and tropical climates: in temperate waters there is a much greater yearly range of temperature. Conditions at the cold and warm ends of this range are liable to have profound effects on the lives of fishes. Scrutiny of the fine book by Henry B. Bigelow and William C. Schroeder on the fishes in the Gulf of Maine shows this very well [177]. For instance, a good many species such as the mackerel, silver-hake, scup and weak-fish, migrate coastwards to their spawning grounds in spring and summer. Before the colder season advances, when shallow inshore waters become chillier than those offshore, these species retire seaward to warmer, deeper waters. On the other hand, the cold-loving fishes of temperate waters, such as haddock and pollack, may have to retreat in summer from warm, inshore waters – where temperatures may be too high for their comfort – to cooler, deeper reaches. Fishes of temperate, coastal waters can also be subject to quick changes in temperature. A cold snap, for instance, may cause great mortality among the species that prefer warm, temperate seas.

Compared to tropical seas, temperate waters are intemperate in another respect. Though there may be times when it is more abundant, tropical plankton grows and reproduces throughout the year. In temperate seas, the plankton flourishes in spring and summer, but declines in winter, when, in concert with other conditions, the amount of light every day is not adequate for plant photosynthesis. Temperate fishes, particularly those of the colder waters, thus have less food at their disposal during the winter. A good many have to live on their own fat. This drastic change is reflected in the annual growth rings on their scales, bones and ear-stones.

In these respects and others, temperate fishes face conditions that are immoderate beside those that play on the lives of polar and tropical fishes.

In brief, the coastal faunas of temperate regions contain some of the most adaptable fishes in the sea.

Chapter 19

Freshwater Habitats

Conditions of life

Most kinds of freshwater fishes live in land-waters that are out of reach of the sea. Few tolerate brackish waters; and most of these must seek fresh waters when ready to spawn. Still fewer kinds, such as carp (*Cyprinus carpio*), bream (*Abramis brama*), roach (*Rutilus rutilus*) and barbel (*Barbus brachycephalus*), are more adaptable. Populations of these cyprinids live in the Caspian and Aral Seas, where they lay eggs that will develop at salinities as high as eight to ten parts per thousand (‰). It will be recalled that the mean salinity of sea water is 35‰.

Some fishes leave fresh waters to spawn in the sea. Freshwater eels (*Anguilla*), comprising some sixteen species that are widely but patchily distributed in land waters that flow into the three oceans, are the best known instances. They all breed in the sea, but spend most of their life in rivers and lakes, the females tending to remain longer than the males. Certain of the 'trout' (Galaxiidae) of south temperate lands also enter the sea or estuaries to breed. But more species make the reverse migration. They feed and grow fat in the sea and enter fresh waters when spawning time is near. Lampreys (*Petromyzon*, etc.), sturgeon (*Acipenser*, etc.), salmon, sea trout, steelhead trout, char, smelt (*Osmerus*), shad (*Alosa*, *Ilisha*, etc.) and alewives are among such fishes.

Some of these genera contain species, or have near relatives, that spend their entire lives in land-locked waters: there are no purely marine species, which suggests that they are primarily freshwater fishes. Perhaps the salmon and its allies took to feeding in the sea during the Ice Ages, when the melting of the ice caps led to a great freshening of the waters of northern seas. On the other hand, freshwater eels, like their marine relatives, lay eggs that produce leptocephalus larvae; and no *Anguilla* is known to pass its entire life history in lakes or rivers. This would imply that the ancestral *Anguilla* was a marine fish, for its descendants must still return to the sea to find the right subtropical conditions for spawning. Did the young of the ancestral species, after changing to elvers, become established in fresh waters, where they grew to maturity? It would be fascinating to know the past history of freshwater eels, but until we know more of the life of the Indo-Pacific species, further conjecture is hardly profitable.

There are no freshwater fishes in antarctic lands. On Antarctica, or on the surrounding islands, fresh waters, whether in the form of snow-melt streams or lakes, are only available for a brief time in the summer. The only animals that can live under these conditions are such forms as protozoans, nematode worms, tardigrades and rotifers. They spend most of the year as resting eggs, which hatch when snow and ice begin to melt. The successful ones reach maturity and produce eggs before their habitat is frozen again. But the Arctic, much of which has a less severe climate than the Antarctic, is accessible to some freshwater fishes. For instance, the pike (*Esox lucius*) lives in nearly all the main arctic river systems of Eurasis, while the blackfish (*Dallia pectoralis*), which exists in shallow ponds on the tundra, also extends north of the Arctic Circle. The arctic char (*Salvelinus alpinus*) runs into many arctic river systems to spawn. On Novaya Zemlya, mature and immature char spend the winter in lakes near the upper reaches of rivers. In American regions, arctic char extend even farther north, for they have been taken in fresh waters on northern Ellesmere Island at a latitude of 82° or more. The southernmost strictly freshwater fish would appear to be a South American pygidiid cat-fish, which extends to within 500 miles of Cape Horn. Southern trout (*Galaxias*, etc.), some of which spawn in brackish or salt waters, are found on Tierra del Fuego, the Falkland Islands, on certain sub-antarctic islands south of New Zealand, and in Australia and southern South Africa.

Arctic freshwater fishes live in freezing or near freezing waters, which is true of many temperate species during the winter months. At the other extreme, a few fishes, like certain species of the topminnow genus, *Cyprinodon*, can live in hot springs having a temperature of nearly 40°C. But most freshwater fishes live in waters of intermediate temperatures, particularly in those of warm temperate and tropical regions.

Like their marine relatives, freshwater fishes range into the depths of their environment. Lake Baikal is the deepest stretch of fresh water, where the maximum depth is 1,741 metres. During the winter, bullheads (of the subfamily Abyssocottinae) live on the bottom of this lake at depths down to 1,000 metres or more. However, most fresh waters are relatively shallow and nearly all of their fishes dwell at depths of 10 metres or less.

The deepest parts of Lake Baikal must be quite dark, though the fish that winter here move upward to lighted waters in the spring. Most cave fishes spend their entire lives in sunless waters and are perfectly able to make a living, even though virtually losing their eyes. A good many tropical fishes live in turbid waters but have developed other than visual means of finding each other and their food. Numerous others live in moderately to brilliantly lit surroundings and doubtless use their eyes to good effect.

Concerning their means of acquiring oxygen in diverse freshwater habitats, reference may be made to Chapter 6. One observation should be added. None of the carp-like fishes (Cyprinidae), which are found in temperate and tropical regions and number some two thousand species, have evolved air breathing organs. This is a remarkable omission, so to speak, in a remarkably successful group of fishes. But cyprinids are rare in stagnant swampy waters of the tropics.

Study of the fish fauna at different levels of a river system gives some insight to the interplay of these physical conditions of life. Let us consider a European system [187]. Starting in the hills, the stream is quick and turbulent as it rushes downwards. The water is clear and cold, being rarely over a temperature of 10°C, even in summer. Here one finds brown trout and bullheads. The first, as fishermen know, is an active, muscular fish with splendid lines and endowed with great stamina. At the same time, a trout has a great 'feel' for the more sheltered, quieter flowing parts of a lively stream. It needs much oxygen and its eggs develop in well-aerated redds during the autumn and winter. The bullhead will flatten itself against the gravel or wriggle under a stone. It looks a sluggish fish but it can seize its food in a quick dart. It lurks, strikes and dives for cover.

Proceeding down the slope, we reach the grayling (or minnow) reach. Here the current is brisk while the water is clear and cool. Summer temperatures are rarely above 15°C and the oxygen content is high. Besides grayling and minnow, one may see salmon, brown trout, brook trout, dace, chub and stone-loach. All but the loach, which stays close to the bottom, are active, muscular fishes with a shapely fusiform body. They are fishes with good staying powers.

The slope becomes gentler in the barbel region, where the current is moderate and the waters are often rather turbid. In summer the temperature is usually above 15°C. The oxygen content is high near the surface but may be lower at the bottom. As well as the barbel, there may be roach, rudd, bleak and perch; chub and dace may also be found. Here there are more fishes with deeper bodies and more leisurely ways of moving, such as the roach, rudd and perch.

Moving still more downstream, the river becomes wider and is bordered with dense vegetation. At this level the slope is slight, the current being moderate to slow. The water is often cloudy with detritus and is very warm in summer, reaching temperatures of 20°C or more. There is adequate oxygen near the surface; nearer the bottom the supply may be meagre. This is the bream region, where live bream, carp, tench, crucian carp and eels. All of these are able to live in poorly aerated waters. Again, we see stately-moving, deep-bodied fishes. Bream, carp, tench and crucian carp are certainly this kind of fish.

The river ends in the estuarine reach, where sea and land waters meet. Here the brackish waters carry much detritus. The temperature and oxygen content of the water are much as in the bream region. Characteristic fishes are flounders and sticklebacks (three- and ten-spined). Smelt and shad migrate here to spawn.

From beginning to end, the river thus unfolds, as it were, a succession of fishes. Such unfoldings, involving different species but a like fitness of form and function to surroundings, may be seen in many parts of the world. Fitness for life in torrents is strikingly displayed by small loach-like fishes (Homalopteridae), which live in the hill streams of India, Southern China, Malaya and Indonesia. Except for the tail, the underside of a homalopterid is flat and smooth. Much of this surface, which forms a sucker, is provided by lateral extensions of the paired fins. The forward pectoral rays and most of the pelvic rays are thick and flat and are fashioned below into adhesive pads (figure 77).

Using the sucker to cling and creep, these loaches move slowly over the surface of stones, grazing on the algal covering. Fish has almost become a limpet. Concerning suckers of the genus *Gastromyzon* in Bornean streams, here are some apt observations by R. F. Inger and C. H. Kong:

> In the swift, clean water, one first sees *Gastromyzon*, dark greenish or blackish fishes, 50 to 100 millimetres long, darting from rock to rock, or moving slowly over the surface of the boulders. They are not active fishes

but spend most of the time clinging to rocks. These fishes graze on algae growing on the substrate, and if they are not disturbed, one can observe them moving slowly over the surface of boulders leaving narrow naked trails from which they have scraped the algae [210].

In another hill stream fish, *Gyrinocheilus*, the lips form a suction disc, which means that the gills cannot be irrigated in the usual way. Water is drawn in through an opening at the top of each gill cover, which is separated from a lower outlet by a septum of skin. Some of the South American cat-fishes (Loricariidae) also have widened lips that unite to form a sucker. When clinging to stones – and certain

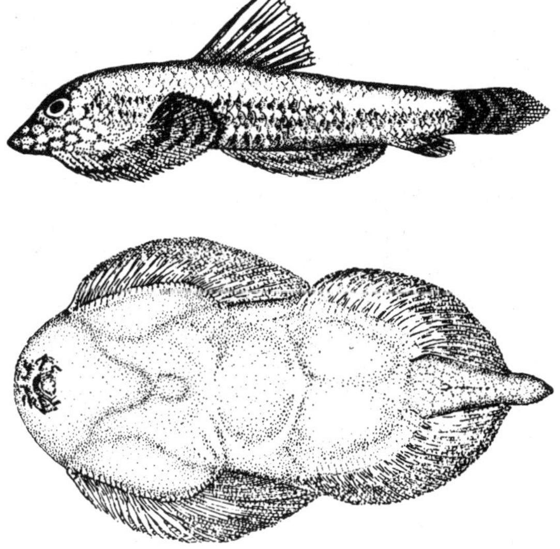

Figure 77. A homalopterid fish (*Sewellia lineolata*) from Asian torrential streams. The lower figure shows the suction disc formed by the expanded pectoral and pelvic fins and the flattened underparts. (After Hora.)

species live in torrential Andean streams – they breathe in much the same way as a *Gyrinocheilus*.

Where the gradient becomes gentler, torrents merge with less turbulent but briskly flowing waters. Here, as in the grayling reach of European rivers, one finds active fusiform fishes such as diverse characins and cyprinids. Spindle-shaped fishes are also found in quieter regions including lakes, but again, as in Europe, there are also deep-bodied, bream-like fishes. Such fishes, which move in precise and stately ways, are able to weave their way through the dense vegetation that grows in calmer freshwater habitats. The South American angel-fishes (*Pterophyllum*) are extreme instances of this type of fish. A deep-bodied fish would be out of place in quickly flowing waters, for it would expose too great a surface to the press of water when not aligned with the current.

The form of a fish is not only hydrodynamically apt, but conforms to its way of making a living. Deep-bodied fishes are usually mid-water feeders or browsers, though a few, such as the cyprinids *Chela* of Thailand and *Oxygaster* of Indonesia, have upturned mouths and feed near the surface. Many of the top-minnows (Cyprinodontidae) have such jaws and feeding habits. Freshwater fishes that feed along the bottom generally show some flattening of the underparts and bear sensitive lips. Barbels are commonly developed. Many catfishes, gudgeons and loaches have acquired these adaptations for a bottom-dwelling life. Moreover, some of these fishes, particularly species of loaches and gudgeons, live in quickly flowing waters and have somewhat flattened heads, followed by a cylindrical trunk and tail. They keep close to the bottom for much of the time, but are able to dart rapidly and briefly from place to place.

Many of the smaller freshwater fishes, particularly cyprinids and characins, depend a great deal on larval insects. The larger, predatory types, those that actively pursue large invertebrates, fishes and amphibians, have long jaws and very often an elongated body. The pikes are paradigms of this kind of freshwater fish. There are also pike-like cyprinids such as *Elopichthys* of China and the Amur River), characins (*Sarcodaces* of Africa) and perches (for example the pike-perch *Lucioperca* and *Luciolates*). But a proper survey of the body shapes and feeding habits of freshwater fishes could easily fill a book of this size.

Distribution

While most kinds of freshwater fishes live in tropical regions, north temperate lands contain quite a diverse fauna, principally of cyprinids, perches and sun-fishes. There are many fewer freshwater fishes in the south temperate zone. But whatever may be the reasons for this difference in diversity between north and south, there is the simple fact that the land masses in the south are much less extensive than those in the north. South of the Tropic of Capricorn are the southern parts of Australia, South Africa, South America and a number of small islands, including New Zealand and Tasmania. Representatives of one family of southern 'trout', the Galaxidae, are found in these regions. Of the overwhelmingly dominant group of freshwater fishes, the Ostariophysi, only a few species live in south temperate regions.

Arctic fresh waters are dominated by salmonid fishes, notably by species of char and white-fish. The salmonids range over most of the north temperate zone but are absent in the tropics. Both in temperate Eurasia and North America, cyprinids are the dominant fishes; their allies, the characins, are tropical, which is also true of most cat-fishes. One interesting difference is that North America is richer than Eurasia in perch-like fishes and suckers (Catostomidae). All but one of the eighty-odd species of suckers is confined to North America, which also contains two endemic groups of perches, the darters (Etheostominae) and the sun-fishes and basses (Centrarchidae).

Tropical Africa holds numerous endemic groups of freshwater fishes. The most diverse family is the Mormyridae (about a hundred and fifty species). The bichirs, lung-fishes of the genus *Protopterus* and two species of osteoglossoid fishes (*Pantodon buchholtzi* and *Clupisudis niloticus*) are also confined to Africa. But the dominant fishes are members of the order Ostariophysi, particularly of the family Cyprinidae. Africa is also rich in spiny-finned fishes, the most diverse of which are cichlids. These abound in the western and central tropical regions, notably in the great lakes such as Victoria, Tanganyika and Nyasa. There are also a number of spiny-eels (Mastacembalidae) and labyrinth fishes (Anabantidae), which also occur in oriental regions. Top-minnows also live in Africa. Indeed, they are found in most tropical and warm temperate parts of the world, but not in Australia.

Characins and cat-fishes are much more diverse in South America than in Africa. There are about eight hundred known species of characins and a thousand species of cat-fishes in South America (the corresponding figures for Africa are a hundred and thirty and two hundred and fifty. But South America is without cyprinid fishes. Their ecological place, so to say, is largely taken by the characins (figure 78). Like Africa, South America contains an endemic group of electrical fishes, the knife-fishes and the electric eel. Another endemic is the lung-fish *Lepidosiren paradoxus*. There are also numerous cichlids, which extend into Central America and Mexico. Viviparous top-minnows (Poeciliidae) live in South and Central America, but they also range into the West Indies and southern regions of the USA.

The oriental lands, particularly those of tropical Asia and Indonesia, are extremely rich in cyprinid fishes (figure 78). The related loaches (Cobitidae) are also most diverse in the orient, while the hill-stream loaches (Homalopteridae) are confined to South-east Asia.

There are also numerous cat-fishes. Spiny-finned fishes include the snakeheads, spiny-eels and labyrinth fishes.

There remains Australia, which has an unusual fauna of freshwater fishes. The Australian lung-fish (*Neoceratodus forsteri*) is confined to the Mary and Burnett river systems of south-eastern Queensland. An osteoglossid (*Scleropages leichhardti*) also lives in tropical Queensland, and in southern New Guinea. Perhaps these two are the only true freshwater fishes in Australia. The others, such as the gobies and silversides, have come from the sea, which may well be true of the galaxiid 'trout' and 'smelt' (Retropinnidae). As Dr P. J. Darlington concluded: 'There is no mystery about the origin of most Australian freshwater fishes. Except (perhaps) for the lung-fish and osteoglossid, they have come directly from or through the sea.' [209]

Consideration of the Australian fauna recalls Wallace's Line, which was named by T. H. Huxley after the naturalist, Alfred Russell Wallace, who traced a boundary between the vertebrate faunas of the Oriental and Australian regions (figure 78). Wallace concluded that the Philippine Islands, Borneo and Bali marked the easternmost limits of the Oriental region. Between these islands and the Australian region came the separation between two different faunas of vertebrates. Concerning the fishes, and noting the two possible exceptions mentioned in the preceding paragraph, there are no strictly freshwater species east of Wallace's Line. This is certainly true of Celebes, which is the nearest island east of Borneo, though on Lombok and Sumbawa, east of Bali, there are a few cyprinid fishes. But so far as fishes are concerned, if one shifts the southern part of the line a little to the east, the entire line does mark a barrier.

During Pleistocene times, when the formation of the ice caps led to considerable lowerings of sea level, there may well have been a land bridge extending from Malaya eastward to Wallace's Line. True freshwater fishes were able to spread as far as the Line, but beyond was a barrier of sea. Hence the present discontinuity in the faunas of freshwater fishes. The present distribution of fishes may even reflect the river systems of this Pleistocene bridge. For instance, the fishes of Sumatra and western Borneo are more alike than those of the western and eastern parts of Borneo. According to P. J. Darlington,

> The explanation . . . is that when the islands were connected with each other and the mainland, the great land area they formed was drained by several separate rivers, especially one flowing north between what are

now Sumatra (and the mainland) and Borneo, and one flowing east between what are Java and Borneo, and that the fishes are still distributed according to the old river systems [209].

Here, and elsewhere, one can only begin to understand present distribution by trying to reconstruct the past.

But how is the past to be reconstructed when we consider the fish faunas of South America and Africa? The African fauna is most closely related to that of the Orient; for these two regions share such groups as the featherbacks (Notopteridae), cyprinids, loaches, bagrid and clariid cat-fishes, the anabantids, snake heads and spiny eels. There is a connexion between these two land masses, but how is one to explain the fact that there are characins and cichlids in Africa and South America? (figure 78).

Some geologists and geophysicists consider that there is now good evidence in favour of Wegener's Hypothesis, which postulates that during early Cretaceous times the continents of America and Africa drifted apart. If there were characins and cichlids in the ancient undivided land, this could neatly explain present distributions. Fossil evidence for or against this idea is lacking, but there are other difficulties. If Africa and South America used to be one, why are their modern fish faunas not more alike, even considering that lung-fishes osteoglossids and cat-fishes, though of different families, are common to both? Why are there no cyprinids in South America? or bichirs? or mormyrids? and so forth.

The past can be reconstructed in another way. The first fact to remember is that the South American fish fauna is overwhelmingly dominated by characins and cat-fishes. In fact, 'the limitations and unbalance of the South American freshwater fish fauna suggest that it is a derived one, descended from a few immigrants which somehow reached South America from Africa' [209]. Dr Darlington considers that these migrants came to North America from eastern Asia by way of a Bering Bridge. Developing this idea, he argues that the Ostariophysi originated in the tropics of the Old World during Cretaceous times. These gave rise to the characins and cat-fishes which subsequently spread across Asia and invaded North America. Spreading south, they finally entered South America through Central America. The cyprinids are considered to have evolved from characins. Radiating in their turn, they invaded the whole of Eurasia and reached Africa and North America. In Eurasia and North America they replaced the characins, while in Africa they became much more diverse than their ancestors.

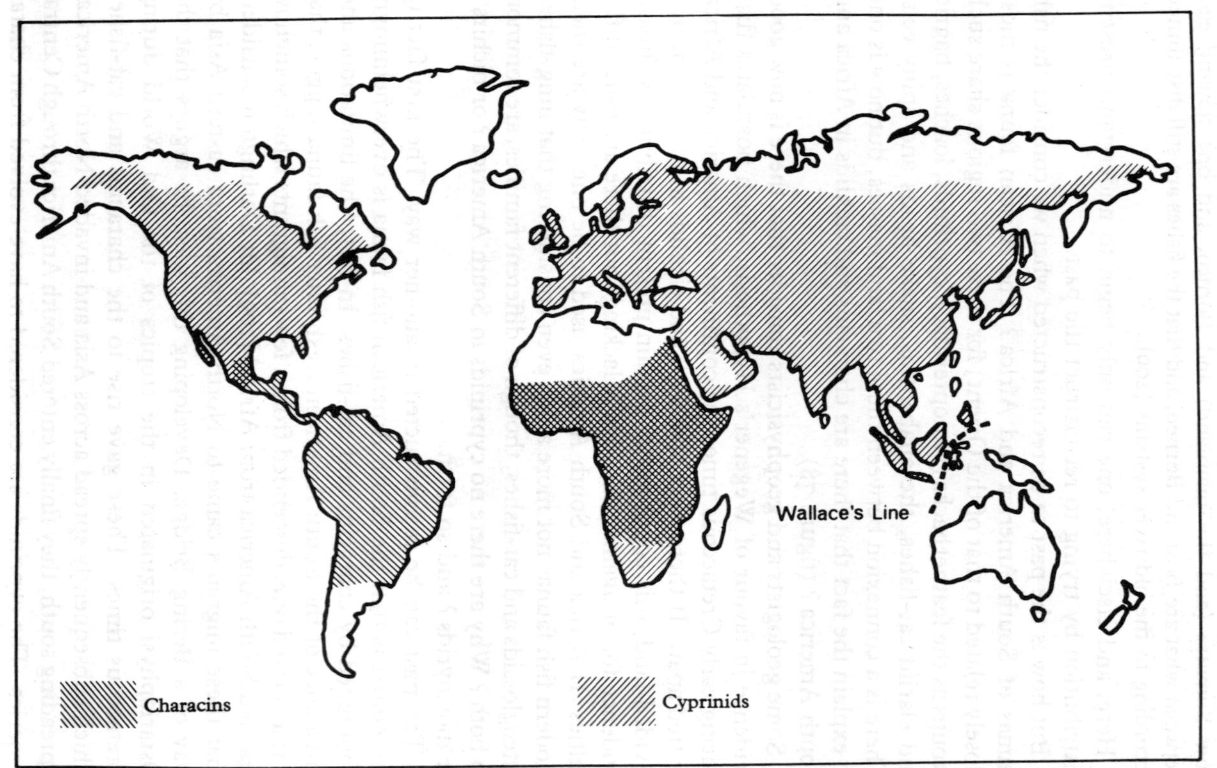

Figure 78. Distribution of the cyprinid fishes and the characins

But why have the cyprinids never crossed Central America? (figure 78). Today, North America cyprinids range no further than Southern Mexico, while their relatives the suckers (Catostomidae) and the ameiurid cat-fishes extend well into Central America, though in decreasing numbers of species. Moreover, a number of South American characins range into Central America. All in all, this region seems unsuitable for cyprinid fishes. Perhaps it was so in the past.

Diversity

Chapter 20

The Diversity of Fishes

PIERRE TEILHARD DE CHARDIN looked on fishes as an 'assemblage of monstrous complexity' [211]. While this is hyperbole, they do form the most diverse and protean group of vertebrates. We know more than twenty thousand living kinds, but our inventory is by no means complete. Each year, as may be seen in the *Zoological Record*, many new species are described. They come from underground streams, rivers, lakes, swamps, inshore waters, coral reefs, the deep sea, and so forth. Each discovery adds to our realization that fishes are masters of many aquatic habitats. Having seen something of their great adaptability, which has taken up most of this book, we may now consider the nature and extent of their diversity.

In the widest sense, fishes are aquatic, gill-breathing, cold-blooded vertebrates, which bear fins that are stayed by an inner skeleton of rod-like fin-rays. If the Pisces are so regarded, the jawless lampreys and hag-fishes are just as much fishes as are the two jaw-bearing groups: the cartilaginous sharks, rays and chimaeras (Chondrichthyes) and the bony fishes (Osteichthyes). The Placodermi, the third main group of jawed fishes, are unknown beyond Permian times.

All fishes have a head and a definite brain, which is protected by a cranium and connected by nerves to elaborate sense organs – the eyes, ears, nasal organs, taste buds, tactile endings and lateral line system. In particular, the ears have one, two or three semicircular canals and contain otoliths or ear-stones that are suspended close to cushions of hair-bearing sense cells, the maculae. The axial framework of the trunk and tail is the vertebral column, which forms the inner attachment of a nested series of muscle segments. Most fishes move by the activity of these muscles. The fins are used as propellors, rudders, stabilizers, brakes, hydroplanes, or even as limbs. The tail fin is the most versatile, for it may serve all of the first four functions.

The jawless fishes

Besides their lack of proper jaws, the lampreys and hag-fishes differ from the jaw-bearing fishes in being without paired fins and true teeth (figure 79). The teeth are thin, horny structures, and in lampreys some are set on the fore-part of a muscular, protrusible tongue, which in the parasitic species is used to draw blood. Hag-fishes feed on dead or dying fish, but they also take worms and crustaceans.

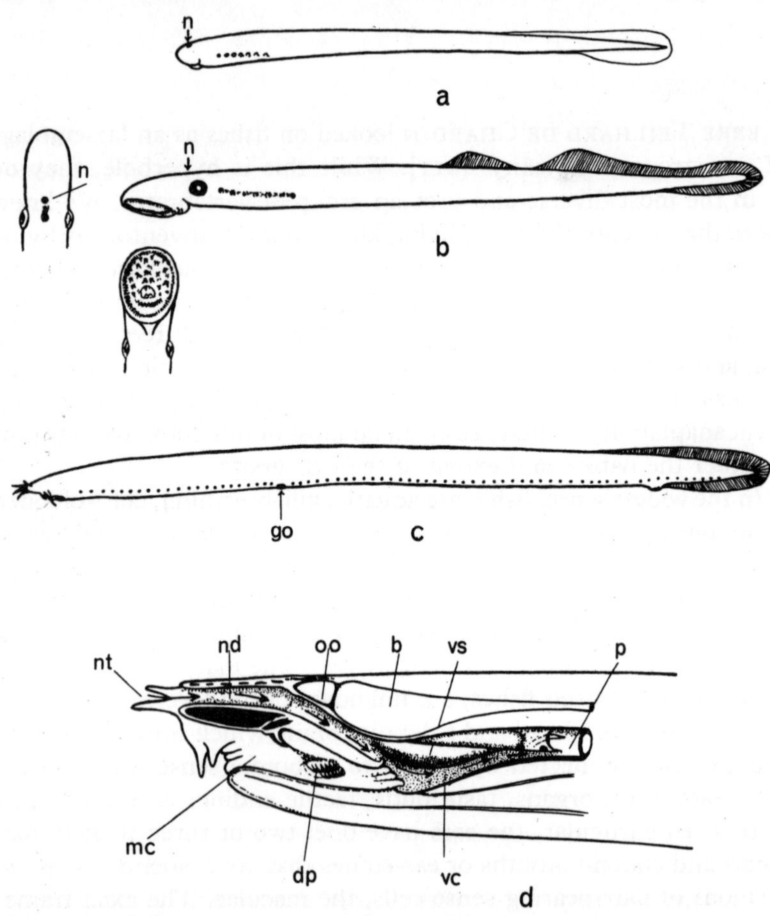

Figure 79. Jawless fishes. The sea lamprey, *Petromyzon marinus* (b) and its ammocoete larva (length 1¼ inches). (a); (c) the hag-fish *Myxine glutinosa* and (d) a longitudinal section through the head. (nd) nasal duct; (n) nostril; (nt) nasal tentacle; (go) gill opening; (oo) olfactory organ; (b) brain; (vs) velar scroll; (vc) velar chamber; (p) pharynx; (mc) mouth cavity; (dp) dental plate. ((a–c) drawn from specimens in the Museum of Comparative Zoology, Harvard University, (d) after R. Strahan.)

Their teeth are fixed to two gristly plates, one on each side of the buccal cavity. Adult lampreys also bear horny teeth on the buccal disc, the circular rim of which forms the adherent part of a suction-cup, a device for clinging to stones, rocks and fish prey (figure 79).

All jawless fishes have a single, median nostril, not one or two on each side of the snout, as in jaw-bearing fishes. The gill system is also different. The vascular, oxygen-catching gill filaments are developed on the inner walls of a paired series of gill pouches in lampreys and hag-fishes and within gill slits in the jawed fishes. The inner ears of jawed fishes and the higher vertebrates have three semicircular canals. In jawless fishes the horizontal canal is absent and there are but two vertical canals in the case of lampreys or a single vertical element in hag-fishes. Lastly, the skeleton of lampreys and hag-fishes, but not of their fossil relatives, is entirely cartilaginous.

Apart from such similarities, lampreys and hag-fishes have an eel-like form and a slimy, scaleless skin. But they are not closely related. Lampreys have the nostril on top of the head and it leads to the nasal organs and a blind hypophysial sac. Hag-fishes bear a terminal nostril, while the hypophysial part opens into the pharynx, the entire canal serving as a respiratory inlet (figure 79). The tip of a hag-fish's snout also carries six fleshy barbels, which are absent in lampreys. The perfectly plain snout of a lamprey overlies the wide sucker-like mouth: hag-fishes have a small non-suctorial mouth. To find their food, hag-fishes rely largely on their well-developed nasal organs, for their eyes are degenerate. Lampreys have perfectly formed eyes, which are certainly put to good use by the parasitic species.

There are some fifteen species of hag-fishes and all but one, which was trawled in deep water off Panama, live in temperate seas off North America, Europe, Japan, New Zealand, Chile, Patagonia and South Africa. They lay large yolky eggs in a horny shell and the young that hatch out of them have much the same organization as the adults.

Lampreys lead a very different and extraordinary life, one that begins in prepared gravel beds in streams, where the female sheds many small eggs that are fertilized by the males. After spawning the fish are so weak that they die. The eggs hatch in about a fortnight and the larvae, known as ammocoetes, are quite unlike their parents. These ammocoetes or prides are worm-like creatures with minute eyes (figure 79). They live in burrows in the sand or mud, emerging at night to feed on the organic detritus on the bottom. This is strained through processes called cirri in the hood-like mouth and passed to the pharynx, where the food particles are caught by the sticky secretions of the endostyle, a homologue of the thyroid gland. After at least three years of this larval life, when the ammocoete has attained a length of 4 to 6 inches, there is a rapid metamorphosis to the adult form. The hood round the mouth is replaced by the buccal disc; the cirri give way to teeth; the eye becomes larger; and the nostril moves from the front to the top of the head. Before very long the sea lampreys, such as *Petromyzon marinus*, become silvery and large-eyed in readiness for their migration to the ocean. The non-parasitic species, which belong to such genera as *Lampetra* and *Ichthyomyzon* and make up about half the family Petromyzonidae, take no more food after metamorphosis, their teeth being small and the gut reduced. They never leave fresh waters, but are provided with adequate food reserves for the ripening of the sexual cells. Their death after spawning is understandable.

The thirty-odd kinds of lampreys live in temperate regions of the Northern and Southern Hemispheres. The southern genera (*Geotria* and *Mordacia*) are found off the coasts of Australia, New Zealand and Chile. Like all lampreys, they spawn in fresh waters. *Petromyzon marinus*, the best-known species, lives on both sides of the North Atlantic, while *Lampetra* is found in North America and Eurasia. The several species of *Ichthyomyzon* are confined to fresh waters in Eastern North America.

The jaw-bearing fishes

During early development, the skeleton of fishes and of other vertebrates consists entirely of cartilage. Young fishes grow fast, and cartilage, being a plastic tissue – one capable of growing rapidly by inner expansion – makes an apt embryonic skeleton. It even forms, as we have seen, the entire adult skeleton of the jawless fishes. But the cartilaginous skeleton of adult sharks, rays and chimaeras is reinforced by mosaics made of bony plates. True bone in these fishes is also found as a component of the dentine of the teeth and scales. More precisely, the fibrous framework of true bone is hardened by apatite, a dense mineral substance which is a complex mixture of calcium phosphates and carbonates. The bony fishes have such a skeleton, which replaces the earlier, cartilaginous template.

There are also large-scale differences in the skeletons of cartilaginous and bony fishes. The upper jaw of sharks, rays and chimaeras is a palatoquadrate cartilage, and to this and the lower jaws of the first two groups is fastened a band of serially replaceable teeth. As the functional teeth wear out those next behind them move forward to take their place. In a few bony fishes tooth-bearing palatoquadrates may also serve as jaws, but in nearly all species the biting, tooth-bearing bones of the upper jaws are premaxillae and maxillae, or simply the former alone. Two other contrasts concern the scales and fin-rays. The scales of cartilaginous fishes have the structure of teeth; those of bony fishes are not so built. The outer fin-webs of cartilaginous fishes are stiffened by hundreds of hair-like, horny rays or ceratotrichia. The soft fins of bony fishes are supported by large, jointed rays, the lepidotrichia, which are made of bone (see also p. 63).

Bony fishes have a swimbladder, which is used to breathe air in the case of lung-fishes, bichirs, reed-fish, garpikes, the bowfin and a few teleosts. In most teleosts it is basically a hydrostatic organ, a float fitted to keep the underwater weight of a fish at vanishing point. Having no swimbladder, cartilaginous fishes are heavier than water and when swimming, they must produce a sustained lift force if they are to keep their level. This contrast in buoyancy between the two groups is reflected in their fin patterns.

Cartilaginous fishes are carnivores. To track their prey, they make great use of their large olfactory organs and, most likely, of their lateral line system. Vision, and nearly all species have nocturnal retinae, is less important. Bony fishes may be herbivores, omnivores or carnivores. Most species are visual feeders and have a sense of colour.

Two other physiological contrasts concern reproduction and water balance. In all cartilaginous fishes fertilization is internal, the sperm being passed to the female by way of grooved claspers, which are formed along the inner margins of the pelvic fins. The eggs of most bony fishes are fertilized as they are laid. When fertilization is internal, the intromittent organ is not derived from the pelvic fins. Concerning water balance, the salt content of the body fluids of fishes is less than that of sea-water, which means that water will tend to pass from a marine fish to its environment. To stop this leakage, sharks, rays and chimaeras make up the difference in salt content by retaining a requisite content of urea in their body fluids. Bony fishes have other ways of keeping an appropriate water balance, which were discussed in Chapter 8. Very few sharks and rays challenge their mastery of freshwater habitats.

The cartilaginous fishes (Chondrichthyes)

Every living cartilaginous fish is matched by more than thirty species of bony fishes and most of these are teleosts. There are some two hundred species of sharks, three hundred and fifty species of rays and twenty-five species of chimaeras.

Sharks and rays (Elasmobranchii) are readily distinguished from the chimaeras (Holocephali). The former have five, six or seven pairs of gills and gill clefts and each cleft has its own opening to the water. The teeth are numerous and the upper jaw is not fused to the cranium. Chimaeras bear four pairs of gills and gill clefts, but the latter lead into left and right gill chambers, each being covered by a fold of skin with a single, exhalant opening at the rear. The teeth consist of six pairs of grinding plates and the upper jaw is fused to the cranium. As well as the pelvic claspers, the males have an erectile clasping device, the tenaculum, in front of each pelvic fin and there is usually another such organ on top of the head. Chimaera is a good group name for such fishes.

It is easy to tell a shark from a ray. Sharks (Selachii) have laterally opening gill clefts, before which the edges of the pectoral fins are free from the sides of the head (plate 10). The gill clefts of rays (Batoidei) open on to the underside of the body and the edges of the pectoral fins, which are usually very large, are fused to the sides of the head before the gills (plate 3). Rays with smaller pectoral fins, notably the saw-fishes (*Pristidae*) have a shark-like appearance. On the other hand the angel-fishes (Squatinidae), which belong to the shark group, have large pectoral fins and a flattened ray-like form.

Sharks (Selachii)

Sharks with a single dorsal fin, an anal fin, and six or seven pairs of gill clefts include the frilled-shark (*Chlamydoselachus*), the six-gilled sharks (*Hexanchus*) and the seven-gilled shark (*Heptranchias*). The first, which has six pairs of gill clefts, is known from rather deep waters in the Atlantic and off California and Japan. It is an elongated, brown-coloured shark with a terminal mouth and tricuspid teeth in the jaws (in most sharks the mouth is on the underside of the head). The six-gilled shark *Hexanchus griseus*, which reaches a length of seventeen feet (occasionally 25 feet), has a world-wide distribution in warm temperate and tropical seas. All six- or seven-gilled sharks are ovoviviparous. The young, which soon emerge from large, yolky eggs, develop within the oviducts of the mother until they are fully formed.

The galeoid sharks and Port Jackson sharks (*Heterodontus*) have two dorsal fins, an anal fin and five pairs of gill clefts. Each dorsal fin of Port Jackson sharks is headed by a stout spine. They live in parts of the Indian Ocean and Pacific Ocean but not, it seems, in the Atlantic and the Mediterranean. Port Jackson sharks lay cylindrical eggs with a spiral frill around the horny case. In captivity the Californian species (*Heterodontus californicus*) lays single eggs during February or March. After about seven months incubation, the young that emerge are about 8 inches long.

The Galeoidea, which form the most diverse group of sharks, have spineless dorsal fins. The whale-shark (*Rhincodon typus*), the dog-fishes (*Scyliorhinus*) and certain of the carpet-sharks such as *Chiloscyllium*, produce large eggs enclosed in horny cases; but the rest are either ovoviviparous, or, less commonly, viviparous. The viviparous species include certain hammerhead-sharks (*Sphyrna*), smooth dog-fishes (*Mustelus*), *Carcharhinus* spp. and the blue shark (*Prionace*). The young develop in special uterine pockets of the oviducts and they are nourished through a yolk-sac placenta which has folds and processes that interdigitate with pleats in the mother's uterine wall.

Besides those already mentioned, the galeoid sharks include the basking-shark (*Cetorhinus maximus*), the sand-sharks (Odontaspidae), the mako-sharks (Isuridae), the smooth-sharks (Triakidae) and the threshers (Alopiidae). The whale-shark reaches a length of 60 feet; the basking-shark, when adult, may span 30 feet. They exist entirely or partly on planktonic animals (mostly crustaceans) that are engulfed with seawater. The food is screened by gill rakers set across the inner entrances to the gill clefts. In both species the gill rakers appear to be modified scales, which have become long bristles in the basking-shark and spongy, filter beds in the whale-shark – structures that rest on cartilages spanning the gill arches. The other galeoid sharks, which vary in size from 2 or 3 feet in the case of dog-fishes to over 30 feet in the white-shark, *Carcharodon carcharias*, take prey ranging from small crustaceans to large fishes. Thresher-sharks use their long, flail-like tail fin to herd schools of such fishes as herring, mackerel and bonito. The prey is lashed into a tight school, when they are presumably easier to attack.

Spiny-finned sharks (Squaloidea), angel-fishes (*Squatina*) and the saw-sharks (Pristiophoridae) are also five-gilled but they lack an anal fin. All, it seems, are ovoviviparous. Saw-sharks, which look very like the saw-fish rays (Pristidae), have the snout produced into a long rostrum, armed on each side with sharp teeth. But in bearing lateral gill clefts and pectoral fins that are free from the head, they are true selachians. The saw-sharks, which have been taken in the Western Atlantic and parts of the Indian and Pacific Oceans, reach about 4 feet. Underneath the rostrum is a pair of long barbels, which doubtless help in detecting food on the sea floor.

The ray-like angel-fishes, which have a flattened body and wing-like pectoral fins, are also bottom dwellers. The wide terminal mouth is armed with sharp conical teeth. Angel-fishes lurk on the bottom, usually disguised by a covering of sand or mud. They prey on fishes, crustaceans and molluscs. The several species are found on both sides of the Atlantic, in the Mediterranean and off South Africa, Japan, Australia and the coasts of America.

In most of the squaloid sharks each dorsal fin is headed by a strong spine. The fifty odd kinds of spiny dog-fishes (Squalidae) are found over much of the ocean. Besides the piked dog-fishes (*Squalus*), which live mainly in coastal seas, the family includes dark-skinned, deep-sea genera, such as *Oxynotus*, *Etmopterus* and *Centroscymnus*. Two luminescent deep-sea sharks (*Isistius* and *Euprotomicrus*) belong to the family Dalatiidae, which also includes the Greenland shark (*Somniosus microcephalus*).

Rays (Batoidei)

Saw-fishes (Pristoidea), guitar-fishes (Rhinobatoidea), electric-rays (Torpedinoidea), skates (Rajoidea), sting-rays, eagle-rays, cow-nosed rays, and devil-rays (Myliobatoidea) form the main groups of rays. The saw-fishes live in tropical coastal seas, some reaching a length of 20 feet or more. The snout extends into a flat, narrow blade, which is armed on each edge with a series of sharp teeth. The saw is used to forage along the bottom for food, or it may even be slashed from side to side, so stunning fishes in a school. A saw-fish's tail is robust and shark-like in appearance, the caudal fin and two dorsal fins being well developed. This is also true of the guitar-fishes and torpedo-rays, but in skates and myliobatoid rays the tail is slender to whip-like in form and the median fins are reduced or absent. In the last two groups the tail is of little or no use in locomotion, which is taken over by the muscular and wing-like pectoral fins.

The main skate genus (*Raja*), comprising about a hundred species, is the most diverse genus of cartilaginous fishes. Skates feed on crustaceans, molluscs, worms and fishes. To take active prey such as fishes, they pounce and then cover the victim with their wings.

The myliobatoid rays are close relatives of the skates, though they are subtropical and tropical rather than temperate fishes. They range in width from a few inches to over 20 feet in the devil rays (*Mobula* and *Manta*). The sting-rays (Dasyatidae) have a venomous spine (sometimes two) at the base of the tail. They live at the bottom of shallow waters and feed on molluscs and crustaceans. Eagle-rays have largely abandoned bottom-living habits and swim gracefully in mid-water by bird-like strokes of their pectoral fins. The teeth are broad and flat, well suited to crush hard-shelled molluscs and crustaceans. Devil-fishes (Mobulidae), which have similar ways of swimming, take all or most of their food at mid-water levels. Small, schooling fishes and planktonic animals are guided by the horns into the mouth. The food is retained by sieve-like series of gill plates that are placed parallel to the gill filaments.

Chimaeras (Holocephali)

Chimaeras live near the bottom in coastal to rather deep slope-waters. They have large eyes, a slender tail and mobile dorsal and pectoral fins (figure 80). The first dorsal fin is preceded by a spine, which may be venomous, and from an erect position it can be folded into a groove on the back. The large pectoral fins, unlike those of sharks, play a part in locomotion; as Earl Herald puts it, they can be seen sweeping '... the water gracefully and slowly, making the fish look very fragile, and for some reason mouse-like' [118]. However, the chimaeras are moderately large fishes, ranging from 2 to 6 feet in length. They are all oviparous, and have been seen to lay two eggs at the same time, one from each oviduct. As suggested by their large, highly sensitive eyes, they are more

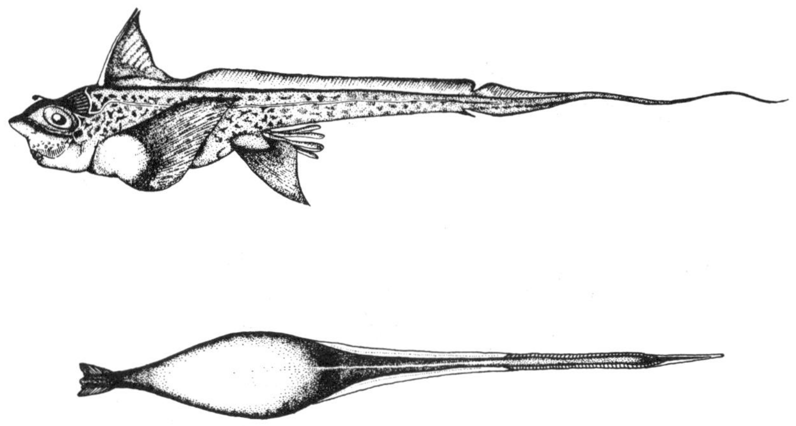

Figure 80. A male *Chimaera monstrosa* and an egg. (After A. P. Andriashev.)

active by night than by day, when they feed on small invertebrates and fishes.

The two most diverse genera, *Chimaera* and *Hydrolagus*, have a rounded or conical snout. The rhinochimaerids, such as *Harriotta*, have a long pointed snout. Most remarkable of all, the elephant-fish (*Callorhinchus*) bears a flexible, hoe-shaped appendage at the end of its snout. It would be interesting to know its use.

Bony fishes (Osteichthyes)

Bony fishes are either tassel-fins (Crossopterygii) or ray-fins (Actinopterygii). Their likenesses point to a mutual ancestry, though the earliest known tassel-fins and ray-fins of Devonian times are perfectly distinct. In the former, the fringes of the tassel are the fin-rays, which spring from a projecting, scale-covered lobe that is braced by an axial skeleton of radial bones. In the ray-finned fishes the entire expanse of each fin, not just the fringe, is stayed by a series of fin rays. The radial skeleton is at the very base of the fins, which look like sails or fans rather than paddles. Ray-fins also differ from tassel-fins in such features as the fine structure of the scales, the pattern of dermal bones on the head, and the course of the lateral line canals. The details of these contrasts need not concern us here.

Tassel-fins

During late Devonian times the osteolepid tassel-fins gave rise to the amphibians, but as fishes they disappear from the fossil record

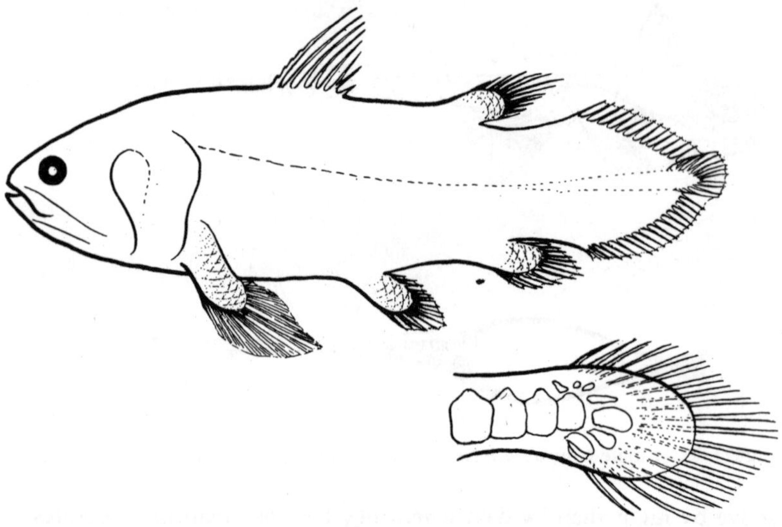

Figure 81. The living coelacanth, *Latimeria chalumnae*. Below: part of a pectoral fin showing the jointed axis of the scale-covered base of the fin.

in the Permian period. The coelacanths, which were once a diverse and widespread group in both marine and freshwater habitats, have one living representative, *Latimeria chalumnae* (figure 81). The third group of tassel-fins, the lung-fishes (Dipnoi), also had an impressive fossil history. After the Devonian they appear to have been confined to fresh waters and today there are seven survivors.

Except for the first described *Latimeria*, which was trawled in 1938 off the mouth of the Chalumna River in south-east African waters, the other specimens of this fish have been taken off the

Comoro Archipelago, a group of islands lying north-west of Madagascar. They were caught on hand lines fishing close to the bottom at depths between 150 and 400 metres.

Latimeria is very like its fossil relatives, which are unknown beyond the Cretaceous period. The rays of the second dorsal fin, the anal fin and the paired fins are inserted on a muscular, scale-covered lobe with a concentrated axial skeleton. The first dorsal fin has no lobe, while the powerful tail fin is symmetrical and formed of upper, lower and terminal sections (figure 81). The lobed fins have a great mobility. Not only can the pectoral fins be moved up and down and to and fro: they can turn on their axes through an angle of 180°. The heavily built body, which is dull grey-blue in colour with irregular white spots, is covered with cosmoid scales. There is a pair of nostrils on either side of the snout. Between the nasal organs is a fibrous sac that is well supplied with nerves. This rostral organ, as it is called, is evidently some kind of sensory device.

The fossil coelacanths contained a large swimbladder with ossified walls, but the swimbladder of *Latimeria*, which originates from the lower side of the gullet, is reduced and copiously invested with fatty tissue. The female produces very large eggs which, on ripening, are likely to be an inch or more in diameter. As the male has no intromittent organ, fertilization may well be external. If so, the incubation period is likely to be long and the young must hatch at an advanced stage. *Latimeria* grows to a length of more than 5 feet and a weight of 180 pounds. By and large, it exists by preying on other fishes.

The lung-fishes (figure 82) have a cellular swimbladder which opens into the lower side of the foregut and is used as an air-breathing organ. In the Australian lung-fish or barramunda (*Neoceratodus forsteri*), the swimbladder is unpaired, but in the South American species (*Lepidosiren paradoxa*) and the African lung-fishes (*Protopterus* spp.) it is divided into right and left lobes. The Australian species, which lives in the Burnett and Mary Rivers in Queensland, is a survivor of a cosmopolitan family (Ceratodidae) that dates back to Triassic times. It has tassel-like paired fins (figure 82), but the dorsal, anal and caudal members are replaced by a continuous median fin, which is also found in the other living lung-fishes. However, the paired fins of the African and South American lung-fishes are reduced to slender, rayless filaments (figure 82). These lung-fishes also differ from the Australian species in having smaller scales and a slimmer body.

The Australian species grows to a length of 6 feet. Like most lung-fishes, it bears grinding tooth plates on the palate and lower jaw. It feeds on water weeds, small crustaceans and molluscs. The eggs are laid among weeds or on the bottom and the larvae that emerge are not provided with external gills, as are those of other lung-fishes, which lay their eggs in some kind of nest.

During aerial respiration, a lung-fish swims to the surface and gulps in air which is passed to the swimbladder. But the Australian lung-fish is much less dependent on its swimbladder than are the

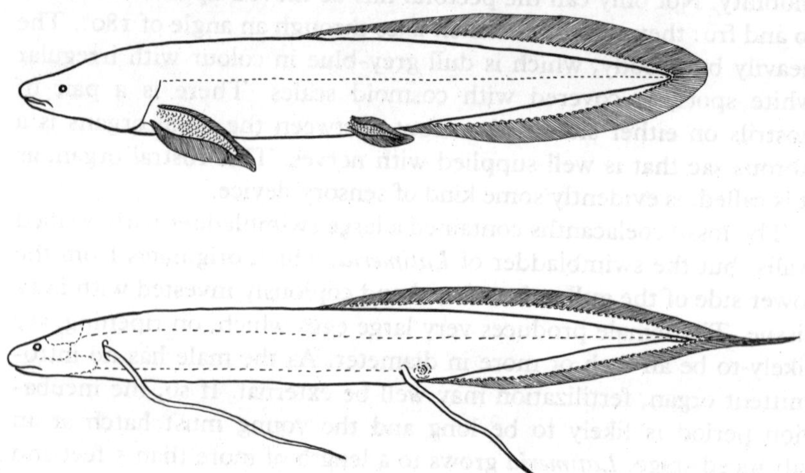

Figure 82. Two lung-fishes. Above: the Australian species *Neoceratodus forsteri* from the Burnett River, Queensland. Note the scaly base of the pectoral and pelvic fins. Below: an African species, *Protopterus aethiopicus*. Note the filamentous pectoral and pelvic fins. (Drawn from specimens in the Museum of Comparative Zoology, Harvard University.)

other species. It has four pairs of complete gills or holobranchs and need only breathe air when the water becomes stagnant. The South American lung-fish, which lives in swamps of the Amazon Basin and North Chaco, has three pairs of complete gills. The gill system of the African lung-fishes, which may also have to live in stagnant waters, is even more reduced, to two pairs of holobranchs. Physiologically, the contrast is still greater, for both the African and South American species obtain more than 95 per cent of the oxygen they need from the air. If prevented from breathing air, they will die, even if they are living in well aerated waters.

The South American lung-fish and three African species (e.g.

Protopterus annectens) have another adaptation to adverse conditions in the form of the drying up of their environment. Before the water disappears they excavate burrows in the mud, where they pass the dry season in a quiescent state. The Australian species is without such means of aestivation.

Ray-fins

There are three main groups of ray-finned fishes: palaeoniscoids, holosteans and teleosteans. The first group flourished from Devonian to Jurassic times and gave rise to the holosteans, which date from the Permian period and attained a considerable diversity during the Mesozoic Era. But during later Mesozoic times, in the Cretaceous, the teleosts which evolved from holostean ancestors became the dominant group of bony fishes. Today, more than 95 per cent of all kinds of fishes are teleosts.

Palaeoniscoid fishes

The modern relicts of the palaeoniscoids consist of some ten species of bichirs and one reed-fish (Polypteridae); twenty or so species of sturgeons (Acipenseridae); and two species of paddle-fishes (Polyodontidae) (figure 83).

The bichirs live in fresh waters of tropical Africa. Apart from certain congruences of skull structure and so on, they resemble their fossil relatives in bearing heavy ganoid scales. But they have many special features. The dorsal fin is represented by a series of free, spine-like finlets; the caudal fin approaches a symmetrical form; and, alone among ray-finned fishes, the pectoral fins have a muscular scale-covered base and fringing fin-rays (however, the radial skeleton is not like that of a tassel-fin). The swimbladder, which is cellular and opens into the floor of the pharynx, has two lobes, the right one being much larger than the left. The young develop a pair of long, feathery, external gills, one above each pectoral fin.

The related reed-fish (*Calamoichthys calabaricus*) is eel-like in form and has no pelvic fins. It lives in tropical West Africa. Of the nine or so species of *Polypterus*, the bichir of the Nile (*Polypterus bichir*) is the best known. It grows to about four feet in length and is a carnivore, feeding on fishes, frogs and crustaceans. The bichirs use the swimbladder to breathe air if the water becomes stagnant. But during most of the time the gills seem to provide adequate means of capturing the oxygen needed.

The sturgeons and paddle-fishes (figure 83), which are closely related, can be traced back to late Cretaceous times. They are unique among bony fishes in retaining an asymmetrical or heterocercal tail fin, which typified their fossil relatives. Moreover, they are unlike all surviving palaeoniscoid and holostean fishes in not developing a lung-like swimbladder.

The body of a sturgeon is armed with five rows of large, bony scutes. The toothless mouth is on the underside of the long pointed snout and before the mouth are two pairs of barbels. As soon as the barbels detect food on the bottom, the protrusible jaws drop down and suck up the prey: gastropod molluscs, crustaceans, insect larvae and small fishes. The largest species is the beluga (*Huso huso*), found in the basins of the Caspian, Adriatic and Black Seas. It reaches a length of 28 feet and a weight of 2,860 pounds. Sturgeons live in cold temperate and temperate parts of Eurasia and North America; they are most diverse and abundant in Eastern Europe and Asia. Except for a few, mainly North American, species, which are confined to fresh waters, sturgeons spend much of their life at sea, particularly on sandy or muddy bottoms along the continental shelf. Here they feed on molluscs, worms and so on. In spring they enter rivers, where they spawn during the early summer. A female beluga weighing a ton may produce about two hundredweights (224 pounds) of eggs.

The paddle-fishes have a depressed, blade-like snout. There are no scutes on the body, which otherwise resembles that of sturgeons in form and fin pattern. The jaws bear small teeth and before the mouth are two barbels. *Psephurus gladius* of Chinese rivers, which is said to grow to a length of 20 feet in the Yangtse-Kiang, has protrusible jaws and probably feeds much like a sturgeon. The paddle-fish or spoon-bill (*Polyodon spathula*) of eastern rivers in the USA reaches six feet in length (figure 83). It has wide, non-protrusible jaws and feeds largely on small planktonic crustaceans, which are screened by fine series of gill rakers.

Holostean fishes

The holostean fishes are now represented by no more than ten species. The garpikes (figure 83) are the descendants of a line of evolution, the semionotoid line, that began in Permian times. A second line has but one survivor, the bowfin (*Amia calva*) (figure 83). But a third evolutionary lineage (or lineages) had great potentialities, for it gave rise to the teleost fishes.

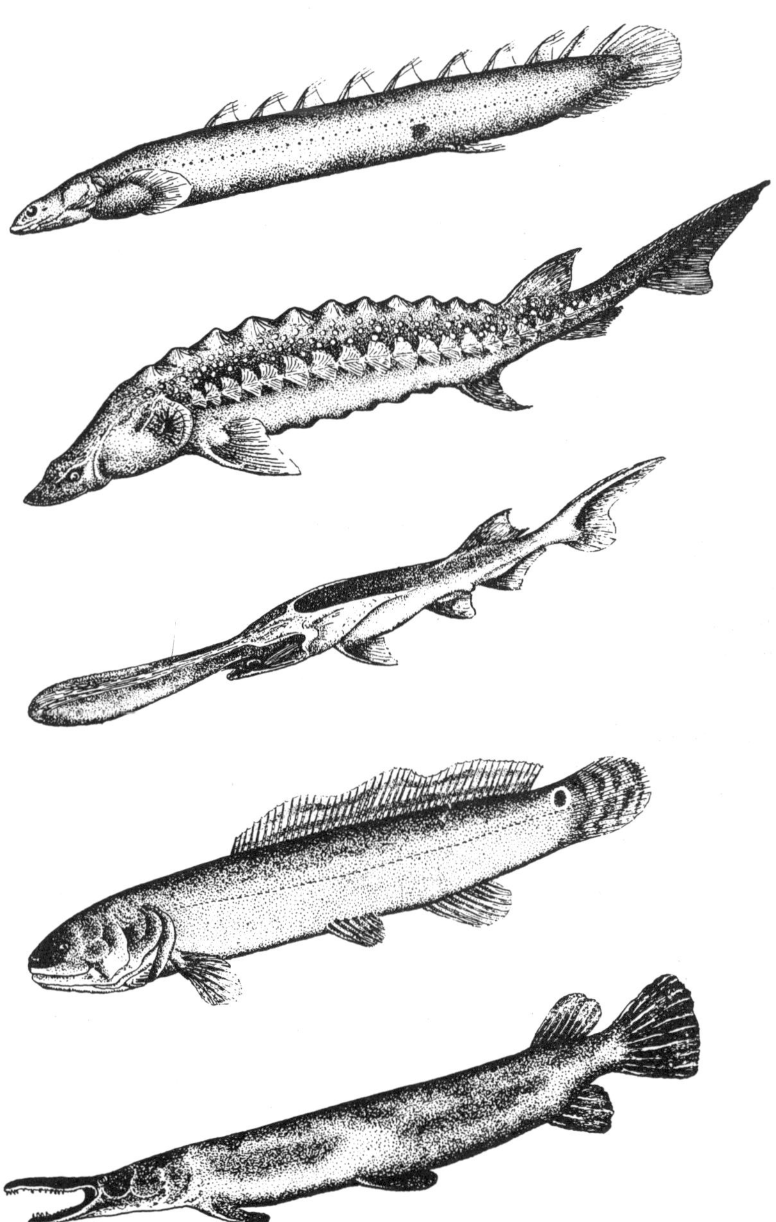

Figure 83. Archaic ray-finned fishes. Top to bottom: bichir (*Polypterus*); sturgeon (*Acipenser güldenstädti*); paddle-fish (*Polyodon spathula*); bowfin (*Amia calva*); garpike (*Lepisosteus tristoechus*).

The garpikes live in fresh and brackish waters of eastern North America and Central America. They have elongated jaws and a slim body, which is protected by a pavement of glassy, ganoid scales. The single dorsal fin is set well to the rear, opposite the anal member. Except when seizing their fish prey, which is taken by a sideways snap of the jaws, garpike are sluggish fishes. The largest of the family is the giant tropical alligator gar (*Lepisosteus tristoechus*), a species that may attain a length of more than 10 feet. Like the bowfin, the garpikes have a lung-like swimbladder, which opens into the roof of the foregut and can be used to breathe air.

The bowfin is found in eastern North America from the Mississippi in Minnesota to the St Lawrence–Champlain Basin in Canada. It has a long dorsal fin that arches like a bow down the greater part of the back. As in the garpikes and many fossil holosteans, the tail fin looks symmetrical, but has an asymmetrical skeleton, the upturned terminal vertebrae and notochord curving over the bases of the caudal rays. The females, which are larger than the males, reach a length of three feet. Bowfins feed on fishes, cray-fishes and other invertebrates. During the spawning season the males clear a circular area in the weeds, where the eggs are laid. The nests are guarded by the males.

Teleostean fishes

There are close to twenty thousand species of teleost fishes. They lead diverse lives in a wide range of habitats. Apart from deep and stagnant waters, such as lie at the bottom of the Black Sea and Lake Tanganyika, the teleosts have radiated into almost every living space in the waters that cover the earth. Some species can even live and move on land (pp. 45–48).

The teleosts have lost the dense ganoin that armoured the dermal bones and scales of their holostean ancestors. Unlike all palaeoniscoids and most holosteans, they bear a symmetrical tail fin, the rays fanning out from enlarged bones, the hypurals, at the end of a well-ossified vertebral column. The tooth-bearing bones of the upper jaw are not simply free from the cheeks, as they are in holosteans, but are attached only at the snout. Through such changes, certain holosteans evolved into teleosteans.

Do such changes give an insight to one outstanding fact, that during the last one hundred million years or so teleosts have become fish-masters of the hydrosphere? During Cretaceous times, birds and mammals also became much more diverse and numerous and we can see how they improved on their reptilian ancestry. How have teleosts become such flourishing fishes?

One salient factor is surely related to their size. The smallest teleost is a Philippine goby (*Pandaka pygmaea*), which grows to a length of 12 millimetres. The largest species, such as the Amazonian arapaima, the blue-fin tunnies and the larger marlins, reach a length of 12 feet or more, but the average span for teleosts is probably less than 6 inches. Now one advantage in being small is a potentiality to exist in relatively restricted living spaces. Many teleosts live, for instance, among the labyrinths and branching heads of coral, or find shelter and food among rocks, stones and vegetation. Small fishes also have modest needs for food. To take but one environment: the mid-waters of the deep ocean contain relatively little food, but they are inhabited by close to a thousand kinds of teleosts, nearly all of which span less than 6 inches. Moreover, there are more kinds of small than large animals, the outstanding group being the insects. A small, newly evolved species must have a better chance of finding a suitable niche than a larger one.

Sharks are much bigger fishes. The smallest living shark (*Squaliolus*) is as large as an average-sized teleost, while the mean size must be close to 6 feet. Most of the rays are over a foot in breadth. A great many ecological niches are clearly debarred to such fishes, which are, moreover, carnivores. Each shark or ray must range over a relatively large living space in order to meet its needs for food.

During Cretaceous times the main competitors of the rapidly evolving teleosts were the holostean fishes; these were mostly small or medium-sized, having about the same size range as the teleosts. All the surviving holosteans use their swimbladder as a lung as do lung-fishes, bichirs and a few teleosts; but in most teleosts the swimbladder is a hydrostatic organ, one fitted to keep the under-water weight of a fish at the vanishing point (pp. 67–80). The capacity of this float need be no more than one twelfth of the entire volume of the fish; for the teleosts have lost much skeletal ballast, both in the form of dense ganoin and in having acquired a light but strong kind of bony architecture. Having thus evolved beyond their holostean ancestors, which had very compact bones, the teleosts not only gained perfect buoyancy but a potentiality for delicate and precise manoeuvre. In this they were greatly aided by inheriting flexible kinds of fin from the holosteans – fins in which each ray is moved by its own set of muscles. Their perfect buoyancy is also reflected in the symmetrical tail fin, which need no longer provide a lift force during motion.

These interrelated advances in buoyancy, bony architecture and fin pattern may well have contributed greatly to the success of the teleosts. And, just as fins became free for more versatile roles, including their use in signalling aggressive or amatory intentions, so did the bones of the upper jaw. Being completely released from the cheek bones, and being connected only to the snout, the premaxilla and maxilla assumed a greater mobility during the opening of the mouth. Prey could be taken more readily. The upper jaw became even more mobile with the evolution of protrusible premaxillae, which could be shot forward from the snout and braced by the maxillae. Well over half the teleosts have such jaws, jaws that can snap up food in awkward places. This versatility is best seen in a coral reef, where most of the fishes, largely belonging to the spiny-finned groups, have jack-in-the-box jaws.

Besides using their fins in a language of signs, many teleosts communicate by sounds, which are elicited by stridulation of certain skeletal parts and by vibration of the swimbladder. Sharks and rays make no more than incidental feeding and swimming sounds. The schooling habit, which is common among teleosts, is rare or absent in other fishes. Lastly, the teleost brain is more advanced and complex that that of other fishes. For instance, if the forebrain is taken from a shark, the fish loses its sense of smell; but there is no obvious interference with its posture swimming or behaviour. A similar operation on a teleost may also result in a loss of initiative. Such tests show that the teleost forebrain is involved in more than simple olfaction. Indeed there is some tendency for the roof of the forebrain to form a 'cortex' of nerve cells recalling that of mammals. This is associated with an increase in the tracts of nerve fibres that connect the forepart with other regions of the brain.

In sum, the teleosts lead a richer life than other kinds of fishes. *Teleos*, meaning complete or perfect, and *osteon*, bone, expresses but a single aspect of their evolutionary achievements.

Turning now to the diversity of teleosts, we can do no more than touch on the main groups. The most primitive forms are found in two orders, the Isospondyli and Ostariophysi, the first containing about nine hundred species, the second about five thousand. The two together form nearly a third of the known kinds of teleosts. The more primitive teleosts – those that retain a good many characters of their holostean ancestors – can be recognized, *inter alia*, by their fin pattern, jaw structure, scaling and swimbladder. There is a single dorsal fin; a symmetrical tail fin typically formed of nineteen principal rays; and a single anal fin. The pectoral fins are set low down on the shoulders and the pectoral girdle is braced on each side by an inner strut of bone called the mesocoracoid. The pelvic fins which have numerous rays, are inserted on the abdomen, well back from the pectorals. There are no fin spines, the fin rays being jointed and soft. The biting part of the upper jaw is formed by a small premaxilla and a long blade-like maxilla. The scales are cycloid, that is without teeth on the free edges. Lastly, the swimbladder opens by way of a pneumatic duct into the roof of the foregut.

The Isospondyli include the tarpons (*Megalops*) and ten-pounders (*Elops*), the herring-like fishes, the deep-sea stomiatoids, the bony tongues (Osteoglossoidea), the salmon-like fishes and the mormyroids (figures 84 and 39). Tarpons and ten-pounders, which live in warm seas, are the most primitive of all living teleosts. Besides the characters listed above, they have numerous rays or branchiostegals supporting the gill cover membranes; a gular plate under the lower jaw; a lateral-line canal across the snout; and fringing fulcral scales on the leading edges of the caudal fin. All such features are found in their ancestors.

There are about three hundred species in the herring family (Clupeidae), most of which live in the coastal seas of the temperate and tropical regions. The herrings have small teeth and most of them feed on planktonic organisms, some developing very fine gill rakers to strain the swallowed food. All the clupeids have a stethoscope-like linkage between the swimbladder and the ears (figure 37), which should enhance their sense of hearing. The family includes the herrings, sprat, pilchards, shads and round herrings (*Dussumeria*, etc.).

The three hundred odd species of stomiatoids live in the mid-waters of the deep ocean. Like the deep-sea salmonoids, the swimbladder (when it is developed) is completely closed, which is a feature of nearly all the deep-sea fishes that have retained this organ. The stomiatoids carry elaborate systems of light organs and in most species there are two rows of lights on each side of the underparts (figure 84).

The bony-tongued fishes (Osteoglossoidea) date back to Cretacous times, but there are only five surviving species: the arapaima (*Arapaima gigas*) and *Osteoglossum bicirrhosum* of the Amazon Basin; *Clupisudis niloticus* and *Pantodon buchholtzi* of tropical Africa, and *Scleropages leichhardti* of Indo-Australian regions. These fishes have retained an archaic character in that two of the palatal bones, the endopterygoids, articulate with the roofing palatal bone, the parasphenoid. The tail fin is also not entirely symmetrical, while in the arapaima, *Clupisudis*, and *Pantodon* the swimbladder is a cellular, air-breathing organ.

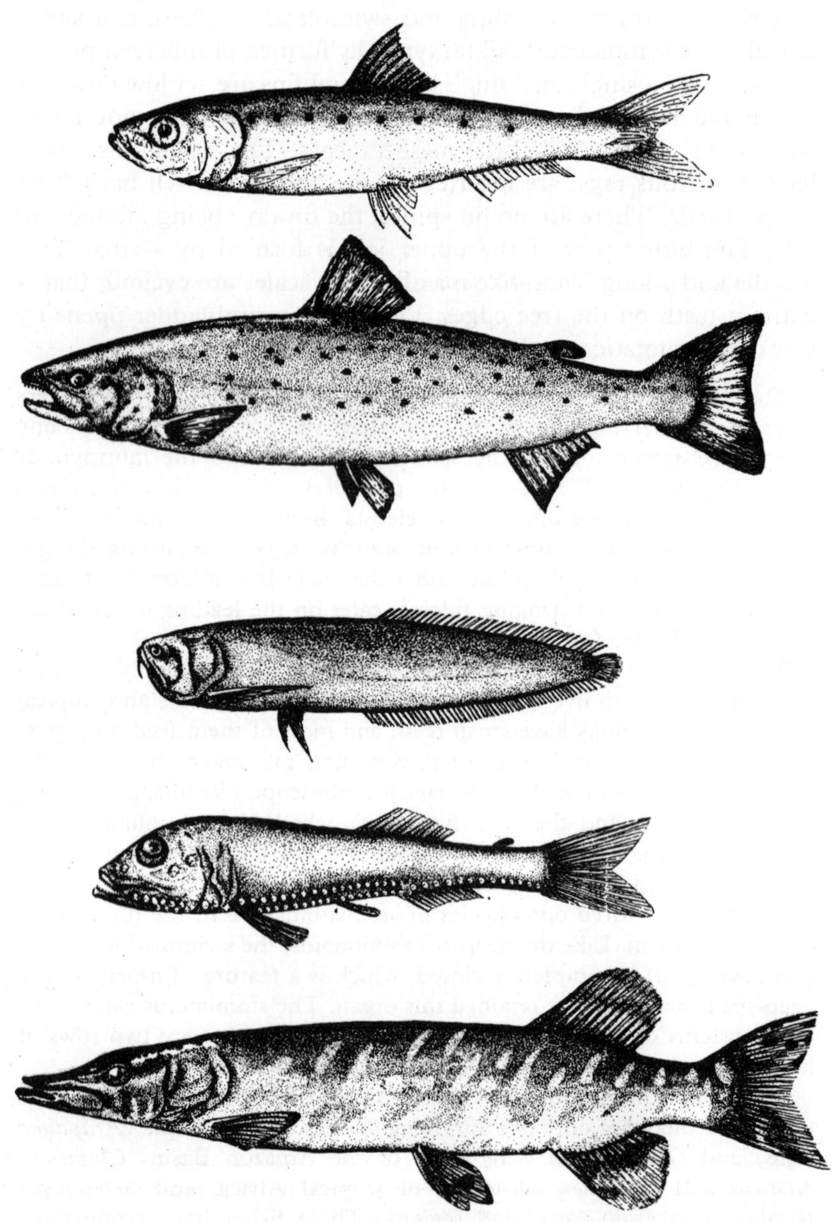

Figure 84. The pike, *Esox lucius* (bottom figure) and members of the order Isospondyli: top to bottom, pilchard (*Sardinops sagax*); salmon (*Salmo salar*); the South American osteoglossid (*Osteoglossum bicirrhosum*); and a stomiatoid (*Vinciguerria*).

Besides the Atlantic salmon, Pacific salmon and trout, the salmonoids include char, smelts and a number of deep-sea species. Most salmonoids have a small, adipose dorsal fin and the females lack true oviducts. Except for the deep-sea species, which are widespread in the ocean, the salmonoids are fishes of arctic and northern regions. Some, like salmon and sea trout, enter rivers to spawn while others, such as graylings and most white-fishes, are confined to fresh waters. The northern char (*Salvelinus alpinus*) lives in arctic seas, but there are a number of land-locked forms of this species, relics of glacial periods, in the Lake District of England, in Ireland, in the Alps and in Eastern North America.

The deep sea salmonoids are large-eyed fishes with small mouths. Certain species, for instance, *Opisthoproctus* (plate 8), have tubular eyes and a light organ close to the anus.

All the mormyroid fishes possess electric organs and a special kind of skin containing sense organs, the mormyromasts, which are probably responsible for sensing electric fields. Each ear is connected with a gas-filled vesicle, detached from the swimbladder during development. The upper jaw bones of the small mouth may depart from the primitive arrangement. The mormyroids, which number about a hundred and fifty species, live in the Nile and other fresh waters of tropical Africa. In some kinds (*Mormyrops*) the snout is moderate in extent: in others (*Petrocephalus*) it is rounded and arches over the mouth. As implied by their name, the snout is greatly elongated in the elephant-fishes (*Gnathonemus*, etc.), which, like pipe-fishes and sea-horses, feed on small crustaceans and so on.

Before turning to the Ostariophysi, we must consider pike (figure 84) (*Esox* spp.), mud-minnows (Umbridae) and the black-fish *Dallia*. Certain authorities place these in the Isospondyli, while others consider them to be in a separate order, the Haplomi. They differ from isospondylous teleosts in having no teeth on the maxillary bones; in lacking a mesocoracoid in the pectoral girdle; and in bearing a pair of large dermal bones, the proethmoids, on top of the snout. The pikes live in the fresh waters of North America and Eurasia. The black-fish is confined to Alaska and north-eastern Siberia. Dr Leo Berg, the eminent Russian ichthyologist, concluded that the Haplomi evolved from the smelt-like osmeroid fishes at the end of the Cretaceous period. The pikes also seem to be related to the 'salmonoid' fishes of south temperate fresh waters, to fishes such as the trout-like galaxiids and the 'smelts' (Aplochitonidae and Retropinnidae).

Except for certain marine cat-fishes, the Ostariophysi (plates 4 and 5) are freshwater fishes. They are widely distributed in Asia, Africa, the Americas and Europe. The ostariophysans are readily distinguished from the Isospondyli in that each inner ear is linked to the swimbladder by a chain of ossicles, the Weberian apparatus. In the characins, gymnotoids and cyprinoids – carp-like fishes – the body is generally covered with scales: in the cat-fishes the body is armoured with bony plates or naked and the two maxillary bones are nearly always rudimentary, each serving to support a barbel (it is, of course, their whisker-like barbels that have led to their being called cat-fishes).

The most primitive ostariophysans are found among the characins, which live in tropical Africa, Central America and South America. The jaws are toothed, but there are also toothed upper and lower pharyngeal bones in the throat. Characins include the main family, Characidae, which usually have an adipose dorsal fin like that of a salmon; the flying hatchet-fishes (Gasteropelecidae); the xiphostomatids; the anostomids, and the citharinids. There is a wide variation in form and feeding habits among these fishes, which range from about an inch to 5 feet in length (plate 4).

The cyprinoid fishes have toothless jaws, but there are sickle-shaped, tooth-bearing, lower pharnygeal bones that bite against a projection from the skull. By far the most diverse is the carp family (Cyprinidae), which is the largest of all fish families, containing about two hundred and seventy-five genera and about two thousand species. The cyprinids are widely distributed in Africa, Europe, Asia and North America, but there are none in Central and South America, Madagascar and Australia. They have protrusible premaxillae, which form the upper border of the mouth, and a small number of pharyngeal teeth. European species include the minnow, dace, roach, rudd, bream, chub, bleak, barbel, gudgeon, tench and bitterling. The carp itself is a native of China. In Africa and Southern Asia there are hundreds of species of *Barbus* and allied genera. The most diverse North American genus is *Notropis*.

The suckers (Catostomidae) of North America and eastern Asia and the loaches (Cobitidae) of Asia, Africa and Europe are also cyprinoid fishes. Loaches are slender, small-scaled fishes with a number of barbels round the mouth and a reduced swimbladder, the front part of which is enclosed in a bony capsule. The swimbladder is similarly modified in the loach-like homalopterids, which live in torrential streams of south-eastern Asia.

There are more than two thousand species of cat-fishes (Siluroidea), of which about twelve hundred live in South American fresh waters. Two families consist largely of marine species, the Ariidae living in warm seas of both hemispheres, the Plotosidae in parts of the Indian and Pacific Oceans. The most primitive genus is *Diplomystes* of Argentina and Chile, in which the maxillary bones are toothed and included in the gape: in the other cat-fishes, as already stated, these bones are reduced to splints for two barbels. Like the loaches and homalopterids, certain cat-fishes have the front chamber of the swimbladder housed in a bony capsule. Some kinds, such as ariids, mochocids and ameiurids, develop elaborate spine-like rays in the dorsal and pectoral fins. A number of cat-fishes, notably the clariids, have air-breathing organs.

The Ostariophysi are overwhelmingly the dominant group of freshwater fishes and they have one outstanding structural feature. Even when the swimbladder is reduced, they still retain some bony linkage between this organ and the ears (figure 37). The whole mechanism thus acts as a hydrophone, endowing these fishes with very sensitive hearing. Many cat-fishes and some cyprinids produce sounds and the characins are also likely to be sonic fishes. A good many members of the Ostariophysi live in muddy waters where vision is restricted, but where sound signals may well be vital, particularly during the breeding season. Such attainments may well have helped them to master freshwater habitats, an assumption which is just as reasonably extended to another unique feature of these fishes. In the epidermis are cells that secrete alarm substances; when the skin is injured these substances diffuse into the water and on being scented by members of the same or closely related species, there follows a general and hasty retreat. In this way, the young will be protected from cannibalism, or from the attacks of allied kinds of fish.

The Ostariophysi have also evolved apt means for finding and taking food. The most diverse family, the Cyprinidae, have protrusile premaxillae. Cat-fishes greatly depend on their barbels to detect food: their eyes are usually small (plate 5). Certain species are particularly sensitive to electrical disturbances, but the part this plays in their lives is as yet unknown. Lastly, many of the Ostariophysi have a remarkable respiratory adaptability. Apart from special means of breathing air, numerous species are able to exist in poorly aerated waters. Carp and gold-fish, for instance, are particularly hardy in this respect. During wintery conditions crucian carp can live for months in waters containing little or no oxygen. Many other instances could be given.

Just as the fresh waters are largely inhabited by the Ostariophysi, so are the land-fringing waters of the ocean dominated by spiny-finned, perch-like teleosts (Percomorphi and related orders). There may be as many as eight thousand species of perch-like fishes, which have evolved away from the primitive teleosts in a number of striking respects. As implied by the name, some of the fin-rays are spiny. In a typical perch-like fish, each pelvic fin has an outer spine and five soft rays; it has moved forward close under the pectoral fins, the pelvic girdle being linked to the pectoral girdle by ligaments. The pectoral fins are inserted on the sides of the shoulders, not on the undersides, and may now, in concert with the pelvic fins, act as brakes. Perhaps the tie between the two girdles helps to take the strains imposed during braking movements. Moreover, the pectorals can act as brakes because their basal axis is now more or less vertical, which also means that they can be used as oars.

The leading dorsal and anal fin-rays are also spiny, while the caudal fin has seventeen principal rays, two fewer than does this fin in a primitive, soft-rayed teleost. Turning to the jaws, the premaxillae are protrusible and exclude the maxillae from the gape (the mechanism differs from that in a cyprinid fish). The swimbladder is closed and the scales usually have toothed free borders: they are ctenoid. Supporting the gill cover membranes are six to seven branchiostegal rays with a characteristic setting on the hyoid arch.

Perch-like fishes must have appeared early in the evolutionary history of teleosts. Even the first known species, from the Upper Cretaceous, had diverged from their soft-rayed ancestors in at least three advantageous respects. Besides their spiny, protective rays, they had mobile jaws and a versatile fin pattern, a complex of characters of great significance in the life of fishes. Moreover, they evolved an organization that could be expressed in all shapes and sizes (think, for instance, of the great variety of spiny-finned fishes in a coral reef). Though not too wise after the event, we can begin to see why such fishes have flourished so well in coastal and coral seas.

Before considering the perch-like fishes, we must look at a number of orders that come, as it were, between them and the more primitive teleosts. Some have not evolved very far from the primitive condition, while others approach the percomorphs in certain respects.

The less advanced orders include the Iniomi and related orders; the Apodes (eels), Heteromi (spiny-eels, etc.) and Synentognathi (flying-fishes, etc.). The iniomous fishes (figure 85) differ from the Isospondyli in having premaxillae that exclude the maxillae from the gape and in having lost the mesocoracoid strut of the pectoral girdle. There are more than four hundred species, most of them living in the deep ocean. The lantern fishes (Myctophidae) are the most diverse, consisting of more than two hundred and fifty species, most of which live in the mid-water twilight zone (200 to 1,000 metres) of the subtropical and tropical ocean. Each species bears an individual constellation of pearl-button, light organs over the sides of the body (figure 44 and plate 16). Some, such as species of *Diaphus*, also have lights on the snout (plate 17). Males and females of certain species can be distinguished by the arrangement of flattened light organs on the tail stalk.

The tripod-fishes (Bathypteroidae) (figure 85), which range from the continental slopes to the abyssal reaches of the deep-sea floor, and the lizard-fishes (Synodontidae) of warm coastal seas, are placed in the same suborder (Myctophoidea) as the lantern-fishes. The members of both families have the pelvic fins set well forward on the abdomen. Tripod-fishes are so-named because the outer ray of each pelvic fin, together with the lowermost caudal rays, form a three-point undercarriage which is used as a rest when these fishes land on the oozes. Lizard-fishes also rest on their pelvic fins and tail fin, ready to take off and seize their prey of crustaceans, small fishes etc. They also make great use of their pelvic fins to crawl along the bottom.

The other suborder (Alepisauroidea) of iniomous fishes is composed of predatory species, which live at much the same mid-water levels as the lantern-fishes. Except for certain barracudinas (Paralepididae), these fishes are without light organs. The lancet-fishes (*Alepisaurus*, figure 85), javelin-fish (*Anotopterus*) and *Omosudis* (figure 85), are elongated fishes with large jaws that carry formidable stabbing teeth. There is a decided tendency in this group for the scales to be reduced or lost, which goes with a capacity to swallow large prey.

Three small orders of teleosts live in the mid-waters of the deep ocean: the Lyomeri (gulper-eels, plate 7), Giganturoidea and Miripinnati (figure 85). The last are most closely related to the iniomous fishes. Perhaps the other two evolved from isospondylous ancestors, but they have such reduced organ systems – seemingly to conform to their food-poor surroundings – that it is difficult to decide on their nearest relatives.

Most of the hundred and twenty or so members of the order Synentognathi live near the surface of the warm temperate and tropical regions of the ocean. Notable exceptions are certain half-beaks, which are most diverse in fresh and brackish waters of south-eastern Asia; they are also unusual in being viviparous. This order, which consists of four families – half-beaks (Hemirhamphidae, figure 8), flying-fishes (Exocoetidae, figure 13), garfishes (Belonidae) and skippers (Scomberesocidae) – retains certain features of the more primitive teleosts (p. 364): soft rays, cycloid scales, abdominal pelvic fins and numerous branchiostegal rays, nine to fifteen in number. Their more advanced features include the following: a closed swimbladder; the lateral-line runs along the underparts of the trunk and tail; the pectoral fins are set high on the shoulders; and the caudal fin is formed of thirteen branched rays rather than seventeen, as in many soft-rayed teleosts. The half-beaks and some of the flying-fishes live in coastal seas, but numerous flying-fishes are truly oceanic, as are the skippers.

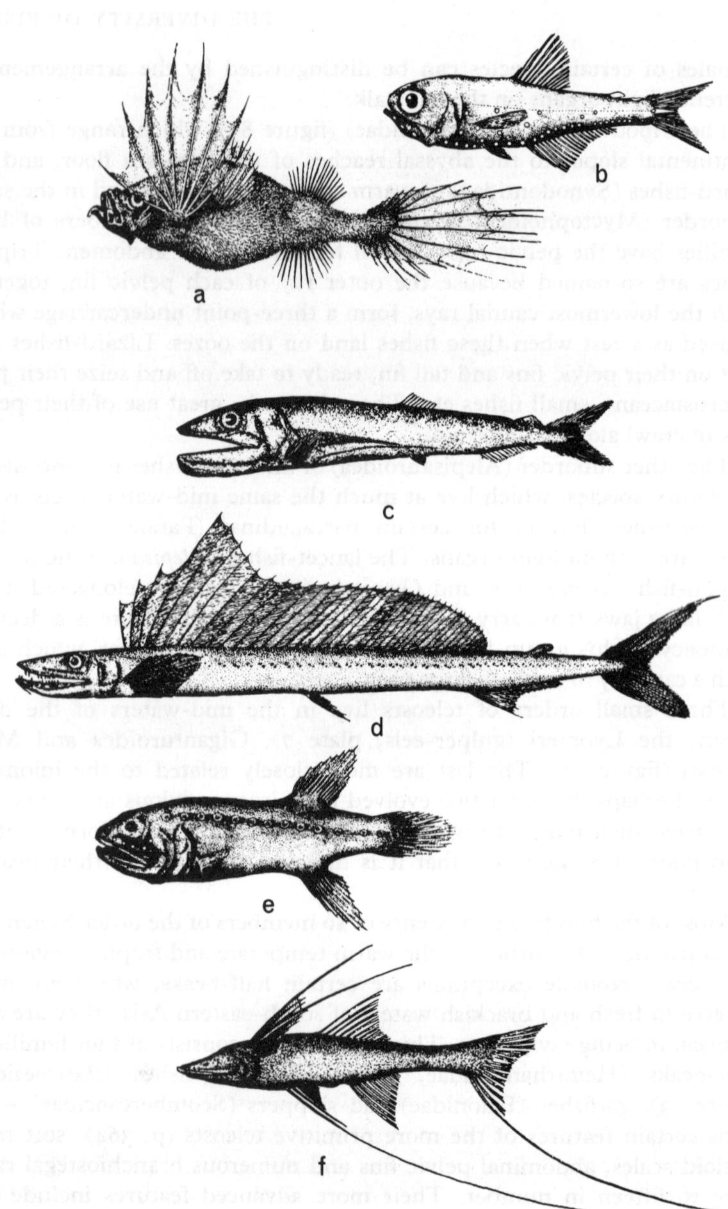

Figure 85. Top left: (a) *Mirapinna esau*, order Miripinnati. The other fishes except (e) belong to the order Iniomi. (b) a lantern-fish (*Hygophum*); (c) *Omosudis lowei*; (d) a lancet-fish (*Alepisaurus ferox*); (e) a whale-fish (*Cetomimus indigator*), order Cetunculi; (f) a tripod-fish (*Benthosaurus*). ((c) and (d) after G. E. Maul; (e) after R. R. Rofen.)

There are several hundred kinds of eels (order Apodes), most of which are entirely marine (plate 33). The notable exceptions are freshwater eels (Anguillidae), which enter rivers as elvers but migrate to oceanic spawning grounds on attaining sexual maturity. Like the primitive teleosts, eels have an open swimbladder, though it is closed in deep-sea species; cycloid scales (when present); soft rays; and a gape that includes the premaxillary and maxillary bones. In particular they resemble the tarpons, ten-pounders and bone-fish in having leptocephalus larvae. Certain families, such as the snipe-eels (Nemichthyidae) and bob-tailed snipe-eels (Cyemidae) which are bathypelagic, and the benthic Synaphobranchidae, live in the deep ocean. One deep-sea eel, *Simenchelys parasiticus*, has a sucker-like mouth and sharp cutting teeth. After fastening itself to another fish, this eel chews pieces out of the flesh of its host. But most eels live in the coastal waters of warm temperate and tropical regions. Moray-eels (Muraenidae) and snake-eels (Ophichthyidae) are particularly diverse in coral reefs.

The halosaurs (plate 13) and notacanths (spiny-eels), which form the order Heteromi, live near the deep-sea floor. In using the pneumatic duct of the swimbladder to eliminate gases, and in having two retia mirabilia that are housed in the wall of the duct, the Heteromi are just like the eels. Moreover, the halosaurs and notacanths have a long body and a backbone made of many vertebrae. These and other similarities suggest that these two groups and the eels came from a common isospondylous ancestor.

The halosaurs are soft-rayed fishes with a single dorsal fin, abdominal pelvic fins and a long, many-rayed anal fin extending to the tip of the tail, which is without a caudal fin. This fin pattern is found in notacanths, but some of the fin-rays have been turned into spines. In both families the mouth is inferior to a snout that is strengthened by a rostral extension of the skull. The eyes, like those of eels, are covered by transparent skin or spectacles, which may well protect the eyes when these fishes forage in the oozes.

Orders containing more advanced fishes than those in the seven that have just been considered include the sticklebacks (Thoracostei), tube-mouths (Solenichthyes), top-minnows or toothed carps (Microcyprini),* cod-fishes and rat-tails (Anacanthini), ribbon-fishes, etc. (Allotriognathi), John Dories, etc. (Zeomorphi) and the squirrel-fishes, etc. (Berycomorphi). In one way or another, they have evolved further away from the organization of a primitive teleost.

The berycomorph fishes were probably derived from a primitive kind of iniomous fish and in turn gave rise to the more advanced spiny-finned fishes (Percomorphi). In retaining nineteen (or eighteen) principal rays in the caudal fin and a sphenoid bone between the orbits, they show traces

* In a recent paper Dr Donn Rosen has shown convincingly that the top-minnows are closely related to the half-beaks and their relatives (Synentognathi).

of their ancestry. Some of the genera in this order, such as *Beryx*, are known from Cretaceous times.

A typical berycomorph has spines in front of the soft rays of the dorsal and anal fins. The pelvic fins, which usually bear an outer spine, have three to thirteen soft rays (a typical percomorph has one spine and five soft rays in each pelvic fin). There are over a hundred species in this order, the majority being contained in the squirrel-fish family (Holocentridae) and the Melamphaidae (plate 6). The squirrel-fishes (plate 34) are large-eyed fishes with strong spines in the dorsal and anal fins. The gill cover bones also bear spines, while the trailing edges of the scales are strongly serrated. Squirrel-fishes are most diverse among coral reefs and tend to be nocturnal in habit. The melamphaids, most of which are dark brown in colour, live in the mid-waters of the deep ocean. They are without light organs and range from one to nine inches in length.

The John Dories and their allies (plate 6) are more advanced than the berycomorph fishes. The forward rays of the dorsal fin and a short forward and anal fin are spiny. Each pelvic fin has an outer spine and from five to nine branched rays. The caudal fin is reduced to eleven to fourteen principal rays. Locomotion is by way of undulations of the opposed, soft dorsal and anal fins. Besides the John Dories (Zeidae) and boarfishes (Caproidae, plate 41), there are a number of deep-sea families in this order, such as the Cyttidae and Grammicolepidae.

The Allotriognathi, meaning 'strange jaws', are so-called because it is the maxillary bones of the upper jaw, not the premaxillae, that are protrusible. In certain species, however, the upper jaw is immovable. The fins are without spines and the pelvic fins, when present, are set near the pectorals and bear from one to seventeen rays. All of the twenty-odd species are fishes that live in the upper waters of the open ocean. The opah or moon-fish is a large deep-bodied fish, up to 6 feet in length, with large pectoral fins and a most handsome colouring (plate 6). Like a typical berycomorph, it has nineteen principal rays in the caudal fin and an orbitosphenoid bone in the neurocranium. The order also includes the ribbon-fishes (Trachypteridae, plate 6), the unicorn-fishes (Lophotidae, figure 50), and the oar-fish (*Regelecus*), all of which have a long, compressed body and a long, many-rayed dorsal fin. One genus (*Stylophorus*), which has tubular eyes, lives in the twilight zone of the deep ocean.

All the sticklebacks, which live in fresh and salt waters of the Northern Hemisphere, have two or more free spines before the soft dorsal fin. The pelvic fins, set a little way back of the pectorals, bear an outer spine and form one to three soft rays. The soft dorsal and anal fins are opposed, while the caudal fin has a reduced number of rays. The hinges of the lower jaw are well in front of the orbits (figure 57).

The sticklebacks may well be most closely related to the tube-mouths, which is suggested by a curious little fish, *Indostomus paradoxus*, that was caught in a freshwater lake in Upper Burma. In many characters *Indostomus* is like a stickleback, but it resembles the pipe-fishes and sea-horses in having about twenty bony rings around the body.

Besides the pipe-fishes (plate 39 and figure 52) and sea-horses (plate 40 and figure 51), the tube-mouths include the trumpet-fishes (*Aulostomus*) (plate 38), cornet-fishes (*Fistularia*), shrimp-fishes (Centriscidae) and the ghost-fishes (Solenostomidae). Most species live in coastal seas, though certain pipe-fishes live in freshwater habitats. The snout is tubular and the mouth relatively small. When present, the first dorsal fin is spinous. The swimbladder, like that of sticklebacks, is divided by a diaphragm into forward gas-producing and gas-resorbing chambers.

The affinities of the Anacanthini and Microcyprini* are by no means evident, though in both orders the rays of the gill membranes are arranged much like those of a spiny-finned fish. Yet it would not be safe to assume that all three groups evolved from a common stock. The order Anacanthini includes cod-fishes (Gadidae) (plate 24), deep-sea cods (Moridae), hake (*Merluccius*) and rat-tails (Macrouridae) (plate 13 and figure 24). They have soft rays; cycloid scales; many rays in the dorsal and anal fins and a very reduced caudal fin, which is altogether absent in the rat-tails. In the cods and hakes most of the caudal fin is formed of dorsal and anal rays. The pelvic fins are close to the pectorals. An oval usually forms the gas-eliminating part of the swimbladder.

Most of the cod-fishes are confined to the seas of the Northern Hemisphere, but the rocklings (*Guidropsarus*) are found both north and south. One genus, *Lota*, the burbot, is the only freshwater fish of the order. The deep-sea cods and the rat-tails live near the ocean floor. They are most diverse in warm temperate and tropical regions. The rat-tail family, comprising some three hundred species, is the most diverse group of anacanthine fishes. There are about seventy species of deep-sea cods and sixty species of cod-fishes.

Nearly all of the Microcyprini (figure 62) have soft-rayed fins. There is a single dorsal fin, and the pelvic fins, with not more than seven rays, are set on the abdomen. The pectorals are on the sides of the shoulder region. The bony structure of the caudal fin is unique. Most of the four hundred-odd species live in fresh and brackish waters of tropical and subtropical regions. The Cyprinodontidae, which lay eggs, include the killifishes (*Fundulus*) and *Orestias* spp., natives of high plateau waters in Chile, Peru and Bolivia. The best known family, Poeciliidae, consists of viviparous species and includes guppies, sword-tails, mollies and platy-fishes.

We have seen that there may be as many as eight thousand species of spiny-finned fishes. One order, the Percomorphi, has already been defined (pp. 369–370). The other order, Scleroparei – mail-cheeked fishes –

* See footnote on p. 373.

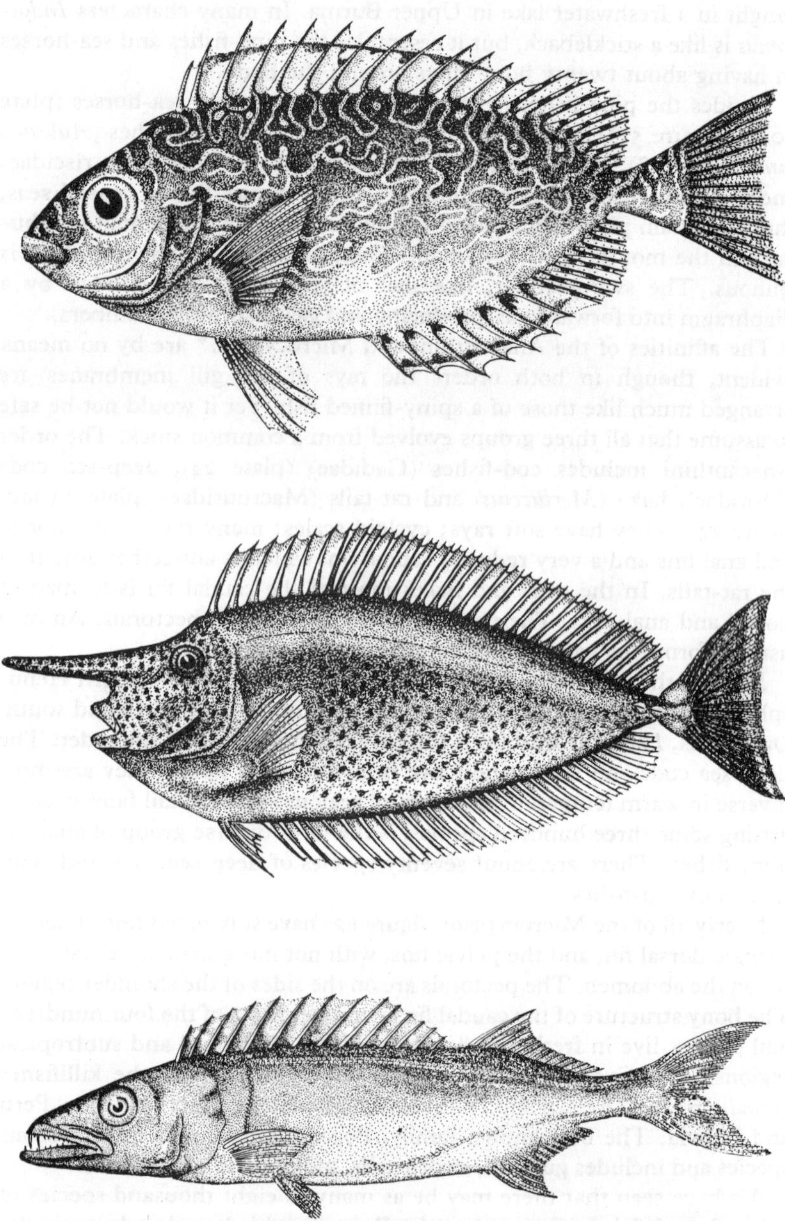

Figure 86. Three kinds of spiny-finned fishes. Above: a rabbit-fish, *Siganus spinus*. Middle: a unicorn surgeon-fish, *Naso brevirostris*. (Both after L. P. Schultz.) Bottom: a trichiuroid fish, *Neoepinnula*.

simply differs from the percomorphs in that the second bone below the orbit, the second suborbital, is fused to the foremost gill-cover bone, the preoperculum. Hence the name, mail-cheeked.

The percomorph fishes include the groupers (figure 74), sea-basses and so on (Serranidae); cardinal-fishes (Apogonidae) (figure 74); jacks and horse mackerels (Carangidae); dolphin-fish (*Coryphaena*); snappers (Lutianidae); drum-fishes (Sciaenidae); sea-breams (Sparidae); red-mullets (Mullidae); butterfly-fishes (Chaetodontidae); damsel-fishes (Pomacentridae); wrasses (Labridae); parrot-fishes (Scaridae) (plate 20); weevers (Trachinidae); nototheniiform fishes; (figure 72) blennies (Blenniidae) (plate 31); sand-eels (Ammodytidae); dragonets (Callionymidae); rabbit-fishes (Siganidae) (figure 86); surgeon-fishes (Acanthuridae) (figure 86); mackerels and tunas (Scombridae); and gobies (Gobiidae) (plate 30). Nearly all of the members of these families are marine. Diverse fresh-water families include the Cichlidae (figure 58), the sunfishes (Centrarchidae), the perches and darters (Percidae) and the labyrinth-fishes (Anabantidae).

The better-known families of mail-cheeked fishes are the scorpion-fishes (Scorpaenidae), gurnards (Triglidae), flatheads (Platycephalidae), bullheads or sculpins (Cottidae) (figure 72) and sea-snails (Liparidae, plate 13). And these are but a few of the families of spiny-finned fishes. No ichthyologist can claim to have an intimate knowledge of more than a small fraction of these two diverse orders of teleosts.

The flat-fishes (Heterosomata), shark suckers (Discocephali), Plectognathi, cling-fishes (Xenopterygii) and angler-fishes (Pediculati) were most probably derived from spiny-finned ancestors.

There are about five hundred species of flat-fishes which are almost entirely confined to coastal seas. A few kinds, such as the European flounder (*Platichthys flesus*) enter estuaries. One genus (*Psettodes*) of eastern Atlantic and Indo-Pacific regions still has spiny rays in the dorsal and anal fins: the others are without spines. The diagnostic feature of adult flatfishes is that the eyes are on one side of the head, which has led to some asymmetry in the skull. Flat-fishes rest on the blind side of the body, which is usually without pigment and is on the right side in the soles (Soleidae) and on the left in the tongue-soles (Cynoglossidae) and flounders (Bothidae). As in all truly bottom-dwelling teleosts, the swim-bladder is absent. Flat-fishes range into polar waters, but most species occur in temperate and tropical regions.

In the shark-suckers (Discocephali), the spinous dorsal fin has been turned into a flat and oval suction disc, which bears a series of transverse, tooth-studded plates (plate 35). By means of this sucker, which is placed on the head and nape, shark-suckers cling to large fishes, turtles and cetaceans. The ten or so species are distributed in tropical and warm temperate seas.

There are some two hundred species of Plectognathi (plate 12). They are small-mouthed fishes – the maxillary bones are united and fused to the premaxillaries – with restricted gill openings. Most are armoured in some way or other. The main families are trigger-fishes (Balistidae), file-fishes (Monacanthidae), box-fishes (Ostraciontidae), puffer-fishes (Tetraodontidae) and porcupine-fishes (Diodontidae). Most members of these families live in tropical seas, while certain species, such as puffers, ascend tropical rivers. The ocean sun-fishes (Molidae) range over the tropical and temperate parts of the ocean.

Cling-fishes (Xenopterygii) have the pelvic fins modified to form a suction disc, which is braced fore and aft by bones of the pectoral girdle. Each pelvic fin has a spine and four branched rays; the other fins are spineless, the dorsal and anal members usually being opposed. As in true bottom-dwelling teleosts, there is no swimbladder. Cling-fishes, most of which live in intertidal regions of tropical and temperate seas, use their suction disc to cling to rocks, weeds and other suitable supports (plate 22). A few species live in swift coastal streams of the American tropics. There are about a hundred species.

In the angler-fishes (Pediculati), the first ray of the spinous dorsal fin is set on the head and transformed into a rod and bait, the illicium, which is used to lure prey. At the base of each pectoral fin there are two to four supporting bones, the radials, the lowermost being enlarged. When present, the pelvic fins are placed on the throat. There are about a hundred and sixty species, all of which are marine. The most diverse group, the deep-sea anglers (Ceratioidea), live in the deeper mid-waters of the ocean (plate 7). There are also numerous frog-fishes (Antennariidae), which hide among sea weeds and in coral reefs (plate 14). The angler-fishes (Lophiidae) have a flattened body and large jaws armed with many sharp teeth. Their usual habit is to lie on the bottom and angle for their prey, which is taken by a sudden expansion of the mouth cavity and a snap of the jaws. The colour pattern can be cunningly varied to suit that of their surroundings.

This concludes a rather cursory survey of the diversity of fishes, though a more detailed one, particularly of the teleosts, could easily fill a book of this size (see also [212] [213] [214]). After all, there are three times as many kinds of fishes as birds, which are the next most diverse of the vertebrates. Fishes are not only the 'birds' of aquatic space, but also dwellers on the interface between water and land. Fishes, as I have tried to show, are masters of their medium.

References

Chapter 1

1 WADDINGTON, C. H. (1957) *The strategy of the genes*. London, Allen and Unwin. 262 pp.
2 MAYR, E. (1963) *Animal species and evolution*. Harvard University Press. 797 pp.
3 COKER, R. E. (1954) *Streams, lakes, ponds*. University of North Carolina Press. 327 pp.

Chapter 2

4 GRAY, J. (1933) The muscular movements of fishes. *Proceedings of the Royal Institution of Great Britain*. **27** 849–74
5 BAINBRIDGE, R. (1963) Caudal fin and body movement in the propulsion of some fish. *Journal of Experimental Biology*. **40** 23–56
6 WELCH, P. S. (1935) *Limnology*. New York, McGraw-Hill, 471 pp.
7 COKER, R. E. (1947) *This great and wide sea*. University of North Carolina Press. 325 pp.
8 GERO, D. R. (1952) The hydrodynamic aspects of fish propulsion. *American Museum Novitiates*. No. 1601 1–32
9 BAINBRIDGE, R. and BROWN, R. H. J. (1958) An apparatus for the study of the locomotion of fish. *Journal of Experimental Biology*. **35** 134–7
10 BAINBRIDGE, R. (1958) The speed of swimming of fish as related to size and to the frequency and amplitude of the tail beat. *Journal of Experimental Biology*. **35** 109–33
11 BLAXTER, J. H. S. and DICKSON, W. (1959) Observations on the swimming speeds of fish. *Journal du Conseil International pour l'exploration de la mer*. **24** 472–9
12 BAINBRIDGE, R. (1960) Speed and stamina in three fish. *Journal of Experimental Biology*. **37** 129–53
13 GRAY, J. (1953) The locomotion of fishes. In *Essays in marine biology*. Edinburgh, Oliver and Boyd

Chapter 3

14 HARRIS, J. E. (1953) Fin patterns and mode of life in fishes. In *Essays in marine biology*. Edinburgh, Oliver and Boyd

15 BREDER, C. M. (1926) The locomotion of fishes. *Zoologica.* **4** 159–297

16 HARRIS, J. E. (1937) The mechanical significance of the position and movements of the paired fins in the Teleostii. *Carnegie Institute of Washington. Publication.* No. 475 171–89

17 HARRIS, J. E. (1938) The role of the fins in the equilibrium of the swimming fish and The rôle of the pelvic fins. *Journal of Experimental Biology.* **15** 32–47

18 RAY, C. and CIAMPI, E. (1958) *The underwater guide to marine life.* London, Nicholas Kaye. 338 pp.

19 BREDER, C. M. and EDGERTON, H. E. (1942) An analysis of the locomotion of the sea-horse, *Hippocampus*, by means of high-speed photography. *Annals of the New York Academy of Sciences.* **42** 145–72

20 LISSMANN, H. W. (1958) On the function and evolution of electric organs in fish. *Journal of Experimental Biology.* **35** 156–91

21 WEITZMAN, S. H. (1945) The osteology and relationships of the South American characid fishes of the subfamily Gasteropelecinae. *Stanford Ichthyological Bulletin.* **4** 211–63

22 GREENWOOD, P. H. and THOMSON, K. S. (1960) The pectoral anatomy of *Pantodon buchholzi* Peters (a freshwater flying-fish) and the related Osteoglossidae. *Proceedings of the Zoological Society, London.* **135** 283–301

23 GREGORY, W. K. and CONRAD, G. M. (1936) The evolution of the pediculate fishes. *American Naturalist.* **70** 193–208

24 SMITH, J. L. B. (1961) *The sea fishes of Southern Africa.* South Africa, Central News Agency Ltd. 590 pp.

25 ARONSON, L. R. (1951) Orientation and jumping behaviour in the gobiid fish, *Bathygobius soporator. American Museum Novitates.* No. 1486 22

26 HARRIS, V. A. (1960) On the locomotion of the mud skipper, *Periophthalmus koelreuteri* (Pallas): Gobiidae. *Proceedings of the Zoological Society, London.* **134** 107–35

Chapter 4

27 VAN OOSTEN, J. (1957) The skin and scales. In *The physiology of fishes.* Vol. 1. (Ed. BROWN, M. E.) New York, Academic Press Inc. 207–44

28 BREDER, C. M. (1947) An analysis of the geometry of symmetry with especial reference to the squamation of fishes. *Bulletin of the American Museum of Natural History, New York.* 321–412

29 HOME, E. (1809) *Philosophical Transactions of the Royal Society, London.* **99** 17–87

30 GOODRICH, E. S. (1930) *Studies on the structure and development of vertebrates.* London, Macmillan. 837 pp. For detailed and comparative account see Chapters 1, 2, 4, 6, 7

31 SCHAEFFER, B. and ROSEN, D. E. (1961) Major adaptive levels in the evolution of actinopterygian feeding mechanism. *American Zoologist.* **1** 187–204

Chapter 5

32 JONES, F. R. H. and MARSHALL, N. B. (1953) The structure and functions of the teleostean swimbladder. *Biological Reviews.* **28** 16–83

33 DENTON, E. J. and MARSHALL, N. B. (1958) The buoyancy of bathypelagic fishes without a gas-filled swimbladder. *Journal of the Marine Biological Association.* **37** 753–67

34 FÄNGE, R. (1953) The mechanism of gas transport in the euphysoclist swimbladder. *Acta Physiologica Scandanivica.* **30** Supplementum 110, pp. 133

35 SCHOLANDER, P. F. (1958) Counter current exchange. *Hvalrådets Skrifter, Oslo.* **44** 1–24

36 KUHN, W., RAMEL, A., KUHN, H. J. and MARTI, E. (1963) The filling mechanism of the swimbladder. *Experientia.* **19** 497–511

37 QUTOB, Z. (1962) The swimbladder of fishes as a pressure receptor. *Archives Néerlandaises de Zoologie.* **15** 1–16

38 MARSHALL, N. B. (1960) Swimbladder structure of deep-sea fishes in relation to their systematics and biology. *Discovery Reports.* **31** 1–122

Chapter 6

39 KLEEREKOPER, H. and VAN ERKEL, G. A. (1960) The olfactory apparatus of *Petromyzon marinus. Canadian Journal of Zoology.* **38** 209–23

40 STRAHAN, R. (1958) The velum and respiratory current of *Myxine. Acta Zoologica.* **39** 227–40

41 HUGHES, G. M. and SHELTON, G. (1958) The mechanism of gill ventilation in three freshwater teleosts. *Journal of Experimental Biology.* **35** 807–23

42 HUGHES, G. M. (1960) A comparative study of gill ventilation in marine teleosts. *Journal of Experimental Biology.* **37** 28–45

43 GOSLINE, W. A. (1959) Mode of life, functional morphology and the classification of modern teleostean fishes. *Systematic Zoology.* **8** 160–4

44 VAN DAM, L. (1938) On the utilization of oxygen and regulation of breathing in some aquatic animals. *Dissertation.* Groningen

45 FRY, F. E. J. (1957) The aquatic respiration of fish. In *The physiology of fishes.* Vol. 1. (Ed. BROWN, M. E.) New York, Academic Press Inc. 1–63

46 HAZELHOFF, E. H. and EVENHUIS, H. H. (1952) Importance of the 'counter current principle' for the oxygen uptake in fishes. *Nature.* **169** 77

47 BLAŽKA, P. (1958) The anaerobic metabolism of fish. *Physiological Zoology.* **31** 117–28

48 CARTER, G. S. and BEADLE, L. C. (1931) Reports of an expedition to Brazil and Paraguay in 1926–27. *Journal of the Linnean Society. Zoology.* **37** 327–68

49 ODUM, H. T. and CALDWELL, D. K. (1955) Fish respiration in the natural oxygen gradient of an anaerobic spring in Florida. *Copeia* No. 2 104–6

50 CARTER, G. S. (1957) Air breathing. In *The physiology of fishes.* Vol. 1. (Ed. BROWN, M. E.) New York, Academic Press Inc. 65–79

51 DAS, B. K. (1936) On ecology and bionomics of an air-breathing loach, *Lepidocephalus guntea* (Ham. Buch.) with a review on air-breathing fishes. *XII Congrès International de Zoologie, Lisbonne, 1935*

52 HUBBS, C. L. (1941) The relation of hydrological conditions to speciation in fishes. In *Symposium on hydrobiology.* University of Wisconsin Press

Chapter 7

53 SHERRINGTON, C. (1955) *Man on his nature.* London, Penguin Books. 312 pp.

54 HICKLING, C. F. (1961) *Tropical inland fisheries*. London, Longmans. 287 pp.
55 MACMAHON, A. F. M. (1946) *Fishlore*. London, Penguin Books. 208 pp.
56 ELLISON, W. A. (1951) The menhaden. In *Survey of marine fisheries of North Carolina*. (Ed. TAYLOR, H. F.) The University of North Carolina Press. 85–107
57 BROWN, M. E. (1957) Experimental studies on growth. In *The physiology of fishes*. Vol. 1. (Ed. BROWN, M. E.) New York, Academic Press. 361–400
58 ROSE, S. M. (1959) Failure of survival of slowly growing members of a population. *Science*. **129** 1026
59 MAGNUSON, J. J. (1962) An analysis of aggressive behaviour, growth and competition for food and space in medaka, *Oryzias latipes*, Pisces (Cyprinodontidae). *Canadian Journal of Zoology*. **40** 313–63
60 WYNNE-EDWARDS, V. C. (1962) *Animal dispersion in relation to social behaviour*. Edinburgh, Oliver and Boyd. 653 pp.
61 KINNE, O. (1960) Growth, food intake and food conversion in euryplastic fish exposed to different temperatures and salinities. *Physiological Zoology*. **33** 288–317
62 MATTHEWS, L. H. (1962) The shark that hibernates. *New Scientist*. **13** 756–59
63 PARKER, R. R. and LARKIN, P. A. (1959) A concept of growth in fishes. *Journal of the Fisheries Research Board of Canada*. **16** 721–45
64 PICKFORD, G. E. and ATZ, J. W. (1957) *The physiology of the pituitary gland of fishes*. The New York Zoological Society. 613 pp.

Chapter 8

65 BLACK, V. S. (1957) Excretion and osmoregulation. In *The physiology of fishes*. Vol. 1. (Ed. BROWN, M. E.) New York, Academic Press. 163–205
66 KROGH, A. (1939) *Osmotic regulation in aquatic animals*. Cambridge University Press. 242 pp.
67 WALFORD, L. A. (1958) *Living resources of the sea*. New York, Ronald Press. 321 pp.
68 BAGGERMAN, B. (1960) Salinity preference, thyroid activity and the seaward migration of four species of Pacific salmon (*Oncorhynchus*). *Journal of the Fisheries Board of Canada*. **17** 295–322
69 MACALLUM, A. B. (1926) The palaeochemistry of the body fluids and tissues. *Physiological Reviews*. **6** 316
70 ROBERTSON, J. D. (1957) The habitat of the early vertebrates. *Biological Reviews*. **32** 156
71 TARLO, L. B. (1962) The earliest vertebrates. *New Scientist*. **14** 151–3

Chapter 9

72 LE GROS CLARK, W. E. (1945) The anatomical basis of sensory experience. In *New biology*. London, Penguin Books. 72–85
73 WALLS, G. L. (1942) The vertebrate eye. *Cranbrook Institute of Science Bulletin*. No. 19 785
74 BAYLOR, E. R. and SHAW, E. (1962) Refractive error and vision in fishes. *Science*. **136** 157–8

75 DIJKGRAAF, S. (1962) The functioning and significance of the lateral-line organs. *Biological Reviews.* **38** 51–105

76 LOWENSTEIN, O. (1957) The sense organs: the acoustico-lateralis system. In *The physiology of fishes.* Vol. 2. (Ed. BROWN, M. E.) New York, Academic Press. 153–186

77 BUDKER, P. (1958) *Whales and whaling.* London, Harrap. 182 pp.

78 NELSON, D. R. and GRUBER, S. H. (1963) Sharks; attraction by low frequency sounds. *Science.* **142** 975–7

79 HASLER, A. D. (1957) The sense organs: olfactory and gustatory senses of fishes. In *The physiology of fishes.* Vol. 2. (Ed. BROWN, M. E.) New York, Academic Press Inc. 187–209

80 STEVEN, G. A. (1930) Bottom fauna and the food of fishes. *Journal of the Marine Biological Association.* **16** 677–700

81 EGAMI, N. and NAMBU, M. (1961) Factors initiating mating behaviour and oviposition in the fish, *Oryzias latipes. Journal of the Faculty of Science, Tokyo. Section 4, Zoology.* **9** 263–78

82 BULL, H. O. (1957) Behavior: Conditioned responses. In *The physiology of fishes.* Vol. 2. (Ed. BROWN, M. E.) New York, Academic Press. 211–228

83 LOTKA, A. J. (1956) *Elements of mathematical biology.* New York, Dover Publications Inc. 465 pp.

Chapter 10

84 GRIFFIN, D. R. (1958) *Listening in the dark.* Yale University Press

85 MORROW, J. E. and MAURO, A. (1950) Body temperatures of some marine fishes. *Copeia.* 108–16

86 LISSMANN, H. W. and MACHIN, K. E. (1958) The mechanism of object location in *Gymnarchus niloticus* and similar fish. *Journal of Experimental Biology.* **35** 451–86

87 GRUNDFEST, H. (1960) Electrical fishes. *Scientific American.* **203** 115–24

88 DARWIN, C. (1956) *The origin of species.* Oxford University Press. 592 pp. (reprint of sixth edition 1872)

89 KEYNES, R. D. (1957) Electric organs. In *Physiology of fishes.* Vol. 2. (Ed. BROWN, M. E.) New York, Academic Press Inc. 323–43

90 MOULTON, J. M. (1960) Swimming sounds and the schooling of fishes. *Biological Bulletin. Woods Hole.* **119** 210

91 DIJKGRAAF, S. (1941) Haben die Lautäusserungen der Elritze ein biologische Bedeutung? *Zoologisches Anzeiger.* **136** 103

92 FISH, M. P. (1954) The character and significance of sound production among fishes of the Western North Atlantic. *Bulletins of the Bingham Oceanographic Collections.* **14** article 3

93 WINN, H. E. and STOUT, J. F. (1960) Sound production by satin fin shiner. *Notropis analostanus*, and related fishes. *Science.* **132** 222

94 DELCO, E. A. (1960) Sound discrimination by males of two cyprinid fishes. *Texas Journal of Science.* **12** 48

95 SØRENSEN, W. (1895) Are the extrinsic muscles in the air bladder of some Siluridae and the 'elastic spring' apparatus of others subordinate to the voluntary production of sounds? What is, according to our present knowledge, the function of the Weberian ossicles? *Journal of Anatomy.* London. **29**

96 REGAN, C. T. (1936) *Natural history (Fishes)*. London, Ward, Lock. 896 pp.
97 JOHNSON, M. W. (1948) Sound as a tool in marine ecology, from data on biological noises and the deep scattering layer. *Journal of Marine Research*. **7** 443
98 TAVOLGA, W. N. (1960) Foghorn sounds beneath the sea. *Natural History*. **69** 44
99 TAVOLGA, W. N. (1958) The significance of underwater sounds produced by males of the gobiid fish, *Bathygobius soporator*. *Physiological Zoology*. **31** 259
100 MARSHALL, N. B. *Aspects of deep sea biology*. London, Hutchinsons. 380 pp.
101 MARSHALL, N. B. (1962) The biology of sound-producing fishes. *Symposia of the Zoological Society of London*. No. 7 45–60
102 CLARKE, G. L. and BACKUS, R. H. (1956) Measurement of light penetration in relation to vertical migration and records of luminescence of deep-sea animals. *Deep-sea Research*. **4** 1–14
103 NICOL, J. A. C. (1958) Observations on luminescence in pelagic animals. *Journal of the Marine Biological Association*. **37** 705–52
104 BERTELSEN, E. (1951) The ceratioid fishes. *Dana Reports*. No. 3 1–276
105 HENEDA, Y. (1951) The luminescence of some deep-sea fishes of the families Gadidae and Macrouridae. *Pacific Science*. **5** 372–8
106 MCELROY, W. D. and SELINGER, H. H. (1962) Biological luminescence. *Scientific American*. **207** 76–89
107 HARVEY, E. N. (1957) The luminous organs of fishes. In *The physiology of fishes*. Vol. 2. (Ed. BROWN, M. E.) New York, Academic Press. 345–436

Chapter 11

108 HOOGLAND, R., MORRIS, D. and TINBERGEN, N. (1957) The spines of sticklebacks (*Gasterosteus* and *Pygosteus*) as a means of defence against predators (*Perca* and *Esox*). *Behaviour*. **10** 205–36
109 TINBERGEN, N. (1951) *The study of instinct*. Oxford University Press. 228 pp.
110 HERTER, K. (1953) *Die Fischdressuren und ihre sinnesphysiologischen Grundlagen*. Berlin, Akademie-Verlag. 326 pp.
111 ATZ, J. W. (1951) It swims upside down. *Animal Kingdom, New York Zoological Society*. **54** 18–21
112 REIGHARD, J. E. (1908) An experimental field-study of warning colouration in coral reef fishes. *Carnegie Institute of Washington*. **2** 257–325
113 SUMNER, F. B. (1934) Does 'protective colouration' protect? – Results of some experiments with fishes and birds. *Proceedings of the National Academy of Sciences, Washington*. **20** 559–64
114 SUMNER, F. B. (1940) Quantitative changes in pigmentation resulting from visual stimuli in fishes and Amphibia. *Biological Reviews*. **15** 351
115 HEALEY, E. G. (1957) The nervous system. In *The physiology of fishes*. Vol. 2. (Ed. BROWN, M. E.) New York, Academic Press. 1–119
116 PARKER, G. H. (1948) *Animal colour changes and their neurohumors*. Cambridge University Press. 377 pp.
117 BUCHSBAUM, R. and MILNE, L. J. (1960) *Living invertebrates of the world*. London, Hamish Hamilton. 303 pp.

118 HERALD, E. S. (1961) *Living fishes of the world.* London, Hamish Hamilton. 304 pp.

119 RANDALL, J. E. and RANDALL, H. E. (1960) Examples of mimicry and protective resemblance in tropical marine fishes. *Bulletin of Marine Science of the Gulf and Caribbean.* **10** 448–80

120 Quoted by BREDER, C. M. (1946) An analysis of the deceptive resemblances of fishes to plant parts, with critical remarks on protective colouration, mimicry and adaptation. *Bulletin of the Bingham Oceanographic Collections.* 1–49

121 MORTENSEN, T. (1917) Observations on protective adaptations and habits, mainly in marine animals. *Videnskdbelige Meddelelser fra Dansk naturhistorisk Forening i Kjøbenhavn.* **69** 57–96

122 LORENZ, K. Z. (1952) *King Solomon's ring.* London, Methuen. 202 pp.

123 BEEBE, W. and TEE-VAN, J. (1928) The fishes of Port-au-Prince Bay, Haiti. *Zoologica.* **10** 1–279

124 HALSTEAD, B. W. (1959) *Dangerous marine animals.* Maryland, Cornell Maritime Press. 146 pp.

125 MEDWAR, P. B. (1957) *The uniqueness of the individual.* London, Methuen. 191 pp.

Chapter 12

126 BREDER, C. M. (1959) Studies on social groupings in fishes. *Bulletin of the American Museum of Natural History.* **117** 393–482

127 HILDEBRAND, S. F. and SCHROEDER, W. C. (1928) Fishes of Chesapeake Bay. *Bulletins of the U.S. Bureau of Fisheries.* **43** 366

128 BREDER, C. M. (1929) Report on synentognath habits and development. *Yearbook of the Carnegie Institute of Washington.* **28** 279–82

129 HIATT, R. W. and BROCK, V. E. (1948) On the herding of prey and the schooling of the black skipjack. *Euthynnus yaito* Kishinouye. *Pacific Science.* **2** 297–8

130 BROCK, V. E. and RIFFENBURGH, R. H. (1960) Fish schooling: a possible factor in reducing predation. *Journal du Conseil International pour l'exploration de la mer.* **25** 307–17

131 LIMBAUGH, C. (1961) Cleaning symbiosis. *Scientific American.* **205** 42–9

132 RANDALL, J. E. (1958) A review of the labrid fish genus. *Labroides,* with descriptions of two new species and notes on ecology. *Pacific Science.* **12** 327–47

133 MACGINITIE, G. E. and MACGINITIE, N. (1949) *Natural history of marine animals.* New York, McGraw-Hill. 473 pp.

134 EIBL-EIBESFELDT, I. (1960) Beobachtungen und Versuche an Anemonenfischen (*Amphiprion*) der Maldiven und der Nicobaren. *Zeitschrift für Tierpsychologie.* **17** 1–10

135 DAVENPORT, D. and NORRIS, K. (1958) Observations on the symbiosis of the sea anemone *Stoichactis* and the pomacentrid fish *Amphiprion percula. Biological Bulletin, Woods Hole.* **115** 397–410

136 VERWEY, J. (1930) Coral reef studies 1. The symbiosis between damsel fishes and the sea anemones in Batavia Bay. *Treubia* **12** 305–66

137 GUDGER, E. W. (1942) Physalia, the fish-eater. *Animal Kingdom, New York Zoological Society.* **65** 62–6

138 PFAFF, J. R. (1942) On a new genus and species of the family Gobiesocidae from the Indian Ocean, with observations on sexual dimorphism in the Gobiesocidae, and on the connection of certain gobiesocids with echinids. *Videnskabelige Meddelelser fra Dansk naturhistorisk Forening i Kjøbenhavn.* **105** 413–21

139 ARNOLD, D. C. (1953) Observations on *Carapus acus* (Brunnich) (Jugulares, Carapidae). *Publicazioni della Stazione Zoologica di Napoli.* **24** 153–67

140 RAPOPORT, A. (1960) *Fights, games and debates.* Ann Arbor, University of Michigan Press. 400 pp.

Chapter 13

141 BRAWN, V. M. (1961) Aggressive behaviour in the cod (*Gadus callarias L.*). *Behaviour.* **18** 107–47

142 TINBERGEN, N. (1953) *Social behaviour in animals.* London, Methuen. 150

143 GERKING, S. D. (1959) The restricted movements of fish populations. *Biological Reviews, Cambridge.* **34** 221–42

144 STUART, T. A. (1957) The migrations and homing behaviour of brown trout (*Salmo trutta. L.*). *Scottish Home Department, Scientific Investigations. Freshwater and salmon fisheries research.* **18** 2–27

145 HASLER, A. D. (1960) Guideposts of migrating fishes. *Science.* **132** 785–92

146 GUNNING, G. E. (1959) The sensory basis for homing in the longear sunfish, *Lepomis lepomis megalotis* (Rafinesque). *Investigations of Indiana Lakes and Streams.* **5** 103–30

147 SAILA, S. B. and SHAPPY, R. A. (1962) Migration by computer. *Discovery.* **23** no. 6

148 EIBL-EIBESFELDT, I. and KRAMER, S. (1958) Ethology, the comparative study of animal behaviour. *Quarterly Review of Biology, Washington.* **33** 181–211

149 NOBLE, G. K. (1938) Sexual selection among fishes. *Biological Reviews, Cambridge.* **13** 133–55

150 TAVOLGA, W. N. (1954) Reproductive behaviour in the gobiid fish *Bathygobius soporator. Bulletin of the American Museum of Natural History.* 427–60

151 RANDALL, J. E. (1961) Observations on the spawning of surgeon-fishes (Acanthuridae) in the Society Islands. *Copeia.* No. 2 237–8

152 BREDER, C. M. (1951) A note on the spawning behaviour of *Caranx sexfasciatus. Copeia.* No. 2 170

153 ROSEN, D. E. and TUCKER, A. (1961) Evolution of secondary sexual characters and sexual behaviour patterns in a family of viviparous fishes (Cyprinodontiformes: Poeciliidae). *Copeia.* No. 2 201–212

154 WYNNE-EDWARDS, V. C. (1962) *Animal dispersion in relation to social behaviour.* Edinburgh, Oliver and Boyd. 653 pp.

155 CLARK, E., ARONSON, L. E. and GORDON, M. (1954) Mating behaviour patterns in two sympatric species of xiphophorin fishes: Their inheritance and significance in sexual isolation. *Bulletin of the American Museum of Natural History.* **103** 135–226

156 De Latil, P. (1954) *The underwater naturalist*. London, Jarrolds. 275 pp.
157 Thorpe, W. H. (1963) *Learning and instinct in animals*. London, Methuen. 558 pp.
158 Russell, E. S. (1934) *The behaviour of animals*. London, Edward Arnold. 184 pp.
159 Koehler, O. (1956) Thinking without words. *XIV International Congress of Zoology, Copenhagen*. 75–88
160 Potter, S. (1960) *Language in the modern world*. London, Penguin Books. 221 pp.
161 Campbell, J. (1960) *The masks of God*. London, Secker and Warburg. 504 pp.

Chapter 14

162 Hoar, W. S. (1957) The gonads and reproduction. In *The physiology of fishes*. Vol. 1. (Ed. Brown, M. E.) New York, Academic Press. 287–321
163 Hela, I. and Laevastu, T. (1962) *Fisheries hydrography*. London, Fishery News (Books). 137 pp.
164 Gunter, G. (1945) Studies on marine fishes of Texas. *Publications of the Institute of Marine Science, University of Texas*. **1** 1–190
165 Talbot, F. H. (1960) Notes on the biology of the Lutjanidae (Pisces) of the East African Coast, with special reference to *L. bohar* (Forskål). *Annals of the South African Museum*. **45** 549–73
166 Smith, H. M. (1945) The fresh-water fishes of Siam, or Thailand. *U.S. National Museum, Bulletin 1 8*. 1–622
167 Smith, C. L. (1958) Hermaphroditism in some serranid fishes from Bermuda. *Papers of the Michigan Academy of Science, Arts and Letters*. **44** 111–8
168 Clark, E. (1959) Functional hermaphroditism and self fertilization in a serranid fish. *Science*. **129** 215–6
169 Harrington, R. W. (1961) Oviparous hermaphroditic fish with internal self-fertilization. *Science*. **13** 1794–950
170 Mead, G. W. (1960) Hermaphroditism in archibenthic and pelagic fishes of the order Iniomi. *Deep Sea Research*. **6** 234–5
171 Matthews, L. H. (1955) The evolution of viviparity in vertebrates. In *Comparative physiology of reproduction and the effects of sex hormones in vertebrates*. (Eds. Chester Jones, I. and Eckstein, P.) Memoirs of the Society for Endocrinology. No. 4 129–48
172 Amoroso, E. C. (1960) Viviparity in fishes. *Symposia of the Zoological Society of London*. Hormones in Fish. No. 1 153–81

Chapter 15

173 Graham, M. (1956) Plaice. In *Sea fisheries: their investigation in the United Kingdom*. (Ed. Graham, M.) London, Edward Arnold. 332–71
174 Shelbourne, J. E. (1956) The effect of water conservation on the structure of marine fish embryos and larvae. *Journal of the Marine Biological Association*. **35** 275–86
175 Parrish, B. B. (1956) The cod, haddock and hake. In *Sea fisheries: their investigation in the United Kingdom*. (Ed. Graham, M.) London, Edward Arnold. 251–331

176 RANDALL, J. E. (1961) A contribution to the biology of the convict surgeonfish of the Hawaiian Islands, *Acanthurus triostegus sandvicensis*. *Pacific Science*. **15** 215–72

177 BIGELOW, H. B. and SCHROEDER, W. C. (1953) Fishes of the Gulf of Maine. *U.S. Department of the Interior. Fish and Wildlife Service. Fishery Bulletin*. **53** 577

178 PEARCEY, W. G. (1962) Ecology of an estuarine population of winter flounder, *Pseudopleuronectes americanus* (Walbaum). *Bulletin of the Bingham Oceanographic Collection, Yale University*. **18** 78

179 HUNTSMAN, A. G. (1918) The effect of the tide on the distribution of the fishes of the Canadian Atlantic Coast. *Transactions of the Royal Society of Canada*. **12** 61–7

180 SETTE, O. E. (1943) Biology of the Atlantic mackerel (*Scomber scombrus*) of North America, Part 1. *U.S. Department of the Interior. Fish and Wildlife Service. Fishery Bulletin*. **50** 149–227

181 BRUUN, A. F. (1937) Contribution to the life histories of the deep sea eels: Synaphobranchidae. *Dana Reports*. No. 9 1–33

182 BREDER, C. M. (1943) The eggs of *Bathygobius soporator* (Cuvier and Valenciennes) with a discussion of other non-spherical teleost eggs. *Bulletin of the Bingham Oceanographic Collection, Yale University*. **8** article 3 1–49

183 STRASBURG, D. W. (1960) The blennies. In *Handbook of Hawaiian fishes*. by GOSLINE, W. A. and BROCK, V. E. Honolulu, University of Hawaii Press. 274–79

184 MOISEEV, P. A. (1961) On the biological basis of fishery in the Western Pacific. *Proceedings of the North Pacific Science Congress of the Pacific Science Association, 1957*. **10** 64–71

185 BREDER, C. M. (1962) On the significance of transparency in osteichthyid fish eggs and larvae. *Copeia*. No. 3 561–7

Chapter 16

186 STUART, T. A. (1953) Spawning migration, reproduction and young stages of loch trout (*Salmo trutta L.*). *Scottish Home Department. Scientific Investigations. Freshwater and salmon fisheries research*. **5** 39

187 SCHINDLER, O. (1957) *Freshwater fishes*. (Translated and Ed. ORKIN, P. A.) London, Thames and Hudson. 243 pp.

188 BUDGETT, J. A. (1901) On the breeding habits of some West African fishes, with an account of the external features in development of *Protopterus annectens*, and a description of the larva of *Polypterus lapradei*. *Transactions of the Zoological Society of London*. **14** 115–36

189 BREDER, C. M. (1936) The reproductive habits of the North American sunfishes (family Centrarchidae). *Zoologica, New York*. **21** 1–48

190 BAERENDS, G. P. and BAERENDS-VAN ROON, J. M. (1950) An introduction to the study of the ethology of cichlid fishes. *Behaviour*. Supplement 1

Chapter 17

191 SPRINGER, S. (1960) Natural history of the sandbar shark, *Eulamia milberti*. *U.S. Department of the Interior. Fish and Wildlife Service*. Fishery Bulletin **178**. 38 pp.

192 HJORT, J. (1912) In *Depths of the ocean.* (Ed. MURRAY, J. and HJORT, J.) London, Macmillan

193 BRIGGS, J. C. (1960) Fishes of worldwide (circumtropical) distribution. *Copeia.* No. 3 171–80

194 BEEBE, W. and CRANE, J. (1939) Deep-sea fishes of the Bermuda Oceanographic Expeditions. Family Melanostomiatidae. *Zoologica, New York.* **24** 65–238

195 CLARKE, R. (1950) (quoting GUNTHER, E. R.) The bathypelagic angler-fish, *Ceratias holbolli. Discovery Reports.* **26** 1–32

196 NICOL, J. A. C. (1960) Studies in luminescence. On the subocular light-organs of stomiatoid fishes. *Journal of the Marine Biological Association.* **39** 529–48

197 NICOL, J. A. C. (1960) Spectral composition of the light of the lantern-fish, *Myctophum punctatum. Journal of the Marine Biological Association.* **39** 27–32

198 NICOL, J. A. C. (1960) Luminescence in marine organisms. *The Times Science Review.* Summer, No. 36. 10–12

199 HANEDA, Y. (1951) The luminescence of some deep-sea fishes of the families Gadidae and Macrouridae. *Pacific Science.* **5** 372–8

Chapter 18

200 SCHOLANDER, P. F., FLAGG, W., WALTERS, V. and IRVING, L. (1953) Climatic adaptation in arctic and tropical poikilotherms. *Physiological Zoology.* **26** 67–92

201 RUUD, J. T. (1959) Vertebrates without blood pigment: a study of the fish family Chaenichthyidae. *Proceedings of the fifteenth International Congress of Zoology, London.* July 1958, 526–8

202 SCHOLANDER, P. F., VAN DAM, L., KANWISHER, J. W., HAMMEL, A. T. and GORDON, M. S. (1957) Supercooling and osmoregulation in arctic fish. *Journal of Cellular and Comparative Physiology.* **49** 5–24

203 GORDON, M. S., AMDUR, B. H. and SCHOLANDER, P. F. (1962) Freezing resistance in some northern fishes. *Biological Bulletin, Woods Hole.* **122** 52–62

204 WELLS, J. W. (1957) Coral reefs. In *Treatise on marine ecology and paleocology.* Vol. 1. (Ed. HEDGPETH, J. W.) Geological Society of America. 1296 pp.

205 ODUM, E. P. (1959) *Fundamentals of ecology.* Philadelphia, W. R. Saunders. 546 pp.

206 HIATT, R. W. and STRASBURG, D. W. (1960) Ecological relationships of the fish fauna on coral reefs of the Marshall Islands. *Ecological Monographs.* **30** 65–127

207 BRIGGS, J. C. (1962) The East Pacific Barrier and the distribution of marine shore fishes. *Evolution.* **15** 545–54

208 WOOSTER, W. S. and REID, J. L. (1963) Eastern boundary currents. In *The sea.* Vol. 2. (Ed. HILL, M. N.) 253–80. Interscience

Chapter 19

209 DARLINGTON, P. J. (1957) *Zoogeography: the geographical distribution of animals.* New York, John Wiley. 675 pp.

210 INGER, R. F. and KONG, G. H. (1961) The Bornean cyprinoid fishes of the genus *Gastromyzon*. *Copeia*. No. 2. 166–76

211 TEILHARD DE CHARDIN, P. (1959) *The phenomenon of man*. London, Collins. 196 pp.

212 GRASSÉ, P. P. (Ed.) (1958) *Traité de Zoologie*. Vol. 13. Poissons, Fascicules 1–3. Paris, Massons

Chapter 20

213 LAGLER, K. F., BARDACH, J. E. and MILLER, R. R. (1962) *Ichthyology*. Ann Arbor, The University of Michigan. 545 pp.

214 NORMAN, J. R. (1963) *A history of fishes*. Revised by GREENWOOD, P. H. London, Ernest Benn. 398 pp.

Index